Environmental Assessment as a Tool for Climate Change Mitigation

Environmental Assessment as a Tool for Climate Change Mitigation

BENOIT MAYER

Professor of Climate Law, University of Reading
with the assistance of Mateusz Slowik

OXFORD
UNIVERSITY PRESS

Great Clarendon Street, Oxford, OX2 6DP,
United Kingdom

Oxford University Press is a department of the University of Oxford.
It furthers the University's objective of excellence in research, scholarship,
and education by publishing worldwide. Oxford is a registered trade mark of
Oxford University Press in the UK and in certain other countries

Published in the United States of America by Oxford University Press
198 Madison Avenue, New York, NY 10016, United States of America

British Library Cataloguing in Publication Data
Data available

Library of Congress Control Number: 2024943017

ISBN 978–0–19–893918–4

DOI: 10.1093/oso/9780198939184.001.0001

Printed and bound by
CPI Group (UK) Ltd, Croydon, CR0 4YY

To Peixi

Acknowledgements

A decade ago, on 16 April 2014, European institutions found it 'appropriate to assess the impact of projects on climate', including their 'greenhouse gas emissions', as part of environmental assessment procedures.[1] When reading about this, I was torn between two diametrically opposed views. On the one hand, it seemed obvious that, as climate change is an environmental issue—and a major one at that—it should be among the considerations weighed before a decision could be made on whether to approve a proposed activity. On the other hand, I kept wondering whether climate assessment could really be an effective and useful tool in relation to impacts that were highly atypical. Unlike most impacts considered in environmental assessment procedures, an activity's greenhouse gas emissions do not directly affect anyone or anything anywhere in the world in any concrete and predictable way.

What convinced me to look further into this issue was the 2017 judgment of the South African High Court in *Earthlife Africa Johannesburg v Minister of Environmental Affairs*,[2] which set aside the environmental approval of a coal-fired power plant on the ground that its GHG emissions had not been properly considered. My research, in the following years, benefited from assistance by Winnie Cheung and from extensive comments and suggestions by Neil Craik. This led to several publications, including an article that received the *International & Comparative Law Quarterly*'s Young Scholar Prize in 2019.[3]

Further discussions with the late Meinhard Doelle led me to realize the potential for a more thorough comparative study on this topic. Developments were taking place at a fast pace across the world but with little interaction across national borders. In this context, it appeared that each national debate could be enriched by a better understanding of how other jurisdictions were addressing similar issues.

The research for this book was conducted mainly at the Faculty of Law of the Chinese University of Hong Kong, but also during fellowships at Tulane Law

[1] Directive 2014/52/EU of the European Parliament and of the Council of 16 April 2014 amending Directive 2011/92/EU on the assessment of the effects of certain public and private projects on the environment, [2014] OJ L 124/1, preamble para 13.

[2] *Earthlife Africa Johannesburg v Minister of Environmental Affairs* [2017] 2 All SA 519 (GP).

[3] Benoit Mayer, 'Climate Assessment as an Emerging Obligation under Customary International Law' (2019) 68 International & Comparative Law Quarterly 271. See also Benoit Mayer, 'Environmental Assessments in the Context of Climate Change: The Role of the UN Economic Commission for Europe' (2019) 28 Review of European, Comparative & International Environmental Law 82; Benoit Mayer and Wu Lan, 'The Environmental Impact Assessment Ordinance: Two Decades, no Change' *Hong Kong Lawyer* (June 2019); Benoit Mayer, 'The Emergence of Climate Assessment as a Customary Law Obligation', in Benoit Mayer and Alexander Zahar (eds), *Debating Climate Law* (CUP 2021) 285.

School and at the British Institute of International and Comparative Law, and while taking a new job at the Law School of the University of Reading. The project was funded by the Hong Kong Research Grants Council.[4] Mateusz Slowik has provided extensive research assistance on a full-time basis for 16 months—he has been extraordinarily dedicated, tenacious, and attentive to detail; has worked tirelessly with, at times, minimal supervision; and has always responded reliably and intelligently to increasingly demanding requests for additional information. Stephanie Du, Roxanne Tsui, and Justin Jin have provided additional help at the late stages of the manuscript preparation.

Early drafts were presented at workshops and conferences held at the Department of Geography and Resource Management and at the Faculty of Law of the Chinese University of Hong Kong; at the annual conference of the International Association of Impact Assessment in Kuching; at the Cambridge Centre for Environment, Energy, and Natural Resource Governance; at Tulane Law School; and at the Research Center ClimLaw at the University of Graz. I benefited immensely from comments and suggestions by many participants to these events. I am also immensely grateful to three anonymous reviewers for their detailed comments and suggestions. At Oxford University Press, this book was commissioned by Fay Gibbons, project-edited by Paulina dos Santos Major, managed by Newgen KnowledgeWorks (P) Ltd, and copy-edited by Christopher Long. The manuscript was finalized on 1 May 2024, with minor updates in August 2024, and all weblinks were last accessed at that time.

[4] General Research Fund No 14602021, 'Environmental Impact Assessment as a Tool for Climate Change Mitigation: Global Comparative Perspectives' (2022–24).

Contents

Table of Cases and Decisions

NB: references are listed in chronological order within each category.

INTERNATIONAL INSTITUTIONS

International Court of Justice

International Tribunal for the Law of the Sea

Arbitral Tribunals

European Court of Human Rights

European Court of Justice

NATIONAL INSTITUTIONS

Austria

California

Canada

Chile

Netherlands

Norway

New South Wales

INDEPENDENT PLANNING COMMISSION

New York State

United States

Table of Instruments

NB: references are listed in chronological order within each category.

UN General Assembly Resolutions

Decisions Adopted by the Parties to Climate Treaties

Documents Adopted by Other International Institutions

FINANCIAL AND DEVELOPMENT INSTITUTIONS

Abbreviations

ABAER	Alberta Energy Regulator
AC	Law Reports: Appeal Cases
ACT	Australian Capital Territory
AEIAR	Approved Environmental Impact Assessment Report (Hong Kong)
AG	Advocate General (European Court of Justice)
ALD	Administrative Law Decisions (Australia)
All ER	All England Law Reports
All SA	All South African Law Reports
Ann	annotated
APEC	Asia-Pacific Economic Cooperation
AUD	Australian dollar
BBNJ	Agreement under the United Nations Convention on the Law of the Sea on the Conservation and Sustainable Use of Marine Biological Diversity of Areas Beyond National Jurisdiction
BC	British Columbia
BC	British Columbia
BC Reg	Consolidated Regulations of British Columbia
BGBl	Bundesgesetzblatt (public gazette of the Federal Republic of Germany/of the Republic of Austria)
BLM	Bureau of Land Management (United States)
BOEM	Bureau of Ocean Energy Management (United States)
CA	climate assessment
CAD	Canadian dollar
Cal 3d	California Reports, Third Series
Cal 4th	California Reports, Fourth Series
Cal 5th	California Reports, Fifth Series
Cal App 4th	California Appellate Reports, Fourth Series
Cal App 5th	California Appellate Reports, Fifth Series
California Pub Res Code	California Code, Public Resources
CarswellAlta	Carswell Alberta
CEQ	Council on Environmental Quality (United States)
CEQA	California Environmental Quality Act
CEQR	City Environmental Quality Review (New York City)
CFR	Code of Federal Regulations (United States)
ch	chapter
Cir	Appellate Circuit (United States)
	CJ Chief Justice

CMLR	Common Market Law Reports
CO_2	carbon dioxide
CO_2e	carbon dioxide equivalent
Colorado Rev Stat	Colorado Revised Statutes
COP	Conference of the Parties to the UN Framework Convention on Climate Change
COSIS	Commission of Small Island States on Climate Change and International Law
CSIH	Court of Session Inner House (Scotland)
Cth	Commonwealth of Australia
CUP	Cambridge University Press
D	District Court (United States)
DC	District of Columbia
DLR	Dominion Law Reports (Canada)
DOF	Diario Oficial de la Federación (Mexico)
EA	environmental assessment
EBRD	European Bank for Reconstruction and Development
ECLI	European Case Law Identifier
ECR	European Court Reports
EIA	environmental impact assessment
EIS	environmental impact statement (United States)
ELRNZ	Environmental Law Reports of New Zealand
EPA	Environmental Protection Agency (US)/Environmental Protection Authority (Northern Territory)
ETS	emission trading scheme
EU	European Union
EUB	Energy and Utilities Board (Alberta)
EUR	Euro
EWCA Civ	England and Wales Court of Appeal (Civil Division)
EWHC (Admin)	England and Wales High Court (Administrative Division)
F 2d	Federal Reporter, Second Series (United States)
F 3d	Federal Reporter, Third Series (United States)
F 4th	Federal Reporter, Fourth Series (United States)
F Supp	Federal Supplement (United States)
F Supp 2d	Federal Supplement, Second Series (United States)
F Supp 3d	Federal Supplement, Third Series (United States)
FC	Federal Court (Canada)
FCA	Federal Court of Australia
FCC	Federal Communications Commission (United States)
FCJ	Federal Court Judgments (Canada)
FCR	Federal Court Reports (Australia/Canada)
FF	Feuille fédérale suisse (Switzerland)
GBP	British pound
GDP	gross domestic product
GG	Government Gazette (South Africa)

GHG	greenhouse gas
GN	Government Notice (South Afrida)
GOQ	Gazette officielle du Québec
GP	Gauteng Division, Pretoria (South Africa)
Gt	gigatonne
GWh	gigawatt hour
GWP100	global warming potential with a time horizon of hundred years
GWP20	global warming potential with a time horizon of twenty years
HKCFAR	Hong Kong Court of Final Appeal Reports
HL	House of Lords (United Kingdom)
ICAO	International Civil Aviation Organization
ICJ	International Court of Justice
ICJ	Rep Reports of Judgments, Advisory Opinions and Orders of the International Court of Justice
IEA	International Energy Agency
IEHC	High Court of Ireland
IEMA	Institute of Environmental Management and Assessment
IESC	Supreme Court of Ireland
ILC	International Law Commission (United Nations)
ILDC	Oxford Reports on International Law in Domestic Courts
ILM	International Legal Materials
IMF	International Monetary Fund
IMO	International Maritime Organization
Inc	incorporated
IPCC	Intergovernmental Panel on Climate Change
IR	Irish Reports
ITLOS	Reports of Judgments, Advisory Opinions and Orders of the International Tribunal for the Law of the Sea
IWG-SCC	Interagency Working Group on the Social Cost of Carbon
IWG-SCGHG	Interagency Working Group on the Social Cost of Greenhouse Gases
J	Justice
JIC	Justis Irish Cases
JJ	Justices
JORF	Journal officiel de la République française (France)
KB	King's Bench division (United Kingdom)
KEELC	Kenya Environment and Land Court
KEPCO	Korea Electric Power Corporation
KLR	Kenya Law Reports
kt	kilotonne
LC	Legislative Council (Hong Kong)
LCA	life cycle assessment
LGERA	Local Government and Environmental Reports of Australia
LLC	limited liability company

LN	Legal notice (Hong Kong)
LNG	liquefied natural gas
Ltd	limited
LULUCF	land use, land-use change and forestry
MEPA	Montana Environmental Policy Act
Minn Ct App	Minnesota Court of Appeals
Minnesota Stat	Minnesota Statutes
MPR	Minnesota Public Radio
Mt	megatonne
NYS 3d	New York Supplement, Third Series
NDC	nationally determined contribution
NEPA	National Environmental Policy Act (United States)
NET	National Environmental Tribunal (Kenya)
NGO	non-governmental organization
NGT	National Green Tribunal (India)
NR	National Reporter (Canada)
NSW	New South Wales
NSWCA	New South Wales Court of Appeal
NSWLEC	New South Wales Land and Environment Court
NSWLR	New South Wales Law Reports
NT	Northern Territory
NY	New York
NZCA	Court of Appeal of New Zealand
NZEnvC	Environment Court of New Zealand
NZHC	High Court of New Zealand
NZLR	New Zealand Law Reports
NZPD	New Zealand Parliamentary Debates
NZRMA	New Zealand Resource Management Appeals
NZSC	Supreme Court of New Zealand
OECD	Organisation for Economic Co-operation and Development
OJ C	Official Journal of the European Union: Courts
OJ L	Official Journal of the European Union: Legislation
OUP	Oxford University Press
PCIJ	Permanent Court of International Justice
ppm	parts per million
PTSR	Public and Third Sector Law Reports
PWh	petawatt hour
QCA	Queensland Court of Appeal
QLC	Queensland Land Court
Qld	Queensland
RIAA	Reports of International Arbitral Awards
RO	Recueil officiel du droit fédéral (Switzerland)
s	section
SA	South Australia
SBC	Statutes of British Columbia

SC	Statutes of Canada
SCC	social cost of carbon/Supreme Court of Canada
SC-GHG	social cost of greenhouse gases
SEA	strategic environmental assessment
SEPP	State Environmental Planning Policy (New South Wales)
SEQR	State Environmental Quality Review (New York)
SOG	Supplement to the Official Gazette (Nigeria)
SOR	Statutory Orders and Regulations (Canada)
SS	single-screw steamship
SSD	State significant development (New South Wales)
STB	Surface Transportation Board
Sup Ct	Supreme Court
t	tonne
Tas	Tasmania
Tt	teratonne
TWh	terawatt hour
UKSC	Supreme Court (United Kingdom)
UN	United Nations
UNCLOS	UN Convention on the Law of the Sea
UNECE	UN Economic Commission for Europe
UNEP	UN Environment Programme
UNFCCC	UN Framework Convention on Climate Change
UNTS	UN Treaty Series
USC	Code of Laws of the United States of America
USD	US dollar
Vic	Victoria
vol	volume
VR	Victorian Reports
WA	Western Australia
WALHI	Wahana Lingkungan Hidup Indonesia (The Indonesia Forum for Environment)
WL	Westlaw
WLR	Weekly Law Reports
ZAGPPHC	South Africa North Gauteng High Court, Pretoria
ZAWT	Water Tribunal of South Africa

I

Introduction

A. Assessing the Climate Impact of Proposed Activities

Most governments have procedures in place to identify, predict, evaluate, and mitigate the effects proposed activities would have on the environment. These environmental assessment (EA) procedures may apply to a wide range of activities, notwithstanding whether they are proposed by public authorities or private companies, including projects, programmes, plans, and policies, or even legislation. An EA procedure generally includes a scientific assessment of the impacts of the activity and some opportunities for the public to express their views about these impacts as a way to inform a decision of a governmental agency. EAs have traditionally focused on localized environmental impacts such as land, air, or water pollution, which may affect surrounding communities, but they have also increasingly been used as a way to address transboundary and global environmental issues, ranging from the pollution of global commons (eg the marine environment or the Antarctic) to the exacerbation of climate change.

This book aims at better understanding whether and how EA can be used as a tool to mitigate climate change—in short, whether and how 'climate assessment' (CA) can be implemented. Climate change mitigation refers to efforts to reduce greenhouse gas (GHG) emissions by sources (eg from the combustion of fossil fuels) and to enhance sinks and reservoirs of GHGs (eg forests). States have cooperated on climate change mitigation for over three decades, but global GHG emissions continue to increase. In this context, EA may appear as a valuable opportunity for governments and society to scrutinize the implications of important decisions for climate change mitigation. For instance, an EA procedure can flag a project's disproportional GHG emissions and identify alterations or alternatives that would avoid or reduce these emissions.

Yet, one faces multiple issues when designing and implementing such CA frameworks. For one, climate change is very different from most localized environmental issues that are typically the focus of EA procedures: GHG emissions have only a global, long-term, multi-causal, cumulative, and diffuse impact. Further, it may be difficult to determine how a potential decision would affect global GHG emissions, given the risk of carbon leakage: preventing the implementation of a project in one jurisdiction could cause a similar project to be implemented elsewhere, with a similar effect on the climate system. Another issue is that, as specific standards are still largely missing for what concerns GHG emissions, there is no obvious

Environmental Assessment as a Tool for Climate Change Mitigation. Benoit Mayer, Oxford University Press.
© Benoit Mayer 2024. DOI: 10.1093/oso/9780198939184.003.0001

threshold to assess the 'significance' of an individual activity, whether this is at the stage of determining the need to conduct a CA or, eventually, when deciding whether the activity should be permitted. And while EA procedures often seek the participation of the affected public, no population is directly affected as a predictable and proximate consequence of an activity's GHG emissions. Altogether, one could wonder whether CA is likely to be an effective tool for climate change mitigation, and how CA frameworks should be designed in order to maximize CA's potential benefits.

These questions have been addressed through an expanding body of case law, legislation, grey literature, and academic publications. Yet these treatments tend to be focused on a jurisdiction taken in isolation of developments unfolding elsewhere. By contrast, this book takes stock of the experience of multiple jurisdictions to enrich the reflection. It identifies common difficulties and innovative solutions, weighs arguments and objections, and considers observations and suggestions that originate from political, administrative, judicial, and academic authorities, in relation to CA procedures established and implemented by multiple national, sub-national, supra-national, or non-governmental institutions across the world. This broad comparison informs a policy-design reflection on whether and, if so, how EA should be implemented as a tool for climate change mitigation. In other words, by taking stock of the way CA is designed and implemented across the world, this book aims at policy recommendations that could help address outstanding issues and, more generally, tap the potentials of EA as a tool for climate change mitigation.

This first chapter provides a general introduction to the book. The three following sections present, respectively, the book's objective, method, and structure.

B. Objective

EA has increasingly been used, across the world, as a tool for climate change mitigation: a means to assess whether a proposed activity might exacerbate climate change, to consider ways to avoid or reduce this climate impact, and thus to inform competent authorities before they make their decision. Various alternatives or alterations of the proposed activity may be considered, and conditions may ultimately be imposed on the proposed activity with a view of limiting its climate impact. While a CA procedure can be conducted in isolation from the assessment of other environmental impacts, it is almost systematically integrated into a comprehensive EA procedure.

For instance, before deciding whether to authorize the construction of a new power plant, most governments would conduct (or require proponents to conduct) an EA. This EA would typically consider the impact that the project would have on the local environment, for instance on the quality of the air and the well-being of the surrounding populations. But this EA may also quantify the GHG emissions resulting from the construction and operation of the power plant. At times, this procedure could lead a government to accept (or force it to concede) that the project's climate impact is unacceptable and that the project should not be authorized. More frequently, CA procedures help governments to identify cost-effective measures that may reduce the power plant's overall climate impact, if only marginally. This could be done, for instance, by reducing methane leakage or by adding features to the project that would allow waste heat to be utilized for a district heating project.

CA aims to allow political communities to reach decisions that, in their own view, are preferable to those they would have reached without this tool. It aims to do so by ensuring the availability of comprehensive and reliable information about the impacts of proposed activities and by providing the opportunity for the public to voice their concerns. Like EA, CA can be approached as primarily a technocratic mechanism aimed at optimizing decisions, as an institutional tool to mediate between competing interests, and (perhaps most promisingly) as a more pluralistic mechanism to facilitate the emergence of a consensus within a political community. One way or another, CA seeks to ensure that, rather than indiscriminately accepting or rejecting all GHG-emitting activities, society and government weigh climate impacts along with other factors before deciding on a case-by-case basis whether a proposed activity is acceptable and under what conditions.

Although CA has been implemented in most jurisdictions across the world, many questions remain about the relevance and effectiveness of this tool. These questions relate in particular to the observation that climate impacts differ in significant ways from the environmental impacts that have traditionally fallen within the scope of EA frameworks. The latter typically apply to the relatively concrete, direct, and foreseeable consequences of proposed activities, which tend to unfold at a local scale and within a relatively short time span. For instance, an activity's air, water, and soil pollution may have proximate consequences for local ecosystems and populations.[1] Even when EA has been applied in a transboundary context, this was often in relation to such concrete and local impacts—with regard to 'normal' projects that just happened to be situated near an international border. Thus, the decision in *Pulp Mills* was concerned with the impact of particular projects 'on the waters of the river' Uruguay and 'other components of the ecosystem of the watercourse such as its flora, fauna, and soil'.[2]

By contrast, an activity's GHG emissions does not cause any such proximate, concrete effects. Climate change is a major global crisis, but it results from the cumulative effect of millions of ordinary activities implemented year after year;

---

[1] See eg Robert Gibson and others, *Sustainability Assessment: Criteria and Processes* (Routledge 2005) 1–13.

[2] *Pulp Mills on the River Uruguay (Argentina v Uruguay), Judgment* [2010] ICJ Rep 2010 14 [188]. See also *Pulp Mills on the River Uruguay (Argentina v Uruguay)*, Memorial of Argentina (15 January 2007), vol I, 139–49.

concrete harm on individuals or ecosystems cannot be attributed to any individual activity.[3] A GHG-emitting activity causes only a very tiny increment in the atmospheric concentrations of GHGs, with vanishingly small effects on, among other things, global average temperature or the likelihood of extreme weather events. On the other hand, even this very tiny effect could potentially be considered significant on the ground that it unfolds on a planetary scale and on the long term. As such, activities emitting large amounts of GHGs can be said to have a significant 'climate impact', if only in this more abstract and statistical sense of the word 'impact'—as an extremely tiny exacerbation of an extraordinarily broad set of risks, rather than as a direct effect on a particular ecosystem or population.

This caveat casts some doubts on the relevance and effectiveness of EA as a tool for climate change mitigation. Tools commonly used to address environmental issues cannot simply be assumed to be as appropriate and effective in addressing climate change.[4] As climate impacts differ significantly from the environmental impacts that have typically been the focus of EA frameworks, one may question whether these frameworks are able to address these impacts in an effective manner. Notwithstanding whether EA is approached as primarily an instrument aimed at enabling a rational decision by competent authorities, to facilitate choices by a political community, or to empower vulnerable populations in processes where they would otherwise be voiceless, one could wonder whether it can fulfil these goals as well, or at all, when it is applied to climate impacts.

These questions have been the object of scholarly and political debates that, so far, have mainly taken place in relative isolation across national jurisdictions.[5] Thus, most publications on the topic are journal articles and book chapters reacting to national developments, in particular in Australia,[6] Canada,[7] the

---

[3] See discussion in Benoit Mayer, 'Attribution Science and the Fate of Climate Litigation' (2022) 13 Global Policy 831.

[4] See *Verein Klimaseniorinnen Schweiz v Switzerland* App no 53600/20 (ECHR, 9 April 2024) [422].

[5] A few exceptions exist where authors of journal articles have looked at CA from a comparative perspective. See eg Brian Preston, 'Contemporary Issues in Environmental Impact Assessment' (2020) 37 Environmental & Planning Law Journal 423; Charles H Eccleston, 'Assessing Cumulative Significance of Greenhouse Gas Emissions: Resolving the Paradox – the Sphinx Solution' (2010) 12 Environmental Practice 105, 111; Philip Byer and others, 'International Best Practice Principles: Climate Change in Impact Assessment' (International Association for Impact Assessment, Special Publication Series No 8, March 2018); Rose Mayembe and others, 'Integrating Climate Change in Environmental Impact Assessment: A Review of Requirements across 19 EIA Regimes' (2023) 869 Science of the Total Environment (article #161850) 1–13, at 7.

[6] Eg Julia Dehm, 'Coal Mines, Carbon Budgets and Human Rights in Australian Climate Litigation: Reflections on *Gloucester Resources Ltd v Minister for Planning and Environment*' (2020) 26 Australian Journal of Human Rights 244; Elena Aydos and others, 'Rocky Hill: A Legal Breakthrough in the Consideration of Climate Change and Social Impacts of Coal Mines' (2020) 14 Carbon & Climate Law Review 98; Victoria McGinness and Murray Raff, 'Coal and Climate Change: A Study of Contemporary Climate Litigation in Australia' (2020) 37 Environmental & Planning Law Journal 87; Kierra Parker, 'Litigating at the Source: Attributing Climate Change Impacts to Coal Mines' (2020) 37 Environmental & Planning Law Journal 67.

[7] Eg Meinhard Doelle, 'Integrating Climate Change Mitigation into the Impact Assessment Act', in Meinhard Doelle and Andrew Sinclair (eds), *Impact Assessment in Transition: A Critical Review of the Canadian Impact Assessment Act* (Irwin Law 2021) 277, 277; Meinhard Doelle, 'Integrating

United Kingdom,[8] and the United States[9]—or, beyond these 'usual suspects',[10] in China,[11] South Africa,[12] and South Korea,[13] among others. While this literature

Climate Change into Environmental Impact Assessment: Key Design Elements', in Francesco Sindico, Stephanie Switzer, and Tianbao Qin (eds), *The Transformation of Environmental Law and Governance* (Edward Elgar 2021) 111, 111; Flavia Castro, 'Canada's Climate Change Mitigation Commitments and the Role of the Federal Impact Assessment Act' (2020) 33 Journal of Environmental Law & Practice 211.

[8] Eg Stephanie Hands and Malcom D Hudson, 'Incorporating Climate Change Mitigation and Adaptation into Environmental Impact Assessment: A Review of Current Practice within Transport Projects in England' (2016) 34 Impact Assessment & Project Appraisal 330; Daria Shapovalova, 'Climate Change and Oil and Gas Production Regulation: An Impossible Reconciliation?' (2023) 26 Journal of International Economic Law 817; Emily Webster, 'The Planning Regime and Climate Change Mitigation' (2022) 4 Conveyancer and Property Lawyer 415; Vicki Elcoate, 'Climate Change Mitigation and the Planning Process – Is There a Gap in the Law Which Allows Mineral Planning Authorities to Ignore the Real Climate Impacts of Onshore Oil and Gas Applications?' *UK Environmental Law Association* (February 2021) <www.wealdactiongroup.org.uk/?mdocs-file=2511>.

[9] Eg Fred Mauhs, 'Cumulative Impact Analysis in NEPA Climate Assessments' (2022) 39 Pace Environmental Law Review 211; Arnold W Reitze, 'Dealing with Climate Change under the National Environmental Policy Act' (2018) 43 William & Mary Environmental Law & Policy Review 173; Nicole Rushovich, 'Climate Change and Environmental Policy: An Analysis of the Final Guidance on Greenhouse Gas Emissions and the Effects of Climate Change in National Environment Policy Act Reviews' (2018) 27 Boston University Public Interest Law Journal 327; Jayni Foley Hein and Natalie Jacewicz, 'Implementing NEPA in the Age of Climate Change' (2020) 10 Michigan Journal of Environmental & Administrative Law 1; Michael Burger, Romany Webb, and Jessica Wentz, *Incorporating Climate Change in NEPA Reviews: Recommendations for Reform* (Sabin Center for Climate Change Law 2022); Lauren Giles Wishnie, 'NEPA for a New Century: Climate Change & the Reform of the National Environmental Policy Act' (2008) 16 New York University Environmental Law Journal 628; Barry Kellman, 'NEPA Review of Climate Change' (2016) 46 Environmental Law Reporter News & Analysis 10378; Maureen O'Dea Brill, 'Assessing the Scope of the National Environmental Policy Act: Recent Attempts by Environmentalists to Add Climate Change Considerations into NEPA Review' (2014) 54 Natural Resources Journal 409; Caleb W Christopher, 'Success by a Thousand Cuts: The Use of Environmental Impact Assessment in Addressing Climate Change' (2008) 9 Vermont Journal of Environmental Law 549.

[10] See Francesco Sindico, Makane Moïse Mbengue, and Kathryn McKenzie, 'Climate Change Litigation and the Individual: An Overview', in Francesco Sindico and Makane Moïse Mbengue (eds), *Comparative Climate Change Litigation: Beyond the Usual Suspects* (Springer 2021) 5, 5 (identifying Australia and the United States as some of the 'usual suspects' in the context of climate litigation). Generally, the academic literature in climate law tends to focus disproportionately on English-speaking developed countries of a common-law tradition. See also Jacqueline Peel and Jolene Lin, 'Transnational Climate Litigation: The Contribution of the Global South' (2019) 113 American Journal of International Law 679, 681; John H Knox and Christina Voigt, 'Introduction to the Symposium on Jacqueline Peel and Jolene Lin: Transnational Climate Litigation: The Contribution of the Global South' (2020) 114 AJIL Unbound 35, 35.

[11] Eg Xiangbai He, 'Integrating Climate Change Factors within China's Environmental Impact Assessment Legislation: New Challenges and Developments' (2013) 9 Law, Environment & Development Journal 50; Qi Gao, 'Mainstreaming Climate Change into the EIA Procedures: A Perspective from China' (2017) 10 International Journal of Climate Change Strategies & Management 342; Xiangbai He, 'Mitigation and Adaptation through Environmental Impact Assessment Litigation: Rethinking the Prospect of Climate Change Litigation in China' (2021) 10 Transnational Environmental Law 413; Yiting Yang and others, 'Integrating Climate Change Factor into Strategic Environmental Assessment in China' (2021) 89 Environmental Impact Assessment Review (article #106585) 1–8.

[12] Eg Melanie Jean Murcott and Clive Vinti, 'The Judge-Made "Duty" to Consider Climate Change in South Africa' (2024) 16 Journal of Human Rights Practice 125.

[13] Eg Jeonghwa Yi and Theo Hacking, 'Incorporating Climate Change into Environmental Impact Assessment: Perspectives from Urban Development Projects in South Korea' (2011) 21 Procedia Engineering 907; Kyeong-Tae Kim and Ik Kim, 'The Significance of Scope 3 GHG Emissions in Construction Projects in Korea: Using EIA and LCA' (2021) 9 Climate 33.

plays important roles, comparative perspectives could help achieve a better understanding of CA. When considering whether and, if so, how to apply EA frameworks as a tool for climate change mitigation, most jurisdictions face similar questions, and any jurisdiction could benefit from considering ideas, developments, and experience from others. By comparison to shorter pieces, a monographic treatment provides room for a more systematic and coherent treatment of the topic.

Thus, this book aims at enhancing our collective understanding of the modalities and, ultimately, the opportunity of applying EA as a tool for climate change mitigation by putting jurisdictional developments and national debates in a global comparative perspective. More specifically, three interrelated tasks are undertaken. First, the book documents the use of EA as a tool for climate change mitigation across the world. Second, it analyses the main conceptual and practical issues facing CA. Third, it evaluates the potential solutions that jurisdictions have contemplated or implemented to address these issues.

Each of the book's chapters contributes to these three tasks. Chapter II provides a general overview of the practice of EA and CA, followed by a preliminary review of the objections that have been raised against the latter. Chapter III considers whether the 'significance' of climate impacts can be appraised in a meaningful fashion, whether at the stage of screening and scoping or, subsequently, to inform a substantive decision on the merits of the proposed activity. Chapter IV considers the difficulty of predicting and assessing the climate impacts of decisions in spite of indirect effects such as market substitution and carbon leakage. Chapter V looks more generally at whether CA can usefully inform agencies' final decisions on whether to approve a proposed activity and whether it can contribute to the mitigation of climate change. Chapter VI concludes by shedding light on some good practices.

The book does not suggest that there is one 'right' or 'best' method to apply EA as a tool for climate change mitigation. There is no denial that CA, like EA in general, 'must be context-specific and tailored to the socio-political and cultural setting where it takes place',[14] and that what works best in one jurisdiction may not necessarily work in another. In some regards, however, the book is able to identify internal inconsistencies in some national approaches and more promising approaches in others. If the latter can be characterized as 'good practices', it is in the sense that they are approaches worth considering by other jurisdictions, and not to suggest that these approaches provide a turn-key solution universally applicable.

While the book focuses on the use of EA as a tool for climate change mitigation, it may also be more broadly relevant to reflections on the role of EAs in addressing other cumulative environmental issues, in particular those of a global nature such as the depletion of the ozone layer, the pollution of low-earth orbital space, or even

---

[14] Angus Morrison-Saunders, *Advanced Introduction to Environmental Impact Assessment* (Edward Elgar 2018) 63.

antibiotic resistance. So far, these other global cumulative environmental issues have attracted little attention in EA law and practice, partly because the activities contributing to these global cumulative environmental issues do not frequently fall within the scope of EA frameworks.[15] This, however, may come to change, in particular as CA law and practice prompt various stakeholders to become more aware of the potential of EA to address such global cumulative environmental issues. Already, some questions have been asked, for instance, on the possibility of applying the US National Environmental Policy Act to satellite licensing as a way to limit pollution in low-earth orbital space.[16]

On the other hand, the theme of this book needs to be distinguished from three other topics on the relations between EA and climate change. First, EA can also be used as a tool to *adapt* to the impacts of climate change—for instance, to avoid urbanization in areas that may become increasingly prone to floodings in the context of a warming climate or to ensure that a project's water consumption is acceptable in light of a future effect of climate change on water scarcity. It is frequent for EA instruments to consider climate change mitigation and adaptation in the same breath,[17] and this often leads to some confusion.[18] However, very different issues arise when EA is used as a tool for climate change adaptation or mitigation. Adaptation calls for an understanding of the changing conditions in which the activity would unfold, such as sea-level rise and the increasing severity of heatwaves.[19] This may raise important methodological questions, for instance regarding the choice of climate model and the assumptions about the implementation and effectiveness of future action on climate change mitigation. Integrating climate change mitigation in EA procedures, by contrast, raises more fundamental

---

[15] But see eg 49 CFR (2024) § 1105.7(e)(5)(iii), applicable to the transportation of ozone-depleting materials.

[16] See eg US Government Accountability Office, 'Satellite Licensing: FCC Should Reexamine its Environmental Review Process for Large Constellations of Satellites' (GAO-23-105005, November 2022).

[17] See Directive 2014/52/EU of the European Parliament and of the Council of 16 April 2014 amending Directive 2011/92/EU on the assessment of the effects of certain public and private projects on the environment, [2014] OJ L 124/1, preamble para 13 (requiring the assessment of 'the impact of projects on climate (for example greenhouse gas emissions) and their vulnerability to climate change'); Meeting of the Parties to the Convention on Environmental Impact Assessment in a Transboundary Context and Meeting of the Parties to the Protocol on Strategic Environmental Assessment, Decision IX/2–V/2, 'Workplan for 2024–2026', ECE/MP.EIA/2023/1–ECE/MP.EIA/SEA/2023/1 (25 September 2023) 8, Annex I § III.B.1 (calling for thematic workshops or seminars on '[c]limate assessment/proofing' as a single topic); Council on Environmental Quality, 'National Environmental Policy Act Implementing Regulations Revisions Phase 2' (1 May 2024) 89 Federal Register 35442, 35575 (to be codified as 42 USC § 1508.1(i)(4)) (clarifying that 'effects' include 'climate change-related effects, including the contribution of a proposed action and its alternatives to climate change, and the reasonably foreseeable effects of climate change on the proposed action and its alternatives').

[18] See eg *AquAlliance v US Bureau of Reclamation* (ED California 2018) 287 F Supp 3d 969, 1030 (noting that an attempt of the defendants to link mitigation with adaptation requirements 'adds nothing but confusion to the discussion').

[19] See eg EU Commission, 'Guidance on Integrating Climate Change and Biodiversity into Environmental Impact Assessment' (2013) 30–31.

questions about the nature and function of EA and its potential to address global impacts, one activity at a time.

Second, EA as a tool for climate change mitigation needs to be distinguished from the application of EA as a tool to avoid or reduce the adverse impacts of climate action.[20] Measures and policies aimed at mitigating climate change or at adapting to its impacts may cause environmental harm, and EA can help to avoid or reduce this harm.[21] States have committed to 'employ appropriate methods, for example impact assessments', to prevent or reduce the adverse impacts of 'projects and measures undertaken ... to mitigate or adapt to climate change'.[22] In many cases, however, this commitment only points to what states are already doing: most relevant measures and policies already fall within the scope of EA frameworks. A hydroelectric dam project, for instance, must undergo the same EA procedure applicable to any other large project notwithstanding whether the project aims at mitigating climate change by producing clean energy, at adapting to it by ensuring a better management of freshwater, or merely at producing cheap electricity to foster economic development. At times, lawmakers or governments have had to add to existing list of activities subject to EA requirements some new types of projects implemented to mitigate climate change, such as wind power plants[23] and facilities for the geological injection and storage of carbon dioxide,[24] but these additions do not affect any structural feature of EA frameworks.

Third, this book does not systematically explore the *use of exemption* from EA requirements as a tool for climate change mitigation. Concerns have emerged that existing EA requirements and the litigation they result in may delay urgent climate action. In a context of a perceived 'climate emergency',[25] some governments

---

[20] See eg Chiara Tea Antoniazzi, 'Strengthening the Complaint Mechanisms of Multilateral Climate Funds and Carbon Markets: A Critical Step Towards a Human Rights-Based Green Transition' (2023) 32 Review of European, Comparative & International Environmental Law 173.

[21] See Elizabeth Fisher, 'Law and Energy Transitions: Wind Turbines and Planning Law in the UK' (2018) 38 Oxford Journal of Legal Studies 528. See also Conseil d'État (State Council) (6ème–1ère chambres réunies), 18 December 2017, 397923, ECLI:FR:CECHR:2017:397923.20171218 (holding that a regional plan on climate change mitigation had to be subjected to an EA procedure); Conseil d'État (State Council) (6ème chambre), 16 May 2018, 408887, ECLI:FR:CECHS:2018:408887.20180516 (reaffirming the same finding).

[22] UN Framework Convention on Climate Change (opened for signature 9 May 1992, entered into force 21 March 1994) 1771 UNTS 107 (UNFCCC), art 4(1)(f).

[23] See eg Environmental Impact Assessment Ordinance (Amendment of Schedules 2 and 3) Order (2023) LN 77/2023, s 3(20) (Hong Kong).

[24] See eg Environmental Planning and Assessment Regulation 2021 (NSW) s 28; Environmental Protection Act 1994 (Qld) s 107(b); Environmental Protection Regulations 2021 (Vic) Sch 1 item 71; Directive 2011/92/EU of the European Parliament and of the Council of 13 December 2011 on the assessment of the effects of certain public and private projects on the environment, [2012] L 26/1, consolidated as of 15 May 2014, Annex I para 22.

[25] On the proclamation of a 'climate emergency', see eg 'Climate Emergency Declarations in 2,351 Jurisdictions and Local Governments Cover 1 Billion Citizens' (*Climate Emergency Declaration*, 22 February 2024) <https://climateemergencydeclaration.org/climate-emergency-declarations-cover-15-million-citizens/>. For a critical view on this political discourse, see Mike Hulme, 'Climate Emergency Politics Is Dangerous' (2019) 36 Issues in Science & Technology 23.

have decided to expedite the approval of some categories of projects. For instance, New York State has sought to 'take appropriate action to ensure that … new renewable energy generation projects can be sited in a timely and cost-effective manner',[26] including by setting aside the application of environmental review procedures that would be 'unreasonably burdensome in view of' the state's goals on climate change mitigation.[27] Similarly, France adopted a series of statutory provisions aimed at accelerating the approval of renewable energy projects, including by establishing a presumption that the benefits of these projects justify some environmental impacts,[28] shortening the period for the review of EA studies,[29] and limiting the circumstances in which courts can set aside the approval of a project.[30] These measures may facilitate climate change mitigation, not by applying EA procedures but, to the contrary, by exempting some projects from some requirements. As such, the questions this raises are different from those discussed in the book.

C. Method

This book reviews EA law and practice from a global comparative perspective with the view of understanding what issues are faced when applying EA as a tool for climate change mitigation, determining what solutions are contemplated to address these issues, and making policy recommendations. The research relies on a broad global survey of CA law and practice supplemented by multiple case studies and interdisciplinary policy analysis.

The global survey identifies statutory and regulatory reforms, judicial decisions, and guidance documents addressing the application of EA as a tool for climate change mitigation. This survey aspires to identify relevant developments wherever they occur, including beyond the jurisdictions most often considered in the climate law literature (eg the European Union (EU), the United States, and Australia). However, such a survey is inevitably limited by practical constraints, including language barriers, the difficulty of accessing copyright materials, and the difficulty of understanding developments taking place in unfamiliar legal systems.

Multiple tools have been used to surmount these difficulties. In addition to commercial databases and institutional websites, the survey has relied on online

---

[26] 2020 New York Sess Laws ch 58 (S 7508-B) (McKinney), part JJJ, § 2(2)(a).

[27] New York Exec Law (McKinney 2024) § 94-c(5)(2). See also *Town of Copake v New York State Office of Renewable Energy Siting* (NY App Div 2023) 191 NY S 3d 181.

[28] Loi 2023-175 du 10 mars 2023 relative à l'accélération de la production d'énergies renouvelables, JORF 11 mars 2023, art 19.

[29] Ibid, art 7(2).

[30] Ibid, art 23.

databases listing national decisions and instruments from multiple jurisdictions,[31] secondary sources, scholarly literature, press releases, and news reports to identify relevant developments. Automatic translation tools have also been used, in particular Google Translate, supplemented at times by bilingual dictionaries. As a whole, relevant developments could be identified in about a hundred jurisdictions, including national and subnational governments as well as a few international organizations. Nonetheless, these practical constraints have prevented a comprehensive documentation of developments in some parts of the world, perhaps most obviously in Latin American and Russian-speaking countries.

This broad preliminary survey has been supplemented by more thorough research into particular developments. These case studies seek a better understanding of policy arguments that were being made by government agencies, lawmakers, courts, and other stakeholders, so as to enrich the evaluation of potential solutions. While the global survey largely focuses on the present state of CA law and practice, some case studies add a historical perspective. Others look further into the motivation of relevant stakeholders, for instance by considering party submissions in court proceedings or parliamentary deliberations in relation to statutory developments. Yet other case studies review EA studies, submissions by the public, the decisions that followed, and any other evidence of the real-world outcome of individual procedures.

These case studies focus on developments that are remarkable, one way or another. These include innovative policy solutions, such as the use of the social cost of carbon to assess the significance of climate impacts, but also developments that are out of the ordinary in any other way, as is the case of the few decisions by national lawmakers to exclude the application of EA as a tool for climate change mitigation.

For the same practical reasons explained above, the selection and depth of case studies depends in part on practical constraints. Most of the case studies are in developed country jurisdictions, such as Australia (including New South Wales, Queensland, and Victoria), Austria, Canada, Czechia, France, New Zealand, Norway, the United Kingdom, and the United States (including California, Montana, New York State, and New York City), although important developments were also be documented in developing countries such as China (including Hong Kong), India, Indonesia, Kazakhstan, Kenya, and South Africa.

All in all, this book implements a classical, 'functionalist' approach to comparative legal methodology: it considers the solutions that various jurisdictions have adopted to the common problems they encounter when applying EA as a tool for

---

[31] See Climate Change Litigation Databases (Sabin Center for Climate Change Law 2024) <https://climatecasechart.com/>; Climate Change Laws of the World (Grantham Research Institute on Climate Change and the Environment 2024) <https://climate-laws.org/>; EcoLex: The Gateway to Environmental Law (International Union for the Conservation of Nature, UNEP, and Food and Agricultural Organization 2024) <www.ecolex.org/>; FAOLEX Database (Food and Agriculture Organization 2024) <www.fao.org/faolex/en/>.

climate change mitigation.[32] Functionalist comparative studies can be conducted on matters in relation to which 'the legal system of every society faces essentially the same problems, and solves these problems by quite different means'.[33] The topic of this book lends itself particularly well to such functionalist comparison. This is because EA is a nearly universal mechanism which, despite important differences in national designs, can be (and has often been)[34] approached from a comparative perspective. Further, most EA frameworks have been applied as tools for climate change mitigation, prompting many lawmakers, courts, and agencies throughout the world to seek solutions to similar issues. Yet, in multiple jurisdictions, CA law and practice remain very rudimentary, and no jurisdiction seems to have found a perfectly satisfactory solution to all issues.

In such circumstances, it is well recognized that functionalist comparative analysis can shed light on 'a reservoir of different solutions' to common issues,[35] thus providing inspiration to lawmakers and decision-makers struggling with these issues. Ralf Michaels, for instance, pointed out that functionalist comparative analysis 'can show alternatives and provide some information and thereby greatly improve [a] policy decision'.[36] This type of research is particularly important in a field of law and policy that is rapidly expanding and where fundamental decisions remain to be made, such as climate change.[37] A premise of this book is that such functionalist comparative analysis can enrich academic and political debates on the subject by prompting a better understanding of the prospects of applying EA as a tool for climate change mitigation.

This potential can be illustrated with a brief example. When applying EA as a tool for climate change mitigation, many national decision-makers have had difficulties defining a consistent way to appraise the significance of the climate impacts of a proposed activity.[38] A few jurisdictions, however, have made important breakthroughs in this regard, for instance by relying on local mitigation targets[39]

---

[32] Mark Van Hoecke, 'Methodology of Comparative Legal Research' [2015] Law & Method 11 ('In its most common understanding, the functional method doesn't compare primarily rules, but solutions to practical problems with conflicting interests').

[33] Konrad Zweigert, *Introduction to Comparative Law* (3rd revised edn, Clarendon 1998) 34.

[34] Thus, comparative perspectives are frequent in the literature on EA law and governance. See eg Neil Craik, 'The Assessment of Environmental Impact', in Emma Lees and Jorge Vinuales (eds), *The Oxford Handbook of Comparative Environmental Law* (OUP 2019) 876, 877; Chris Wood, *Environmental Impact Assessment: A Comparative Review* (2nd edn, Routledge 2013); John Glasson and Riki Therivel, *Introduction to Environmental Impact Assessment* (5th edn, Routledge 2019); Morrison-Saunders (n 14); Kevin Hanna (ed), *Routledge Handbook of Environmental Impact Assessment* (Routledge 2022).

[35] Jonathan Hill, 'Comparative Law, Law Reform and Legal Theory' (1989) 9 Oxford Journal of Legal Studies 101, 102.

[36] Ralf Michaels, 'The Functional Method of Comparative Law', in Mathias Reimann and Reinhard Zimmermann (eds), *The Oxford Handbook of Comparative Law* (2nd edn, OUP 2019) 345, 381.

[37] See Lisa Vanhala, 'The Comparative Politics of Courts and Climate Change' (2013) 22 Environmental Politics 447; Michael Mehling, 'The Comparative Law of Climate Change: A Research Agenda' (2015) 24 Review of European, Comparative and International Environmental Law 341.

[38] See generally Chapter III.

[39] See eg CEQ, 'National Environmental Policy Act Guidance on Consideration of Greenhouse Gas Emissions and Climate Change' (9 January 2023), 88 Federal Register 1196, 1203; European

or on economic valuations of the adverse social value of an additional tonne of GHG emissions.[40] Transplanting either of these approaches to every jurisdiction may not be desirable or even possible—what works best in relation to a particular EA instrument and in a specific legal, political, and cultural context may not work as well, or at all, elsewhere. Nonetheless, these solutions are at least worth contemplating in some other jurisdictions.

A difficult question regards the potential for such functionalist comparative research to lead to normative conclusions, for instance through the identification of good practices or the formulation of policy recommendations. Some authors have suggested that functionalist comparative legal analysis can contribute to singling out 'the better of several laws'[41] or, more precisely, the law that is 'better ... at performing a function'.[42] At times, one solution appears 'clearly superior' to the others.[43] Yet, most authors have also recognized that these evaluations risk relying on questionable assumptions about the function—or functions—that the laws should pursue.[44] For instance, Ralf Michaels observed that laws on car accidents aim both at ensuring the compensation of victims and at incentivizing careful driving, and that fulfilling one function often comes at the expense of another.[45]

These concerns are well-founded, but they do not categorically exclude the possibility of drawing normative conclusions from functionalist comparative analysis. Whether such normative conclusions can be drawn depends largely on the question that is being contemplated. By contrast to laws on car accidents, the application of EA as a tool for climate change mitigation pursues a relatively distinct principal goal in relative isolation from other policy considerations—it seeks to enable better decisions, understood as decisions that balance climate change mitigation with other policy goals in a way that society finds acceptable.[46] Overall,

---

Commission, 'Environmental Impact Assessment of Projects: Guidance on the Preparation of the Environmental Impact Assessment Report' (2017) 39.

[40] See eg CEQ (n 39) 1203; Environment and Climate Change Canada, 'Social Cost of Greenhouse Gas Estimates: Interim Updated Guidance for the Government of Canada' (April 2023) <www.canada.ca/en/environment-climate-change/services/climate-change/science-research-data/social-cost-ghg.html>.

[41] Michaels (n 36) 348.

[42] Ibid, 380. See also Jorge Vinuales, 'Comparative Environmental Law: Structuring a Field', in Lees and Vinuales (eds) (n 34) 3, at 13; James Gordley, 'The Functional Method', in Pier Giuseppe Monateri (ed), *Methods of Comparative Law* (Edward Elgar 2012) 107, 109; Konrad Zweigert, 'Méthodologie du droit comparé', in *Mélanges offerts à Jacques Maury* (Dalloz 1960) 579, 580–83.

[43] Zweigert, *Introduction to Comparative Law* (n 33) 47.

[44] Michaels (n 36) 381. See also Hill (n 35) 104; Manolis Kotzampasakis and Edwin Woerdman, 'The Legal Objectives of the EU Emissions Trading System: An Evaluation Framework' *Transnational Environmental Law* (forthcoming).

[45] Michaels (n 36) 380–81.

[46] Another relevant policy goal is to allow developments to take place as freely as possible with the view of protecting individual freedoms and foster economic development. This policy goal can justify arguments against EA generally, but not obviously against CA in particular, assuming that CA is an effective extension of EA.

because CA law and practice remain in their infancy, stricken by issues that often have not been properly thought through, the contrast between better and worse laws is far sharper and more obvious than in more mature legal fields such as the laws on car accidents.

Thus, at times, functionalist comparative legal analysis can advance academic or political debates on CA by challenging or even refuting certain theories. A somewhat simple example of this relates to the academic proposition that, due to the abstract and diffuse nature of climate impacts, any threshold of significance would be entirely arbitrary.[47] A logical implication of this theory is that, if national EA instruments do define thresholds of significance, these thresholds should be expected to differ extensively from one another. However, comparative analysis reveals that various jurisdictions have independently defined relatively similar thresholds of significance:[48] virtually every jurisdiction would find that an ordinary coal-fired power plant would cause a climate impact significant enough to justify a CA procedure, but that installing a new streetlight would not.[49] These observations confirm that, despite the absence of a clear line between impacts that are significant and those that are not, in practice, the decisions that have been made independently in various jurisdictions are broadly consistent and, therefore, do not appear to be entirely arbitrary.

More often, however, evaluating approaches to CA calls for further policy analysis. Policy analysis refers to a set of methodologies that aim, in particular, at weighing the policy options or underlying policy arguments with the view of formulating policy recommendations. Policy analysis can build upon comparisons between legal systems as a way to identify policy options that can then be evaluated.[50] To determine what policy option works best, as William Dunn observes, policy analysis 'is productively eclectic', in the sense that researchers can 'choose among a wide range of scientific methods ... so long as these yield reliable knowledge'.[51] Within the field of inquiry of this book, policy analysis may need to be informed by disciplines such as climate science, emission accounting, economics, and political science. Importantly, what works in a given context may or may not work in another context. Considering in what context a given good practice could realistically and effectively be transplanted requires further analysis beyond the scope of this book.[52]

Policy analysis is used in this book to weigh alternative policy options that are identified through comparative legal analysis. For instance, comparative legal

---

[47] See Alexander Zahar, 'Environmental Impact Assessment for Greenhouse Gas Emissions is Pie in the Sky', in Benoit Mayer and Alexander Zahar (eds), *Debating Climate Law* (CUP 2021) 297, 300–301.

[48] See Chapter III, subsection B.1 (showing that thresholds of significance are generally defined within the same order of magnitude).

[49] Cf Zahar (n 47) 301.

[50] See Eugene Bardach, *A Practical Guide for Policy Analysis* (7th edn, Sage 2023) 20, 133–37.

[51] William N Dunn, *Public Policy Analysis: An Integrated Approach* (6th edn, Routledge 2017) 3.

[52] See eg Bardach (n 50) 137–39.

analysis shows that, while a few jurisdictions have excluded GHG emissions from the scope of their EA instrument out of concern for a perceived redundancy with national cap-and-trade mechanisms,[53] other jurisdictions such as California and the EU have applied the two instruments simultaneously. Observing this contradiction may push one to question the likelihood of redundancies between EA and cap-and-trade mechanisms. Further analysis reveals that CA procedures routinely consider emissions that would not fall within the scope of most cap-and-trade mechanisms, including emissions taking place in sectors that are not regulated by these mechanisms and emissions occurring beyond the geographical scope of these mechanisms or beyond the timespan of existing caps, which suggests that concerns for policy redundancies are largely unjustified.[54] In this example, comparative legal analysis plays an important role in illuminating tensions and contradictions across jurisdictions, inviting further analysis that calls shaky assumptions into question.

The comparative methodology used in this book has some important limitations. Like any other comparative legal research project, it is 'necessarily unsystematic' due to 'the immense volume' of relevant materials and practical constraints limiting their accessibility.[55] As noted above, language barriers and difficulties in accessing documents inevitably limited the breadth of the global survey and constrained the selection of case studies, while cultural barriers may have impeded the analysis. While this book purports to identify global trends in the use of EA as a tool for climate change, it does not provide an exhaustive list of every development happening in any jurisdiction. Moreover, the breadth of the global survey has limited its depth, and case studies do not fully make up for this. Lastly, the method certainly involves certain biases, as some developments are more likely to be identified than others. For instance, the book is more likely to identify the application of an EA framework as a tool for climate change mitigation when this application was decided by a court of law in a widely reported decision than when it was made through informal administrative decisions on the scoping of successive EAs.[56]

Lastly, this book does not systematically discuss the context in which EA has been applied as a tool for climate change mitigation or in which the good practices it identifies have emerged. Functionalist analysis does not need to pay as much attention to context as other comparative legal methodologies, as the focus is on the existence of common issues rather than on the different circumstances in which

---

[53] See in particular Law amending the Environmental Code (3 December 2011) No 505-IV, art 34-1.2(13)–(14), modifying the Environmental Code (9 January 2007) No 212-III-ZRK, art 38.2(1). See also Resource Management (Energy and Climate Change) Amendment Act 2004 (New Zealand), Public Act 2004 No 2, s 3(b)(ii); but note that this provision was repealed by Resource Management Amendment Act 2020, ss 17–21, 35, 36.

[54] See discussion in Chapter II, subsection C.2(c).

[55] Hill (n 35) 111.

[56] But see Chapter II, subsection B.2(b) (identifying the evolution of administrative practice in China in the absence of any prominent decision).

these issues are addressed.[57] In turn, to the extent that the 'traditional' method of policy analysis on which this book relies may identify desirable ways forward, it does not 'provide information about how governments make use of such evidence or how political will for interventions that may be opposed might be generated'.[58] While acknowledging these limitations, this book hopes to inspire others to consider why certain developments take place in some jurisdictions and to assess how the good practices it identifies might be transposed to other contexts.

D. Structure and Key Findings

This section highlights the main focus and the key findings of each of the next chapters.

1. The concept of climate assessment

Chapter II engages with the question of *whether* EA has been used as a tool for climate change mitigation and starts considering whether it should be used in such a way. It does so by taking stock of CA law and practice and by analysing the main objections that have been made to this law and practice.

In particular, this chapter establishes that most jurisdictions have applied EA frameworks as a tool for climate change mitigation by approaching climate impacts like any other environmental impacts subjected to these procedural frameworks. Yet it also shows that the debate is not necessarily over. Many of the decisions applying EA as a tool for climate change mitigation are unsystematic, leaving fundamental issues open about the scope and modalities of CA.

The chapter further documents the objections that have been made to the use of EA as a tool for climate change mitigation. Several EA frameworks have not been applied as a tool for climate change mitigation, sometimes merely because of administrative inertia or based on constitutional competence-sharing provisions (eg EA frameworks implemented by subnational governments without competence to address climate change), but also at times in application of statutory provisions intended to exclude GHG emissions from the scope of EA frameworks. These decisions have sometimes been based on concerns for policy overlaps between EA and other climate policies, in particular cap-and-trade mechanisms, although, as

---

[57] See Van Hoecke (n 32) 11 (that 'functionalism ... requires a less thorough analysis of the broader cultural context, if any', than other types of comparative legal analyses).

[58] Jennifer Browne and others, 'A Guide to Policy Analysis as a Research Method' (2019) 34 Health Promotion International 1032, 1035.

mentioned above, these concerns appear overblown because such overlaps are bound to be fairly limited.

Overall, more fundamental questions have been asked about the analogy between climate and environmental impacts that often underpins the application of EAs as a tool for climate change mitigation.[59] This analogy is most evident in cases where courts characterize GHG emissions as a type of 'pollution',[60] but it is also at work when GHG emissions are considered more generally as one among the 'environmental impacts' that EA instruments were designed to prevent.[61] If EA is a useful tool to prevent environmental impacts, and if climate impacts can indeed be approached as just another type of environmental impact, then surely EA can also be a useful tool to prevent climate impacts. Yet, the particular nature of climate impacts may cast doubts about this analogy: while EAs have generally assessed impacts that were local, concrete, and relatively foreseeable, climate impacts are global, diffuse, and partly unpredictable. However, the following chapters document ingenious ways to apply EA frameworks to climate change in spite of the peculiar nature of climate impacts.

2. Significance

Appraising the significance of environmental impacts involves not only a prediction of their magnitude, but also 'a value judgement':[62] whether an impact is deemed significant depends on the importance one attaches to what is impacted. This subjective element does not justify arbitrariness: an appraisal of significance as part of an EA procedure is generally expected to be informed by evidence and to be based on 'cogent reasoning'.[63] Chapter III notes that agreement on such value

[59] See eg *Wildlife Preservation Society of Queensland Proserpine v Minister for the Environment and Heritage* (2006) 232 ALR 510 [72]; *Genesis Power Ltd v Greenpeace New Zealand Inc* [2007] NZCA 569, [2008] 1 NZLR 803 [17]; *350 Montana v Haaland* (9th Cir 2022) 50 F 4th 1254, 1281 (dissenting opinion by R Nelson CJ); Montana Code Ann (West 2024) § 75-1-201(2)(a), as amended by Montana Laws 2023, ch 450, § 1, effective 10 May 2023.

[60] See eg *Border Power Plant Working Group v Department of Energy* (SD California 2003) 260 F Supp 2d 997, 1028; *Earthlife Africa Johannesburg v Minister of Environmental Affairs* [2017] 2 All SA 519 (GP) [78].

[61] *Australian Conservation Foundation v Latrobe City Council* (2004) 140 LGERA 100 [43] ('the generation of greenhouse gases from a brown coal power station clearly has the potential to give rise to "significant" environmental effects'); *Save Lamu v National Environmental Management Authority*, NET 196/2016, judgment (26 June 2019) <https://climatecasechart.com/wp-content/uploads/non-us-case-documents/2019/20190626_Tribunal-Appeal-No.-Net-196-of-2016_decision.pdf> (Kenya) [138] ('[c]limate [c]hange issues are pertinent in projects of this nature'); *Pandey v Union of India* [2019] NGT 843 (India) [2] ('[t]he issue of climate change is certainly a matter covered in the process of impact assessment').

[62] Morrison-Saunders (n 14) 58. See also Alan Ehrlich and William Ross, 'The Significance Spectrum and EIA Significance Determinations' (2015) 33 Impact Assessment & Project Appraisal 87, 89; Jane Holder, *Environmental Assessment: The Regulation of Decision Making* (OUP 2005) 24.

[63] Ehrlich and Ross (n 62). See also Department for Communities and Local Government, Circular 2/99, 'Environmental Impact Assessment' (12 March 1999) s 12 (UK) (noting that an EA is 'not discretionary'). This circular was replaced by a new one, which omits any explicit mention of administrative

judgements is more likely in relation to typical environmental impacts, which are concrete and localized, than with regard to climate impacts, which are diffuse and overly abstract in nature.

This issue of determining significance is faced in two different ways in a typical EA: first, at the stages of screening and scoping; and then again, at the decision-making stages. At the stages of screening and scoping, decision-makers often rely on 'common sense' and 'good judgement' to determine the need for an EA and those environmental impacts that must be assessed as part of this procedure.[64] Yet, it may be more difficult to define a threshold beyond which the climate impact of a proposed activity is 'significant' enough to justify a CA procedure, given that this impact will not have any tangible effect on individuals and ecosystems.

At the substantive decision-making stages, likewise, the economic benefits of a proposed activity—that tend to be limited in time and space—can more easily be weighed against typical environmental impacts, which typically unfold in a similar time span and at a similar scale—than against climate impacts that unfold on a global scale over millennia. For instance, when deciding on the merit of a power plant, a decision-maker may forge an informed opinion about the increase in air and water pollution that is acceptable as the 'cost' for cheaper electricity, but she may find it more difficult to determine the acceptable climate impact of the same project.

The lack of obvious criteria to determine the significance of climate impacts has occasionally been used as a reason to exclude such impacts from the scope of an EA procedure[65] or to reach a cursory finding that an activity's climate impact would not be significant.[66] Rather, Chapter III suggests that the objectives of EA are best served by distinguishing between activities whose climate impacts deserve more or less scrutiny, or by differentiating between activities whose climate impacts is or is not acceptable. In this regard, a parallel can be drawn between climate impacts and other cumulative environmental impacts. Yet, this parallel is of limited use in defining approaches to appraising the significance of climate impacts, considering

---

discretion in determining significance. See Department for Levelling Up, Housing and Communities and Ministry of Housing, Communities & Local Government, 'Guidance: Environmental Impact Assessment' (13 May 2020).

[64] See Glasson and Therivel (n 34).

[65] See eg *Anvil Hill Project Watch Association Inc v Minister for the Environment and Water Resources* (2007) 243 ALR 784 398 (Cth) [40]; *Genesis Power Ltd v Greenpeace New Zealand Inc* (n 59); *Greenpeace New Zealand Inc v Genesis Power Ltd* [2008] NZSC 112, [2009] 1 NZLR 730.

[66] See eg *Kearl Oil Sands Project*, Joint Review Panel (Canadian Environmental Assessment Agency and Alberta Energy and Utilities Board), Decision and Recommendations 2007-013 (27 February 2007) 2007 CarswellAlta 704 [327]; Madeleine Siegel and Alexander Loznak, 'Survey of Greenhouse Gas Considerations in Federal Environmental Impact Statements and Environmental Assessments for Fossil Fuel-Related Projects 2017–2018' (Sabin Centre for Climate Change Law 2019) 1.

that cumulative impact assessment is largely viewed as 'one of the most challenging and least successful components of impact assessment'.[67]

To move the debate forward, Chapter III identifies three approaches that have emerged from EA law and practice across the world. The first approach relies on the magnitude of GHG emissions as a proxy for significance. Thus, a CA is generally considered justified when the activity under consideration is likely to exceed a threshold of annual GHG emissions ranging from 20 to 200 kilotonne carbon dioxide equivalent.[68] The second approach consists in benchmarking the activity's climate impact against the emissions of the region where the project is implemented or the mitigation goals applicable to that region. Thus, the merits of a project should be called into question if the project is expected to cause a sizeable portion of the region's emission budget without offering commensurate social and economic benefit to that region. The third approach involves an economic valuation of climate impacts, based either on marginal emission abatement cost (eg the shadow price of carbon)[69] or, more commonly, on the value of the harm caused by additional GHG emissions (ie the social cost of GHG emissions).[70]

Each of these approaches has its advantages and disadvantages but, as a whole, they generally offer effective ways to appraise the significance of climate impacts. The magnitude-based approach is most relevant at the stages of screening and scoping as it provides simple ways to identify proposed activities whose climate impacts would deserve further scrutiny. By contrast, the second and third approaches are more relevant to the substantive decision-making stages, where they can be used either alternatively or in combination. The benchmark-based approach is contingent on the availability of relevant data on GHG emissions or on the existence of a relevant mitigation target at the scale of the project. The valuation-based approach may appear easy to implement, but the public and the competent authorities need to be informed about, and take into account of, major uncertainties related to the valuation of climate impacts. Overall, the aim of these significance appraisal tools can only be to inform decision-making processes, not to replace political judgements.

---

[67] Chris Joseph and others, 'Improving Cumulative Effects Assessment: Alternative Approaches Based upon an Expert Survey and Literature Review' (2023) 41 Impact Assessment & Project Appraisal 162. See also Jill Blakley and Jessica Russell, 'International Progress in Cumulative Effects Assessment: A Review of Academic Literature 2008–2018' (2022) 65 Journal of Environmental Planning & Management 186; Rebecca Nelson and LM Shirley, 'The Latent Potential of Cumulative Effects Concepts in National and International Environmental Impact Assessment Regimes' (2022) 12 Transnational Environmental Law 150; Elizabeth A Masden and others, 'Cumulative Impact Assessments and Bird/Wind Farm Interactions: Developing a Conceptual Framework' (2010) 30 Environmental Impact Assessment Review 1.

[68] See Chapter III, subsection B.1.

[69] Richard Price, Simeon Thornton, and Stephen Nelson, 'The Social Cost of Carbon and the Shadow Price of Carbon: What They Are, and How to Use Them in Economic Appraisal in the UK' (Department for Environment, Food & Rural Affairs 2007) 14.

[70] CEQ (n 39) 1203.

3. Indirect climate impacts

One issue with applying EA to climate change mitigation concerns the very possibility of predicting the outcome of an agency decision. Typically, an activity's environmental impacts depend on the activity's location. Thus, an agency can avoid water pollution affecting a fragile ecosystem by preventing the construction of a coal mine near this ecosystem. By contrast, climate impacts do not depend on the location of the activity because, once GHGs are emitted, they mix into the atmosphere in a matter of weeks and exacerbate climate change on a global scale.[71] While preventing the construction of the coal mine within its jurisdiction, an agency may nevertheless be unable to prevent another coal mine from operating elsewhere with a similar climate impact. In other words, decisions aimed at limiting climate impacts are prone to carbon leakage, in particular due to market substitution effects.

In this regard, Chapter IV discusses how EA frameworks have sought to predict and address indirect climate impacts. In particular, this chapter engages with national debates about the downstream emissions of fossil-fuel projects. An argument that has frequently been made in this context is that, in a competitive market, an agency's decision to reject such a project would only prompt other suppliers to increase their production so as to satisfy existing demand. Yet, basic economic analysis shows that such market substitution, while real, is normally incomplete: constraining the supply of fossil fuel increases prices and diminishes consumption, thus reducing GHG emissions, in proportion to the price elasticities of supply and demand.

Other frequent objections to the assessment of indirect climate impacts concern the extraterritorial reach of EA procedures when they consider downstream GHG emissions occurring overseas. Chapter IV argues that these objections confuse different dimensions of extraterritoriality. Even when EA instruments are only applicable to activities that are implemented within the state's territory, national agencies generally have the authority to assess these activities in the light of their extraterritorial effects, including their effects on other activities implemented overseas.

As a whole, Chapter IV argues that a comprehensive assessment of indirect climate impacts furthers the objectives of EAs: it helps to identify potential measures to reduce climate impacts and, thus, to provide relevant information to decision-makers. The uncertainties associated with indirect climate impacts do not justify excluding these impacts from the scope of EA procedures; to the contrary, they vindicate the need for a methodical assessment of the predictable effects and of

---

[71] By contrast, some short-lived climate pollutants may have a more localized impact, in particular black carbon. See eg V Ramanathan and G Carmichael, 'Global and Regional Climate Changes Due to Black Carbon' (2008) 1 Nature Geoscience 221; Shichang Kang and others, 'A Review of Black Carbon in Snow and Ice and Its Impact on the Cryosphere' (2020) 210 Earth-Science Reviews (article #103346) 1–12.

the irreducible uncertainties surrounding them. Further, the chapter submits that these indirect climate impacts ought to be considered not only at the scoping and decision-making stages, but also in screening decisions. For instance, these indirect impacts would justify the application of EA requirements to energy-intensive projects without tail-end emissions, such as data centres and crypto-currency mining facilities.

4. Better decisions?

Chapter V considers more generally whether EA can achieve its goal of enhancing decision-making processes, and decisions themselves, when applied to climate impacts. The three main sections of the chapter discuss, respectively, the role and influence of different stakeholders, the content of the substantive decisions, and the effect these decisions may have on climate change.

First, the nature of climate impacts affects the roles that stakeholders can play in the EA. Since the climate impact of an activity does not directly affect any individual or community, public participation in CA cannot be justified as a way of empowering affected communities, although it can still be justified on other grounds. Further, the lack of clear benchmarks to appraise the significance of climate impacts may prompt courts to treat agency decisions with more deference. As a result, CA is a higher risk than EA to unfold as a formalistic and technocratic process controlled by agencies.

Second, Chapter V considers the content of the substantive decisions reached following CA procedures. As in other aspects of EA practice, agencies seldom reject proposed activities, but they do regularly impose conditions aimed at reducing their impacts. These conditions should follow a hierarchy, whereby priority is given to measures aimed at avoiding climate impacts entirely over those aimed at minimising these impacts or at rectifying or compensating them. Some of the most common conditions impose the use of technologies that could limit GHG emissions at a limited cost: for instance, a new urban development may be required to provide space for cleaner modes of transportation.[72] To be effective, these conditions should be defined in clear and specific terms, they should include reporting requirements, and appropriate monitoring mechanisms should be put in place. Some conditions are best imposed through general and abstract regulation, but CA can help governments to identify these conditions and experiment with their implementation.[73]

---

[72] Cour administrative d'appel de Paris (Administrative Court of Appeal of Paris), 23 June 2021, 20PA02347 [32].

[73] See for instance *Hunter Environment Lobby Inc v Minister for Planning* [2011] NSWLEC 221; *Hunter Environment Lobby Inc v Minister for Planning (No 2)* [2012] NSWLEC 40; *Gray v Minister for Planning* [2006] 152 LGERA 258 (NSWLEC); State Environmental Planning Policy (Mining, Petroleum Production and Extractive Industries) 2007 (NSW) s 14(2), reg 65.

Third, Chapter V assesses the real-world effect of CA procedures. It shows that, in some cases, CA has led to better-informed and more carefully considered decisions. In particular, CAs have led to changes in projects or to the imposition of conditions that are likely to reduce climate impacts. Newhall Ranch, a large urban development project, is a case in point: following the 2015 decision of the Supreme Court of California,[74] the project proponents adopted 13 mitigation measures to avoid or offset all GHG emissions associated with the project.[75] Some other projects have not survived the increased scrutiny they were subjected to as a result of CA procedures and related adjudication.[76] CA may also have more diffuse effects, including on the initial design of proposed activities—for instance, by discouraging some of the most unsustainable projects from being submitted at all[77]—and on the public's and agencies' awareness of climate change and ways to mitigate it. Ultimately, how effective EA can be as a tool for climate change mitigation will likely depend on how administrative, political, and judicial authorities engage with this tool and on the institutional and cultural context in which it is used.

5. Tapping the potential of CA

Chapter VI concludes the book with some prospective reflections and policy recommendations. It suggests that EA should be used as a tool for climate change mitigation as it has the potential to enhance decision-making processes. For one, EA ensures that climate change is among the factors that stakeholders consider when appraising the merits of a proposed activity. Further, it helps to gather and circulate reliable information about the climate impact of proposed activities and ways to avoid or reduce it. However, Chapter VI also highlights that climate impacts differ from other environmental impacts in important ways. Merely deciding to expand EA frameworks to climate impacts is insufficient: tapping the full potential of EA as a tool for climate change mitigation requires an adaptation of EA frameworks.

---

[74] *Center for Biological Diversity v Department of Fish and Wildlife* (2015) 62 Cal 4th 204, 218.

[75] See *Newhall Ranch Resource Management and Development Plan and Spineflower Conservation Plan*, California Department of Fish and Wildlife, Final Actions and Supplemental Findings (14 June 2017) <https://nrm.dfg.ca.gov/FileHandler.ashx?DocumentID=145821&inline>, 15–26.

[76] See eg *Earthlife Africa Johannesburg v Minister of Environmental Affairs* (19 November 2020) <https://perma.cc/UJ4B-5534> [2].

[77] See generally Morrison-Saunders (n 14).

Conclusion

This first chapter has introduced the book's objective, method, and key conclusions. As a whole, the book vindicates the relevance of EA as a tool for climate change mitigation. It does not argue that EA is *the* right instrument for climate change mitigation—addressing climate change requires the simultaneous implementation of multiple policies and measures—but it purports to show that EA is a useful addition to national toolkits for climate change mitigation. CA ensures the availability to all stakeholders of information on how a proposed activity would contribute to climate change and how this climate impact could be avoided or reduced, thus enabling sounder and more productive deliberations. Concretely, CA studies often set aside arguments on a perfect market substitution and can identify relevant benchmarks to appraise the significance of an activity's climate impact. More generally, CA may help to ensure that considerations for climate change mitigation are present in political debates and are considered in final decisions on relevant activities despite the absence of directly affected individuals or communities.

On the other hand, the book sheds light on the need to adapt EA when applying it as a tool for climate change mitigation. Climate impacts differ in significant ways from the environmental impacts that have typically been the focus of EA procedures and these differences raise important issues. Through a global survey of CA law and practice supplemented by case studies and policy analysis, the book identifies and evaluates ingenious solutions to these issues. Yet there remains room for progress even in the most advanced CA regimes, for instance concerning the lack of a coherent approach to indirect climate impacts or the way one could assess the 'significance' of climate impacts.[78] It is hoped that this book will inspire further research, including jurisdiction-specific studies on the opportunity of legal transplant. Overall, the policy recommendations formulated in this book will hopefully inspire policy developments that will further realize the potential of EA as a tool for climate change mitigation.

---

[78] See eg *Dakota Resource Council v US Department of Interior* (D DC, 22 March 2024) 22-cv-1853 (CRC), 2024 WL 1239698, at *20 (noting that the latest CEQ guidance does 'not establish any threshold or method to decide when GHG emissions and their associated social costs become significant').

II

The Concept of Climate Assessment

Introduction

This chapter introduces the use of environmental assessment (EA) as a tool for climate change mitigation—in short, climate assessment (CA). Section A presents the context in which CA emerged, Section B provides an overview of existing CA law and practice, and Section C considers some of the main objections to CA frameworks.

To contextualize CA, Section A gives an overview of existing action on climate change mitigation, reflects the nearly universal recognition of EA as a tool for environmental protection, and illustrates how certain activities can cause large greenhouse gas (GHG) emissions that exacerbate climate change. In this context, the implementation of EA as a tool for climate change mitigation relies on the assumption that the climate impact of a proposed activity can usefully be considered among other impacts subjected to EA.

Turning to a global survey of EA law and practice, Section B shows that many national EA instruments have been reinterpreted or reformed in ways that extended their application to climate impacts and that there are some indications of the emergence of EA as a legal requirement in relation to global environmental impacts. Yet, decisions to apply EA as a tool for climate change mitigation tend to be unsystematic and to leave fundamental questions open with regard to the modalities of CA.[1]

Lastly, Section C identifies factors that have led some jurisdictions to oppose the integration of climate impacts in existing EA frameworks. In particular, it identifies two frequent objections to the use of EA as a tool for climate change mitigation. The first objection questions the need for CA procedures when climate change mitigation policies are already in place. For instance, when a project falls within the scope of a cap-and-trade mechanism, it has been argued that a CA, if it was to reduce emissions from this project, would merely free emission allowances that other activities could use, thus achieving no net mitigation outcomes. However, the chapter shows that emissions assessed in a CA are often beyond the scope of

---

[1] See also Rose Mayembe and others, 'Integrating Climate Change in Environmental Impact Assessment: A Review of Requirements across 19 EIA Regimes' (2023) 869 Science of the Total Environment (article #161850) 1–13, at 7 (qualifying some EA practice as 'perfunctory').

Environmental Assessment as a Tool for Climate Change Mitigation. Benoit Mayer, Oxford University Press.
© Benoit Mayer 2024. DOI: 10.1093/oso/9780198939184.003.0002

cap-and-trade mechanisms, and that reducing demand for emissions may allow public authorities to impose more stringent caps over time.

The second, more fundamental objection to CA casts doubt on the analogy between climate and environmental impacts. While EAs have generally assessed impacts that were local, concrete, and relatively foreseeable, climate impacts are global, diffuse, and often uncertain. The word 'impact' can certainly be defined in a broad sense, extending beyond direct consequences to encompass diffuse effects of a statistical nature such as a very small increase in the likelihood of an extremely broad and far-reaching range of risks. Yet one may question the relevance and usefulness of EA to address this different type of impact. To assess this argument, one needs to consider how EA can approach climate impacts, for instance by predicting climate impacts and appraising their significance in a non-arbitrary way—questions that are further addressed in the following chapters.

A. Contextualising CA

This section provides an overview of the context in which CA has emerged. The first two subsections provide background information, respectively, on climate change mitigation and EA. The third subsection presents a preliminary argument for the application of EA as a tool for climate change mitigation.

1. Climate change mitigation

This subsection provides background information on climate change mitigation. It defines climate change and introduces international cooperation as well as national measures and policies to mitigate it.

a) Climate change

The Intergovernmental Panel on Climate Change (IPCC), an organization aimed at assessing the science on climate change, finds it 'unequivocal that human influence has warmed the atmosphere, ocean and land'.[2] Building on a large body of scientific research, the IPCC estimates that the global average temperature has increased by about 1.1°C since the second half of the nineteenth century,[3] with consequences that are 'unprecedented over many centuries to many thousands of years'.[4] The IPCC reflected the scientific consensus that climate change 'has caused

---

[2] Richard P Allan and others, 'Summary for Policymakers', in Valérie Masson-Delmotte and others (eds), *Climate Change 2021: The Physical Science Basis. Working Group I Contribution to the Sixth Assessment Report of the Intergovernmental Panel on Climate Change* (CUP 2021) 3, 4.

[3] Ibid, 5.

[4] Ibid, 8.

widespread adverse impacts and related losses and damages to nature and people',[5] including through 'hot extremes on land and in the ocean, heavy precipitation events, drought and fire weather'.[6] It also highlighted scientists' concerns, in particular for the impacts of climate change on unique and threatened ecological and human systems, for the intensification of extreme weather events, for the capacity of these impacts to exacerbate inequalities, for the sheer ambit of global aggregate impact, and for the possibility of large-scale singular events ('tipping points').[7]

Climate change results from anthropogenic emissions of GHGs, in particular carbon dioxide, methane, nitrous oxide, and a range of fluorinated gases.[8] Once emitted into the atmosphere, GHGs cause radiative forcing: they 'trap' heat within the Earth's system by reflecting some of the infrared radiations emitted from the Earth. As GHGs may stay in the atmosphere for decades or centuries,[9] continuing emissions result in a gradual increase in GHG concentrations in the atmosphere. Thus, it is estimated that the anthropogenic emission of over two trillion tonnes of carbon dioxide since the mid-nineteenth century has led atmospheric carbon dioxide concentration to increase by half,[10] reaching concentrations not seen in hundreds of thousands of years.[11] GHG emissions result from a large range of activities related to power generation, transport, industrial processes, land-use change, agricultural processes, and waste management.[12] Many (but not all) of these activities involve the combustion of fossil fuels.

The future intensity of climate change depends on future anthropogenic GHG emissions. The IPCC has affirmed the existence of 'a near-linear relationship between cumulative anthropogenic CO_2 emissions and the global warming they cause', noting that each trillion tonnes of carbon dioxide emissions would likely cause an increase in global average temperature by about 0.45°C.[13] By contrast, society and ecosystems are affected in more complex and less linear ways, for

[5] Hans-Otto Pörtner and others, 'Summary for Policymakers', in Hans-Otto Pörtner and others (eds), *Climate Change 2022: Impacts, Adaptation and Vulnerability. Working Group II Contribution to the Sixth Assessment Report of the Intergovernmental Panel on Climate Change* (CUP 2022) 3, 9.

[6] Ibid.

[7] Hans-Otto Pörtner and others, 'Technical Summary', in Pörtner and others (n 5) 35, 69.

[8] Allan and others (n 2) 5; Pörtner and others, 'Summary for Policymakers' (n 5) 9.

[9] See Matthew Collins and others, 'Long-term Climate Change: Projections, Commitments and Irreversibility', in Thomas F Stocker and others (eds), *Climate Change 2013: The Physical Science Basis* (CUP 2013) 1029, 1033.

[10] Allan and others (n 2) 6–7. See also Global Monitoring Laboratory, 'Trends in Atmospheric Carbon Dioxide' (*National Oceanic and Atmospheric Administration*, 5 November 2022) <https://perma.cc/68R9-S5PE>.

[11] Shobhakar Dhakal and others, 'Emissions Trends and Drivers', in Priyadarshi R Shukla and others (eds), *Climate Change 2022: Mitigation of Climate Change. Working Group III Contribution to the Sixth Assessment Report of the Intergovernmental Panel on Climate Change* (CUP 2022) 215, 218.

[12] Pörtner and others, 'Summary for Policymakers' (n 5) 11; M Crippa and others, 'CO_2 Emissions of All World Countries' (JRC/IEA/PBL 2022 report) <https://edgar.jrc.ec.europa.eu/booklet/CO_2_emissions_of_all_world_countries_2022_report.pdf>; Johannes Friedrich, 'World Greenhouse Gas Emissions: 2019', *World Resources Institute* (23 June 2022) <https://perma.cc/NX5W-8PEL>.

[13] Allan and others (n 2) 28 (emphasis removed).

instance as a result of 'tipping points or thresholds in responses'.[14] While major uncertainties prevent a prediction of these tipping points, it is largely expected that 'some components may cancel each other out'.[15] Given these uncertainties, one can reasonably assume, as an approximation, a linear relation between GHG emissions and the risk of social and ecological harm.[16]

b) International cooperation on climate change mitigation

Governments respond to climate change in mainly two ways.[17] First, they mitigate it by addressing its cause: the increase in atmospheric concentrations in GHGs. Climate change mitigation consists in reducing GHG emissions by sources (eg by reducing the combustion of fossil fuels) and enhancing removals by sinks (eg by planting more trees). Second, states seek to reduce the consequences of climate change through adaptation, that is, by promoting the adjustment of societies and ecosystems to actual or expected consequences of climate change.[18] Concrete adaptation actions include, for instance, measures and policies on disaster-risk reduction, food security, and freshwater management. Adaptation is necessary as climate change is already causing adverse impacts, but mitigation is also necessary given the existence of technical and financial limits to adaptation.[19] Adaptation and mitigation raise different types of legal issues, and this book focuses exclusively on mitigation.[20]

---

[14] Rawshan Ara Begum and others, 'Point of Departure and Key Concepts', in Pörtner and others (n 5) 121, 147.

[15] Ibid.

[16] See eg Marshall Burke, Solomon M Hsiang, and Edward Miguel, 'Global Non-Linear Effect of Temperature on Economic Production' (2015) 527 Nature 235, 239 ('our projected global losses are roughly linear—and slightly concave—in temperature, not quadratic or exponential'); Richard SJ Tol, 'The Economic Impacts of Climate Change' (2018) 12 Review of Environmental Economics & Policy 7 (noting that most studies agree on a 'piecewise linear model'). This assumption of approximate linearity informs, for instance, the calculation of a social cost of carbon. See Chapter III, subsection D.2.

[17] See eg United Nations Framework Convention on Climate Change (adopted 9 May 1992, entered into force 21 March 1994) 1771 UNTS 107 (UNFCCC) art 4(1)(b); Paris Agreement (adopted 12 December 2015, entered into force 4 November 2016) 3156 UNTS 79, art 2(1). Finance, the third goal defined by the Paris Agreement, aims at assisting mitigation and adaptation action.

[18] Vincent Möller and others, 'Glossary', in Pörtner and others (n 5) 2897, 2898.

[19] Pörtner and others, 'Summary for Policymakers' (n 5) 26.

[20] On the integration of adaptation in EA procedures, see eg Shardul Agrawala and others, 'Incorporating Climate Change Impacts and Adaptation in Environmental Impact Assessments: Opportunities and Challenges' (2012) 4 Climate & Development 26; Julius Kamau and Francis Mwaura, 'Climate Change Adaptation and EIA Studies in Kenya' (2013) 5 International Journal of Climate Change Strategies & Management 152; Alexandra Jiricka and others, 'Consideration of Climate Change Impacts and Adaptation in EIA Practice: Perspectives of Actors in Austria and Germany' (2016) 57 Environmental Impact Assessment Review 78; Antonio Ledda and others, 'Integrating Adaptation to Climate Change in Regional Plans and Programmes: The Role of Strategic Environmental Assessment' (2021) 91 Environmental Impact Assessment Review (article #106655) 1–9; Alvaro Enriquez-de-Salamanca and others, 'Environmental Impacts of Climate Change Adaptation' (2017) 64 Environmental Impact Assessment Review 87.

States have sought to cooperate in mitigating climate change for over three decades.[21] In 1992, they adopted the UN Framework Convention on Climate Change (UNFCCC). The UNFCCC acknowledges climate change and its adverse effects as a 'common concern of humankind'[22] and adopts an 'ultimate objective' of preventing 'dangerous anthropogenic interference with the climate system'.[23] According to 'equity' and an ambiguous principle of 'common but differentiated responsibilities and respective capabilities', the Convention suggests that 'developed country Parties should take the lead in combating climate change'.[24] Every Party commits to adopt 'programmes containing measures to mitigate climate change',[25] and the developed country parties listed in Annex I further commit to 'adopt national policies and take corresponding measures on the mitigation of climate change'.[26] Yet, the UNFCCC does not specify what these programmes or policies must consist in or how ambitious they must be. All Parties must also develop and publish national emissions inventories and reports on the measures they are taking on climate change mitigation, with more stringent requirements applicable to Annex I Parties.[27]

In 1997, the Conference of the Parties to the UNFCCC (COP) adopted the Kyoto Protocol, which requires Annex I Parties to limit or reduce emissions within their territory by an agreed-upon national percentage indicated in Annex B of the Protocol.[28] The Kyoto Protocol initially applied to a commitment period from 2008 to 2012, but the Doha Amendment extended its application to a second commitment period from 2013 to 2020.[29] However, the Kyoto Protocol was plagued by issues of participation: the United States did not ratify;[30] Canada withdrew before the end of the first commitment period;[31] and Japan, New Zealand, and Russia refused any commitment for the second commitment period.[32] The Kyoto Protocol did not create any new commitments applicable to non-Annex I Parties.[33]

[21] See UN General Assembly Resolution 43/53, 'Protection of Global Climate for Present and Future Generations of Mankind' (6 December 1988).

[22] UNFCCC (n 17) preamble para 2.

[23] Ibid, art 2.

[24] Ibid, art 3(1).

[25] Ibid, art 4(1)(b).

[26] Ibid art 4(2)(a).

[27] Ibid, arts 4(1)(a), 4(2)(b)–(c), 12.

[28] Kyoto Protocol to the United Nations Framework Convention on Climate Change (adopted 11 December 1997, entered into force 16 February 2005) 2303 UNTS 162, art 3(1) and Annex B.

[29] Doha Amendment to the Kyoto Protocol (adopted 8 December 2012, entered into force 31 December 2020), in FCCC/KP/CMP/2012/13/Add.1 (28 February 2013) 7.

[30] Senate Resolution 98, 105th Congress, 143 Cong Rec S8138-39 (25 July 1997).

[31] Kyoto Protocol Compliance Committee, Note by the Secretariat, 'Canada's Withdrawal from the Kyoto Protocol and its Effects on Canada's Reporting Obligations under the Protocol' (20 August 2014) CC/EB/25/2014/2.

[32] Eg Chee Ling, 'Kyoto Protocol Second Commitment Period Remains Elusive' (2012) 264 Third World Resurgence 17 <https://perma.cc/ZD8K-Y6RL>; 'We'll Never Accept 2nd Kyoto Period, Says Japan, Sparking Doubts on KP's Survival' *Third World Network* (30 November 2010); Tim Groser, 'New Zealand Commits to UN Framework Convention' (*New Zealand Government*, 10 November 2012) <https://perma.cc/TK8P-UR55>.

[33] Kyoto Protocol (n 28) art 10.

The Copenhagen Accord of 2009 (an informal, non-binding agreement) and the Cancun Agreements in 2010 (a COP decision) defined an alternative, nationally determined approach to international cooperation on climate change mitigation.[34] All Parties were invited to communicate a mitigation pledge to be implemented by 2020, which should consist in 'quantified economy-wide emission reduction targets' for Annex I Parties and in other 'nationally appropriate mitigation actions' for non-Annex I Parties.[35] The Parties also agreed on the collective objective of holding global warming 'below 2°C above pre-industrial levels.'[36]

This alternative, 'bottom-up' approach was formalized in 2015 with the adoption of the Paris Agreement, a treaty aimed at fostering the implementation of the UNFCCC. The Paris Agreement defines the collective objective of holding global warming 'well below 2°C ... and pursuing efforts to limit [it] to 1.5°C.'[37] To achieve this objective, the Paris Agreement relies on nationally determined contributions (NDCs) which Parties must communicate every five years.[38] The NDCs of 'developed country Parties' should contain 'economy-wide absolute emission reduction targets', whereas 'developing country Parties' are merely 'encouraged to move over time towards economy-wide emission reduction or limitation targets.'[39] Each NDC is expected to 'reflect [the Party's] highest possible ambition' and to 'represent a progression' beyond the Party's previous NDC.[40] Parties must take measures with the view of achieving the mitigation objectives contained in their NDC.[41] Parties are also encouraged to communicate long-term mitigation strategies.[42]

International cooperation on climate change mitigation has yielded limited results so far. While Annex I Parties' GHG emissions have plateaued since 1990, developing countries' emissions have intensified rapidly;[43] as a result, global GHG

---

[34] See Daniel Bodansky, 'The Paris Climate Change Agreement: A New Hope?' (2016) 110 American Journal of International Law 288.

[35] Copenhagen Accord, in annex of Decision 2/CP.15, FCCC/CP/2009/L.7 (18 December 2009) 4 paras 4–5; Decision 1/CP.16, 'The Cancun Agreements: Outcome of the Work of the Ad Hoc Working Group on Long-term Cooperative Action under the Convention', FCCC/CP/2010/7/Add 1 (15 March 2011) 2 paras 36–67.

[36] Decision 1/CP.15, 'Outcome of the Work of the Ad Hoc Working Group on Long-term Cooperative Action under the Convention', FCCC/CP/2009/11/Add 1 (18–19 December 2009) 3 para 4. See also Copenhagen Accord (n 35) para 2.

[37] Paris Agreement (n 17) art 2(1)(a).

[38] Ibid, arts 3, 4(2), 4(9).

[39] Ibid, art 4(4). See Benoit Mayer, 'Progression Requirements Applicable to State Action on Climate Change Mitigation under Nationally Determined Contributions' (2023) 23 International Environmental Agreements: Politics, Law and Economics 293; Benoit Mayer, 'The "Highest Possible Ambition" on Climate Change Mitigation as a Legal Standard' (2024) 73 International and Comparative Law Quarterly 285.

[40] Paris Agreement (n 17) art 4(3).

[41] Ibid, art 4(2). See also Benoit Mayer, 'International Law Obligations Arising in Relation to Nationally Determined Contributions' (2018) 7 Transnational Environmental Law 251.

[42] Paris Agreement (n 17) art 4(19).

[43] See Jan C Minx and others, 'A Comprehensive and Synthetic Dataset for Global, Regional, and National Greenhouse Gas Emissions by Sector 1970–2018 with an Extension to 2019' (2021) 13 Earth System Science Data 5213 (estimating that Annex I parties' emissions decreased from 18 to 17 Gt CO_2e

emissions increased from 38 to 59 Gt CO_2 equivalent (CO_2e) between 1990 and 2019,[44] and atmospheric carbon dioxide concentrations have risen at an accelerating pace.[45] Nonetheless, international mitigation action has not been entirely ineffective.[46] Macroeconomic studies suggest that emissions could have increased faster, were it not for climate treaties[47] and national measures and policies.[48] More precisely, estimates compiled by the IPCC suggest that global efforts have reduced annual global GHG emissions by between 1.8 and 5.9 Gt CO_2e[49]—that is, a reduction in global GHG emissions by 3 to 10 percent.[50] Going forward, states and observers have recognized that current NDCs fall short of the level of ambition consistent with the temperature targets of the Paris Agreement.[51] Far from the global objectives of holding global warming to 1.5 or 2°C, current NDCs seem consistent with a global mitigation pathway resulting in a global warming of about 2.5°C.[52]

c) Measures and policies for climate change mitigation

While climate treaties require states to mitigate climate change, they do not prescribe the particular measures states must take to do so. The UNFCCC affirms

---

between 1990 and 2019, while non-Annex I parties' emissions increased from 14 to 34 Gt during the same period, not including land use, land-use change and forestry (LULUCF) emissions.

[44] Priyadarshi R Shukla and others, 'Summary for Policymakers', in Shukla and others (eds) (n 11) 3, at 21 (figure 1) (including LULUCF emissions).

[45] Global Monitoring Laboratory, 'Annual Mean Growth Rate for Mauna Loa, Hawaii' (National Oceanic and Atmospheric Administration) <https://perma.cc/UM3F-3W2L> (from 0.9 ppm increase per year in the 1960s to 1.5 ppm per year in the 1990s and 2.4 ppm per year in the 2010s).

[46] Cf Steinar Andresen, 'The Paris Agreement and Its Rulebook in a Problem-Solving Perspective' (2019) 9 Climate Law 122.

[47] See Shaikh MSU Eskander and Sam Fankhauser, 'Reduction in Greenhouse Gas Emissions from National Climate Legislation' (2020) 10 Nature Climate Change 750; Nada Maamoun, 'The Kyoto Protocol: Empirical Evidence of a Hidden Success' (2019) 95 Journal of Environmental Economics & Management 227.

[48] See William F Lamb and others, 'A Review of Trends and Drivers of Greenhouse Gas Emissions by Sector from 1990 to 2018' (2021) 16 Environmental Research Letters (article #073005) 1–31; Corinne Le Quéré and others, 'Drivers of Declining CO_2 Emissions in 18 Developed Economies' (2019) 9 Nature Climate Change 213; Sebastian Schäfer, 'Decoupling the EU ETS from Subsidized Renewables and Other Demand Side Effects: Lessons from the Impact of the EU ETS on CO_2 Emissions in the German Electricity Sector' (2019) 133 Energy Policy (article #110858) 1–12; Jonathan Colmer and others, 'Does Pricing Carbon Mitigate Climate Change? Firm-Level Evidence from the European Union Emissions Trading Scheme' (The Centre for Economic Performance 2023). But see William Nordhaus, 'Climate Change: The Ultimate Challenge for Economics' (2019) 109 American Economic Review 1991, 2008–10.

[49] Anthony Patt and others, 'International Cooperation', in Shukla and others (eds) (n 11) 1451, 1480 (cross-chapter box 10, figure 1).

[50] Calculated based on Shukla and others (n 44) 21 (figure 1).

[51] See eg Decision 1/CMA.5, 'Outcome of the First Global Stocktake', FCCC/PA/CMA/2023/16/ Add.1 (15 March 2024) 2 para 21; UNFCCC Secretariat, 'Summary Report Following the Second Meeting of the Technical Dialogue of the First Global Stocktake under the Paris Agreement' (31 March 2023) <https://unfccc.int/sites/default/files/resource/TD1.2_GST_SummaryReport.pdf>, 10.

[52] UNEP, 'The Closing Window: Climate Crisis Calls for Rapid Transformation of Societies' (27 October 2022) 35.

'the principle of sovereignty of States in international cooperation to address climate change',[53] thus leaving it for each state to determine what measures and policies they wish to take with the view of limiting and reducing their GHG emissions. Consistently, the Kyoto Protocol required each Annex I Party to 'implement ... policies and measures in accordance with its national circumstances'.[54] As an exception, states have agreed on international policies and measures to limit GHG emissions in international shipping[55] and international civil aviation,[56] on the ground that uncoordinated national measures cannot regulate these sectors efficiently. At times, states have emphasized the need for certain policies to be implemented nationally, for instance the phasing out of fossil-fuel subsidies,[57] but they have done so without adopting any binding commitments.

Thus, it generally belongs to each state to determine *how* to mitigate climate change. Some of the most prominent measures that governments have taken seek to impose an economic incentive against GHG emissions, in particular in the form of a 'carbon price'.[58] Cap-and-trade mechanisms such as the EU Emissions Trading Scheme require economic actors to surrender allowances in proportion of GHG emissions from specified sources.[59] Notwithstanding whether emission allowances are freely allocated ('grandfathered') or whether they are auctioned, they can subsequently be traded, thus creating a marginal economic incentive for every actor to reduce its emissions. A carbon price can also be imposed more directly through tax,[60] either directly on emissions or, more conveniently, on emission-intensive goods and services (eg fuel and airline tickets).[61] Lastly, subsidies can be established to support more sustainable practices, for instance by guaranteeing feed-in

---

[53] UNFCCC (n 17) preamble para 10.

[54] Kyoto Protocol (n 28) art 2(1)(a).

[55] Eg International Maritime Organization, Marine Environment Protection Committee, Resolution MEPC.304(72), 'Initial IMO Strategy on Reduction of GHG Emissions from Ships' (13 April 2018), Annex 11. See also Beatriz Garcia, Anita Foerster, and Jolene Lin, 'Net Zero for the International Shipping Sector? An Analysis of the Implementation and Regulatory Challenges of the IMO Strategy on Reduction of GHG Emissions' (2021) 33 Journal of Environmental Law 85.

[56] Eg International Civil Aviation Organization, Resolution A41-21, 'Consolidated statement of continuing ICAO policies and practices related to environmental protection – Climate change' (October 2022), in ICAO Doc 10184, Assembly Resolutions in Force (as of 7 October 2022), I-79. See also Benoit Mayer and Zhuoqi Ding, 'Climate Change Mitigation in the Aviation Sector: A Critical Overview of National and International Initiatives' (2023) 12 Transnational Environmental Law 14.

[57] G20, Pittsburgh Summit, Leaders' Statement (September 2009) para 29; G20, Bali Summit, Leaders' Declaration (November 2022) para 12; APEC, Singapore Summit, Leaders' Declaration (November 2009); APEC, Bangkok Summit, Leaders' Declaration (November 2022) para 15; G7, Climate and Environment Ministers' Communiqué (21 May 2021); Decision 1/CMA.3, 'Glasgow Climate Pact', FCCC/PA/CMA/2021/10/Add.1 (8 March 2022) 2 para 36; Decision 1/CMA.5 (n 51) para 28.

[58] See generally World Bank, 'State and Trends of Carbon Pricing 2022' (24 April 2022).

[59] Directive 2003/87/EC of the European Parliament and of the Council of 13 October 2003 establishing a system for greenhouse gas emission allowance trading within the Union and amending Council Directive 96/61/EC, [2003] OJ L 275/32.

[60] Carbon Tax Regulation, BC Reg 125/2008, s 12.

[61] Mayer and Ding (n 56) 18–19.

tariffs on renewable energy.[62] Governments have also relied on more traditional types of regulation. They can, for instance, impose performance standards (eg on new road vehicles),[63] regulate fuel supply (eg biofuel-content requirements),[64] or even ban certain technologies (eg internal combustion engines)[65] with the view of limiting and reducing GHG emissions. Governments have also relied on industrial policies, for instance through economic incentives to the development and deployment of new technologies[66] or by directing state-owned enterprises towards lower-emission activities.[67]

Rather than relying on a single policy, most governments have taken multiple measures and policies—some of which are applicable to multiple sectors; others, only to very specific activities.[68] Such a multipronged approach is essential because no measure can address effectively emissions from all sectors at once and on all relevant time scales. For instance, carbon pricing tools can only apply to sectors where it is relatively convenient to monitor, report, and verify emissions—even a relatively comprehensive measure such as the EU Emissions Trading Scheme applies to less than half of the EU's emissions,[69] and the economic incentive it creates for emission reduction is stronger on the short term than on the longer term, where the cap remains malleable and uncertain.[70] In this regard, many governments have come to see EA as yet one policy in their toolkit on climate change mitigation.

---

[62] Eg European Commission, Report to the European Parliament and the Council on the Performance of Support for Electricity from Renewable Sources Granted by Means of Tendering Procedures in the Union (15 November 2022) COM(2022) 638 final.

[63] See eg Regulation (EU) 2019/631 of the European Parliament and of the Council of 17 April 2019 setting CO_2 emission performance standards for new passenger cars and for new light commercial vehicles, and repealing Regulations (EC) No 443/2009 and (EU) No 510/2011, [2019] OJ L 111/13; Environmental Protection Agency, 'Revised 2023 and Later Model Year Light-Duty Vehicle Greenhouse Gas Emissions Standards' (30 December 2021) 86 Federal Register 74434.

[64] Eg 42 USC (2024) § 7545; Directive (EU) 2018/2001 of the European Parliament and of the Council of 11 December 2018 on the promotion of the use of energy from renewable sources, [2018] OJ L 328/82; Clean Fuel Regulations, SOR/2022-140 (Canada).

[65] See Jonathan Gitlin, 'California Calls Time on Internal Combustion Engines from 2035', *ArsTechnica* (25 August 2022) <https://perma.cc/T4FG-Q5Q3>; 'EU Ban on the Sale of New Petrol and Diesel Cars from 2035 Explained', *News EU Parliament* (3 November 2022).

[66] See generally Jeffrey Rissman and others, 'Technologies and Policies to Decarbonize Global Industry: Review and Assessment of Mitigation Drivers through 2070' (2020) 266 Applied Energy (article #114848) 1–34.

[67] See eg Philippe Benoit, 'Engaging State-Owned Enterprises in Climate Action' (Columbia Center on Global Energy Policy 2019).

[68] See generally Navroz K Dubash and others, 'National and Sub-National Policies and Institutions', in Shukla and others (eds) (n 11) 1355, 1394–99.

[69] See EU, Eighth National Communication and Fifth Biennial Report under the UNFCCC (2022) <https://unfccc.int/documents/624694>, 199 ('The EU ETS covers approximately 40% of the EU's greenhouse gas emissions').

[70] The number of emission allowances auctioned can be adjusted in response to demand for allowances during each phase of the ETS. See Directive 2003/87/EC (n 59), consolidated as of 5 June 2023, art 10a(5). The amount of emission allowances to be issued beyond the current phase (ie beyond 2030) is yet to be determined as of 2024.

2. Environmental assessment

This subsection defines EA as a tool for environmental protection. It first defines the concept and the rationale for EA before describing the main stages of a typical EA procedure and retracing the widespread adoption of this instrument throughout the world.

a) The concept

EA is a process that aims at identifying, predicting, evaluating, and mitigating the effects of projects, plans, programmes, and policies on the environment before a final decision liable to cause environmental harm is made.[71] Most EA frameworks apply notwithstanding whether the decision relates to an activity that would be implemented by a public or private entity, provided only that an administrative approval of the activity is required.[72] EA seeks to 'ensure ... that environmental considerations are properly taken into account when project decisions are made'.[73] The emphasis of EA law is on procedural requirements. As Supreme Court Justice John Stevens put it, EA procedures 'merely prohibit ... uninformed—rather than unwise—agency action'.[74] Yet, the hope and expectation clearly are that these procedural requirements will eventually improve the substance of decisions.

Different theories suggest ways EA may improve decisions. In particular, Jane Holder distinguishes between information theories and culture theories. Information theories approach EA as 'a means of informing decision makers of the possible environmental consequences of a proposed project or action'.[75] These theories suggest that 'environmentally sound and sustainable development' is more likely when decisions are explicitly informed by scientific research and public deliberations.[76] By contrast, culture theories 'propose, more fundamentally, that environmental assessment inculcates environmental protection values amongst those taking decisions'.[77] These theories suggest that EA procedures may prompt people to 'move beyond strict self-interest, or administratively determined interests, to seeing their private or individual interest linked with other, shared

---

[71]   See John Glasson and Riki Therivel, *Introduction to Environmental Impact Assessment* (5th edn, Routledge 2019) 3; Eric L Hyman and others, *Combining Facts and Values in Environmental Impact Assessment: Theories and Techniques* (Routledge 2018) 5; Hannah Kevin and Lauren Arnold (eds), *An Introduction to Environmental Impact Assessment* (Routledge 2020) 3.

[72]   Angus Morrison-Saunders, *Advanced Introduction to Environmental Impact Assessment* (Edward Elgar 2018) 7.

[73]   European Commission, '35 Years of EU Environmental Impact Assessment' (Publications Office of the European Union 2021) 1.

[74]   *Robertson v Methow Valley Citizens Council* (1989) 490 US 332, 351.

[75]   Jane Holder, *Environmental Assessment: The Regulation of Decision Making* (OUP 2005) 22.

[76]   UNEP, 'Goals and Principles of Environmental Impact Assessment' (16 January 1987) UNEP/GC.14/17 Annex III, preamble. This document was subsequently endorsed by UNEP, Governing Council resolution 14/25 (17 June 1987), A/42/25, 77 para 1.

[77]   Holder (n 75) 22. See also Serge Taylor, *Making Bureaucracies Think: The Environmental Impact Statement Strategy of Administrative Reform* (Stanford University Press 1984).

interests, and thus make decisions based on the common good'.[78] In particular, Robert Bartlett has argued that EA procedures can act like a 'worm in the head' of national agencies.[79]

Elaborating among other things on Holder's research, Neil Craik further identifies three theoretical approaches that are situated on a continuum ranging from information to culture theories. The first such approach is a 'comprehensive rationality model', which 'views the environmental goals that underlie EIA as being uncontested' and suggests that 'trade-offs between environmental and development goals can be objectively determined'.[80] The second, 'more pluralistic ... approach accepts that EIA processes must necessarily mediate between competing sets of interests that are not easily reconcilable'.[81] As Holder put it, EA may thus provide 'a forum for negotiation and bargaining about the design of a project, and mitigating measures to be taken, between interest groups and within and between agencies'.[82] The third approach 'views the role of EIA as developing and inculcating shared values around environmental decisions', based on the assumption that 'interests [are] capable of change'.[83]

Scholars have also identified pitfalls that may limit the effectiveness of EA. In particular, a recurrent concern is that EA may be 'reduced to a formality, resulting in the generation of detailed reports that few read and that have no effect on decision-making process'.[84] As Craik notes, the ability of EA to influence decisions will depend on factors such as 'the degree of transparency' within EA processes, 'the presence of shared values to root justification in a common goal', the political culture in which the EA procedure is implemented, and the 'institutional infrastructure to support transparency, participation, and access to justice'.[85] Theorists generally suggest that an EA framework that is too technocratic would 'have little relevance to the public's perception of environmental harm'[86] and, thus, little effect on decisions.[87] Some also raise concerns that EA could 'lend ... an air of neutrality

---

[78] Holder (n 75) 28.

[79] Robert Bartlett, 'Ecological Reason in Administration: Environmental Impact Assessment and Green Politics', in Robert Paehlke and Douglas Torgerson (eds), *Managing Leviathan: Environmental Politics and the Administrative State* (2nd edn, University of Toronto Press 2019) 47, 48. See also Megan Jones and Angus Morrison-Saunders, 'Understanding the Long-Term Influence of EIA on Organisational Learning and Transformation' (2017) 64 Environmental Impact Assessment Review 131.

[80] Neil Craik, 'The Assessment of Environmental Impact', in Emma Lees and Jorge Vinuales (eds), *The Oxford Handbook of Comparative Environmental Law* (OUP 2019) 876, 881; Chris Wood, *Environmental Impact Assessment: A Comparative Review* (2nd edn, Routledge 2013) 881.

[81] Craik (n 80) 882; Wood (n 80) 882.

[82] Holder (n 75) 19. See also Juan Palerm and Carla Aceves, 'Environmental Impact Assessment in Mexico: An Analysis from a "Consolidating Democracy" Perspective' (2004) 22 Impact Assessment & Project Appraisal 99.

[83] Craik (n 80) 882; Wood (n 80) 882.

[84] Robert V Bartlett and Priya A Kurian, 'The Theory of Environmental Impact Assessment: Implicit Models of Policy Making' (1999) 27 Policy & Politics 415, 418.

[85] Craik (n 80) 883; Wood (n 80) 882–83.

[86] Holder (n 75) 18.

[87] See Bartlett and Kurian (n 84) 417.

to decision making', in particular if these procedures are 'used to convert political and normative issues into bogus scientific and technical ones'.[88]

Hannah Keven and Lauren Arnold note that EA is both 'a planning and decision support tool'.[89] As a decision-making tool, EA helps decision-makers to determine *whether* to allow an activity. But, as a planning tool, it also helps decision-makers to determine *under which conditions* to allow the activity.[90] By contrast to the binary nature of the decision-making function, the planning decision implies the need to consider various alternatives to the proposed activity with the view of maximizing its benefits while minimizing its impacts.[91] While the decision-making function highlights the need to assess the activity's overall impact, the planning function puts more emphasis on environmental management.[92]

Eric Human and colleagues draw a useful distinction between the objective and subjective components of EA procedures. The objective component 'deals with facts and falls within the domain of the natural and physical sciences'.[93] It focuses on ascertaining the impacts of the decision with the view of identifying the probability and magnitude of the risks of environmental impacts. By contrast, the subjective component 'deals with values'.[94] As Holder noted, '[t]he absence of clear, positive, environmental standards means that ... the ultimate decision whether or not to proceed with a development project will depend on economic judgments and political perspectives'.[95] Thus, the subjective component of any EA 'involves comparing diverse effects ... and weighing the environmental impact against impacts on other societal objectives'.[96] This subjective component implies, more or less explicitly, a comparison of the cost and benefit of the decision, either qualitatively or quantitatively, for instance through a pecuniary valuation. In a typical EA procedure, the comparison is between the environmental cost of the activity under consideration and its economic benefit, thus involving the juxtaposition of 'two contradictory value systems: the ethics of the market and environmental ethics'.[97] While the objective component is usually carried out by experts, the subjective component is normally conducted by political or administrative authorities following some consultations with the public or specific stakeholders.

---

[88] Holder (n 75) 24, citing Francis Sandbach, *Environment, Ideology, and Policy* (Allanheld, Osmun & Co 1980) 96–100.

[89] Kevin and Arnold (eds) (n 71) 3. See also *R (Finch) v Surrey County Council* [2024] UKSC 20, [2024] All ER (D) 71 (Jun) [105] (asserting that, while '[i]dentifying mitigating measures, where they are available, may be a valuable result of the EIA process ... it is not its sole ... purpose').

[90] Eg G Werner, 'Environmental Impact Assessment in Asia: Lessons from the Past Decade', in Asit K Biswas and SBC Agarwal (eds), *Environmental Impact Assessment for Developing Countries* (Elsevier 1992) 16, 19; Glasson and Therivel (n 71) 7–8.

[91] Glasson and Therivel (n 71).

[92] Werner (n 90) 19.

[93] Hyman and others (n 71) 23.

[94] Ibid.

[95] Holder (n 75) 24.

[96] Hyman and others (n 71).

[97] Holder (n 75) 25.

EAs may be conducted in relation to activities defined at different levels of abstractness, or 'tiers',[98] ranging from national policies to their implementation through particular projects. In particular, a distinction can be drawn between the environmental impact assessment (EIA) of a project and the strategic environmental assessment (SEA) of a more general policy, plan, or programme.[99] The conclusions of higher-tier EAs (eg SEA of a national policy on electricity production) address overarching strategic issues, which allows lower-tier EAs (eg an EIA for a power plant) to focus on project-specific questions. Further, there can be several tiers of SEAs, where 'a policy is … defined as an inspiration and guidance for action …, a plan as a set of coordinated and timed objectives for the implementation of the policy …, and a programme as a set of projects in a particular area'.[100] Some jurisdictions have distinct instruments for EIA and SEA (eg the EU),[101] while others apply the same instrument but, often, with some difference in the modalities of implementation (eg the US).[102]

The scope and content of EAs varies across jurisdictions. While all EA frameworks involve an assessment of the direct adverse environmental effects of the proposed activity, some EAs extend to social, cultural, and archaeological impacts,[103] or consider indirect and cumulative effects.[104] Beside the adverse effects, some EAs are interested in assessing the proposed activity's overall sustainability or its merits by documenting anticipated benefits.[105] To the extent that EAs seek to identify 'the best options rather than merely acceptable proposals',[106] they identify and compare potential alterations of the proposed activity, including additional measures that could mitigate adverse impacts.[107]

---

[98] Timothy O'Riordan, 'Beyond Environmental Impact Assessment', in Timothy O'Riordan and Richard Hey (eds), *Environmental Impact Assessment* (Saxon House 1976) 202, 209.

[99] Christopher Wood and Mohammed Dejeddour, 'Strategic Environmental Assessment: EA of Policies, Plans and Programmes' (1992) 10 Impact Assessment 3; Riki Therivel and Maria do Rosário Partidário, *The Practice of Strategic Environmental Assessment* (Earthscan 1996). Note that some of the literature and legislation use different terminologies, referring for instance to SEA as 'EIA of planning' or 'programmatic' EIA. See, respectively, Yang Yang, 'Reformed Environmental Impact Assessment in China: An Evaluation of its Effectiveness' (2020) 11 Journal of Environmental Protection 889, 894; Michael Boots, 'Memorandum for Heads of Federal Departments and Agencies' (CEQ, 18 December 2014).

[100] See Glasson and Therivel (n 71) 313.

[101] Directive 2011/92/EU of the European Parliament and of the Council of 13 December 2011 on the assessment of the effects of certain public and private projects on the environment, [2012] OJ L 26/1; Directive 2001/42/EC of the European Parliament and of the Council of 27 June 2001 on the assessment of the effects of certain plans and programmes on the environment, [2001] OJ L 197/30.

[102] 40 CFR (2024) §1508.1(q)(2)–(3). See discussion in Riki Therivel *Strategic Environmental Assessment in Action* (2nd edn, Routledge 2010) 45–65.

[103] Glasson and Therivel (n 71) 23.

[104] Ibid, 89.

[105] Ibid, 131–36.

[106] Morrison-Saunders (n 72) 53. See generally Robert Gibson and others, *Sustainability Assessment: Criteria and Processes* (Routledge 2005).

[107] Glasson and Therivel (n 71) 141–43.

Likewise, the legal force of EA instruments varies between jurisdictions. EA frameworks can be established by statutory law[108] or by regulation,[109] at times in application of an international[110] or constitutional[111] obligation, and they are often complemented by guidance or other technical documents.[112] While most EA procedures are presented as mandatory, their enforcement may be incomplete, in particular in the EA frameworks adopted by developing countries, international organizations, and non-state actors.

EA differs from decision-making and planning procedures that pursue different goals. For instance, land-planning procedures may invite public participation in response to a research report, but their primary focus is on promoting the optimal use of the land rather than environmental protection.[113] Likewise, law-making procedures may include impact assessments aimed at ensuring the effectiveness of new laws.[114] In practice, environmental considerations have sometimes been inserted in procedures that were primarily designed in the pursuance of other goals, such as development consent[115] and land-planning procedures.[116] Moreover, different assessment procedures may be implemented jointly, or merged into a unique procedure, as in the case of social and environmental assessment procedures[117] and some all-encompassing 'sustainability' assessments.[118] As such, there is not always a clear line between EA and other assessment procedures.

b) A typical EA procedure

No two EA frameworks are identical, but most share some common features. In particular, most EA procedures include six main stages.[119]

---

[108] Eg Impact Assessment Act, SC 2019, c 28; Gesetz über die Umweltverträglichkeitsprüfung (Environmental Impact Assessment Act), 18 March 2021, BGBl.I 540, revised 22 December 2023, BGBl.I Nr. 409 (Ger).

[109] Eg Environmental Impact Assessment Regulations, GN R.984 of GG 38282 (4 December 2014) as amended by GN 325 of GG 40772 and GN 517 of GG 44701 (SA); Ministry of Environment and Forests, 'Environmental Impact Assessment Notification' (14 September 2006) (India).

[110] Eg Directive 2011/92/EU (n 101); Directive 2001/42/EC (n 101); United Nations Convention on the Law of the Sea (adopted 10 December 1982, entered into force 16 November 1994) 1833 UNTS 397 (UNCLOS) art 206.

[111] Eg Constitution (2010) (Kenya) art 69(f).

[112] Eg CEQ, 'National Environmental Policy Act Guidance on Consideration of Greenhouse Gas Emissions and Climate Change' (9 January 2023) 88 Federal Register 1196.

[113] Eg Town and Country Planning Act 1990 (UK) Pt II. See Hyman and others (n 71) 9–13.

[114] See eg Loi organique 2009-403 du 15 avril 2009 relative à l'application des articles 34-1, 39 et 44 de la Constitution, JORF 16 April 2009, art 8 (France).

[115] Environmental Planning and Assessment Act 1979 (NSW), Pts 4 and 5. See also Gerard Maxwell Bates, *Environmental Law in Australia* (LexisNexis Butterworths 2016) 408–9.

[116] Eg Town and Country Planning (Environmental Impact Assessment) Regulations 2017 (UK).

[117] Eg Interinstitutional Agreements between the European Parliament, the Council of the European Union and the European Commission on Better Law-Making [2016] 59 OJ L 123 paras 12–18 (describing an 'impact assessment' procedure applicable to EU law-making process aimed at 'assessing the economic, environmental and social impacts in an integrated and balanced way').

[118] Eg Morrison-Saunders (n 72) 42.

[119] Eg Owen Harrop and Ashley Nixon, *Environmental Assessment in Practice* (Taylor & Francis 2005).

1. *Screening.* The first stage of a typical EA procedure seeks to ensure that competent authorities 'focus on those projects with potentially significant adverse environmental impacts'.[120] A competent authority screens out the proposed activities that do not risk causing any significant environmental impacts, and it may impose a simplified EA procedure when the risk of such impact is limited.[121] Screening can be conducted by application of predefined thresholds (eg the project's size),[122] on a case-by-case basis (ie against a list of criteria),[123] or—as is most frequently the case—through a hybrid approach.[124]

2. *Scoping.* After deciding that an EA procedure must be conducted, a competent authority typically defines the scope of the procedure and determines how it should be conducted. In particular, the authority identifies the impacts that are to be considered and determines the information needed to assess them.[125] Further, the authority defines the boundaries of the assessment, for instance the geographical and temporal span of the risks that are to be assessed.[126] It can define the methodologies that the assessment must follow and the assumption it should build upon, for instance regarding the evolution of ecological resources in a baseline scenario without the proposed activity.[127] Lastly, it decides what alternatives and harm-mitigating measures must be considered.[128] Thus, scoping helps to concentrate the next phases of the EA procedure on the most important issues.

3. *EA study.* The EA study is typically carried out by consultants, which can be engaged either by the project proponent or by the competent authority.[129] It involves a prediction of the impacts, their evaluation, and an assessment of the possibility of mitigating these impacts.[130] The impacts considered may include direct and indirect effects unfolding at different scales and over different periods of time. These effects may or may not be quantifiable, and may be associated with different levels of probability.[131] Many methods can be used, including trend analysis, analogies, scenarios, and expert judgement.[132] EA studies may facilitate an assessment of the significance of the

---

[120] Glasson and Therivel (n 71) 86.
[121] Wood (n 80) 157–58.
[122] Eg Directive 2011/92/EU (n 101), consolidated as of 15 May 2014, art 4(1) and Annex I.
[123] Eg California Pub Res Code (2024) § 21080(d).
[124] Eg Environment Act 2000 (Papua New Guinea), s 50(1) (predefined thresholds for 'level 3' activities) and (2) (criteria for 'level 2' activities). See Glasson and Therivel (n 71) 86–87; Craik (n 80) 885; Wood (n 80) 140.
[125] Morrison-Saunders (n 72) 50. See also Glasson and Therivel (n 71) 88; Neil Craik, *The International Law of Environmental Impact Assessment: Process, Substance and Integration* (CUP 2008) 888–89; Wood (n 80) 159.
[126] Morrison-Saunders (n 72) 50.
[127] Kevin and Arnold (eds) (n 71) 10.
[128] Glasson and Therivel (n 71) 89–95.
[129] Wood (n 80) 183.
[130] Kevin and Arnold (eds) (n 71) 10–12; Morrison-Saunders (n 72) 54–62.
[131] Glasson and Therivel (n 71) 115–16.
[132] Ibid, 116–26.

predicted impacts, for instance by conducting a cost-benefit analysis based on a monetary valuation of environmental impacts[133] or by relying on any available benchmarks such as political objectives and legal standards.[134] Lastly, EA studies consider alternatives and mitigation measures that may avoid, minimize, or offset significant adverse effects.[135] The EA study is made available to the competent agency and, generally, to the public. While the EA study is an extremely long and technical document, it must generally include a non-technical summary.[136]

4. *Review of the EA study.* Once completed, the EA study must be reviewed so that it can inform the final decision. Typically, this starts with a quality control whereby the competent authority verifies that the EA study conforms with the scoping decision and other applicable requirements.[137] This is followed by an appraisal of the findings of the EA study.[138] The public—in particular, the individuals and communities more likely to be affected by the proposed activity—is invited to participate, for instance by submitting written comments and attending public meetings.[139] While public participation is often facilitated at the screening and scoping stages, it is during the review of the EA study that the public tends to play a more central role.[140] In addition to the public, relevant national agencies may be invited to share their views on the project.[141] The project proponent may be invited to respond to the comments received at this stage.

5. *Decision-making.* The EA study and its review may lead to a formal recommendation by an independent agency before a final decision is made.[142] The final decision is typically reached by either a political or administrative authority. The decision may be to approve the proposed activity as it standards, to reject it, or (most frequently) to approve it with conditions.[143] It is generally required that the decision, the reasons for it, and the conditions it includes be made available to the public.[144] The decision must take into account the EA study and its review.[145] The decision may be challenged

---

[133] Ibid, 126–36.

[134] Kevin and Arnold (eds) (n 71) 11.

[135] Eg Directive 2011/92/EU (n 101) Annex IV para 6. See also Kevin and Arnold (eds) (n 71) 11; Glasson and Therivel (n 71) 136–41; Morrison-Saunders (n 72) 86–97.

[136] Wood (n 80) 176.

[137] Glasson and Therivel (n 71) 161–63.

[138] See Kevin and Arnold (eds) (n 71) 12–13; Glasson and Therivel (n 71) 149–67.

[139] Morrison-Saunders (n 72) 64. See also Craik (n 125) 890–2.

[140] Wood (n 80) 204.

[141] Eg Directive 2011/92/EU (n 101), consolidated as of 15 May 2014, art 6(1).

[142] Kevin and Arnold (eds) (n 71) 13.

[143] Morrison-Saunders (n 72) 66; Kevin and Arnold (eds) (n 71) 14; Glasson and Therivel (n 71) 163–64.

[144] Wood (n 80) 224.

[145] See eg *Greater Boston Television Corporation v Federal Communications Commission* (DC Cir 1970) 444 F 2d 841, 851; Directive 2011/92/EU (n 101), consolidated as of 15 May 2014, art 9(1)(b); Environmental Assessment Act (环境影响评价法) of 28 October 2002 (China) arts 13–14;

through administrative or judicial appeal procedures,[146] including broad 'merits review' in some Australian states[147] and the narrower judicial reviews in most other jurisdictions,[148] subject to conditions of standing[149] and various applicable standards of review.[150]

6. *Monitoring.* EA procedures generally extend to a post-decision stage that involves a monitoring of the proponent's compliance with the decision and the impacts of the activity on the environment. Non-compliance can lead to enforcement proceedings.[151] In some cases, the competent authority may also be able to alter the conditions imposed by the decision approving the activity, in particular when some uncertainty was initially attached to the environmental impacts of the activity.[152] Some agencies also conduct post-EA auditing aimed at evaluating the effectiveness of these procedures.[153]

c) Widespread implementation

Governments have long implemented mechanisms aimed at assessing and reducing the environmental harm of proposed activities.[154] Yet, the invention of EA in its modern form is often credited to the US National Environmental Policy Act (NEPA) of 1969.[155] NEPA imposes a procedural requirement on every major federal action.[156] It is 'a generally worded law',[157] whose interpretation has relied on case law[158] and on regulations and guidance adopted by the

Convention on Access to Information, Public Participation in Decision-Making and Access to Justice in Environmental Matters (adopted 25 June 1998, entered into force 30 October 2001) 2161 UNTS 447 (Aarhus Convention) art 6(9). See Morrison-Saunders (n 72) 65–66.

[146] See Wood (n 80) 224; Glasson and Therivel (n 71) 164.

[147] Elizabeth Fisher, 'Administrative Law, Pluralism and the Legal Construction of Merits Review in Australian Environmental Courts and Tribunals', in Linda Pearson, Carol Harlow, and Michael Taggart (eds), *Administrative Law in a Changing State: Essays in Honour of Mark Aronson* (Hart 2008) 325.

[148] Eg Directive 2011/92/EU (n 110), consolidated as of 15 May 2014, art 11(1). See Glasson and Therivel (n 71) 165–67.

[149] See eg Elizabeth Fisher, 'Environmental Impact Assessment: "Setting the Law Ablaze"', in Douglas Fisher (ed), *Research Handbook on Fundamental Concepts of Environmental Law* (Edward Elgar 2016) 422, 430–33.

[150] Eg Jin Zining, 'Environmental Impact Assessment Law in China's Courts: A Study of 107 Judicial Decisions' (2015) 55 Environmental Impact Assessment Review 35; Áine Ryall, 'Enforcing the Environmental Impact Assessment Directive in Ireland: Evolution of the Standard of Judicial Review' (2018) 7 Transnational Environmental Law 515.

[151] Kevin and Arnold (eds) (n 71) 14–15; Glasson and Therivel (n 71) 175–82.

[152] Craik (n 125) 895. See eg Environmental Assessment Act (China) (n 145) art 15; Environment Protection and Biodiversity Conservation Act 1999 (Cth), consolidated as of 15 December 2023, s 143.

[153] Kevin and Arnold (eds) (n 71) 15; Glasson and Therivel (n 71) 182–84. See generally Wood (n 80) 240–46.

[154] See eg Holder (n 75) 3–4 (describing the ancient writ of *ad quod damnum* as an early form of EA in English law).

[155] See eg Glasson and Therivel (n 71) 32–42.

[156] 42 USC (2024) § 4332.

[157] Glasson and Therivel (n 71) 33.

[158] Eg *Kleppe v Sierra Club* (1976) 427 US 390, 411. See Neil Orloff and George Brooks, *The National Environmental Policy Act: Cases and Materials* (Bureau of National Affairs, 1980); Bradley C Karkkainen, 'NEPA and the Curious Evolution of Environmental Impact Assessment in the United

Council on Environmental Quality (CEQ).[159] In addition to NEPA, several US states (eg California,[160] Montana,[161] and New York),[162] territories (eg District of Columbia)[163] and even cities (eg New York City[164] and Rochester, NY)[165] have adopted their own EA procedures, often referred to as 'little NEPAs'.[166]

Other developed countries adopted comparable instruments from the 1970s onwards. Rather than a one-off reform, national EA legislations often emerged through incremental steps, from a simple, discretionary procedure, to a more robust, mandatory one. For instance, the Government of Canada first adopted an EA procedure as merely a cabinet policy in 1973,[167] then as an order in Council in 1984;[168] the latter was treated as a discretionary procedure until 1989, when courts found that it was mandatory.[169] Canada then enacted its first statutory EA framework in 1992,[170] and recodified it in 2012[171] and again in 2019.[172] Australia's EA procedure followed a similar trend: the first federal EA, adopted in 1974, was discretionary and seldom implemented,[173] until it was replaced by a more robust procedure in 1999.[174] All Australian states and territories have also adopted EA procedures applicable within their constitutional competences.[175] Similarly, New

States', in Jane Holder and Donald McGillivray (eds), *Taking Stock of Environmental Assessment: Law, Policy and Practice* (Routledge 2007) 45, 55–58.

[159] Karkkainen (n 158) 45, 58–60; Glasson and Therivel (n 71) 33; Wood (n 80) 20–21. See also 42 USC (2024) §§ 4332, 4344; 40 CFR (2024) §§ 1500–08.

[160] California Environmental Quality Act, 1970 California Stat 2780.

[161] Montana Laws 1971, ch 238 (MEPA). See Montana Code Ann. (West 2024) § 75-1-101.

[162] New York Env't Conserv Law (McKinney 2024) § 8–0101 (New York State Environmental Quality Review Act).

[163] District of Columbia Environmental Policy Act of 1989, 36 DC Register 5741. See DC Code (West 2024) § 8-103.01.

[164] New York City Executive Order 91 of 1977; New York City Executive Order 149 of 2011.

[165] Rochester, NY, City Council Order 78-600 of 1978, as amended.

[166] See Glasson and Therivel (n 71) 38–39; Kenneth S Weiner, 'NEPA and State NEPAs: Learning from the Past, Foresight for the Future' (2009) 39 Environmental Law Reporter News & Analysis (article #10675). See generally CEQ, 'States and Local Jurisdictions with NEPA-like Environmental Planning Requirements' (*National Environmental Policy Act*, n.d.) <https://ceq.doe.gov/laws-regulations/states.html> (listing 20 state or local instruments similar to NEPA).

[167] See Jamie Benidickson, *Environmental Law* (Irwin Law 2013) 257–65.

[168] Environmental Assessment and Review Process Guidelines Order, SOR/84-467 (Canada).

[169] *Canadian Wildlife Federation Inc v Canada (Minister of the Environment)* [1989] FCJ 530, (1989) 99 NR 72.

[170] Canadian Environmental Assessment Act, SC 1992, c 37 (Canada).

[171] Canadian Environmental Assessment Act, SC 2012, c 19 (Canada).

[172] Impact Assessment Act 2019 (n 108).

[173] Environmental Protection (Impact of Proposals) Act 1974 (Cth). See John Formby, 'The Australian Government's Experience with Environmental Impact Assessment' (1987) 7 Environmental Impact Assessment Review 207, 208; Robert J Fowler, 'Environmental Impact Assessment: What Role for the Commonwealth? – An Overview' (1996) 13 Environmental & Planning Law Journal 246, 247.

[174] Environment Protection and Biodiversity Conservation Act 1999 (Cth). See Bates (n 115) 181.

[175] Environmental Planning and Assessment Act 1979 (NSW), Pts 4 and 5; Environmental Protection Act 1994 (Qld); Planning and Development Act 2007 (ACT); Environmental Assessment Act 1982 (NT); Environmental Effects Act 1978 (Vic); Development Act 1993 (SA) s 46; Environmental Protection Act 1986 (WA), pt 4; Land Use Planning and Approvals Act 1993 (Tas).

Zealand's 1973 EA framework[176] was replaced by a more stringent EA procedure under the 1991 Resource Management Act.[177] And the Cabinet of Japan adopted non-binding guidelines on EA for specified projects from 1972 onwards,[178] while local governments established their own EA procedures. In 1984, the Cabinet adopted a general non-binding EA guideline applicable to all major projects.[179] Eventually, the National Diet adopted a general statutory EA in 1997.[180]

Several European states, such as France[181] and the German Federal Republic,[182] established national EA frameworks in the 1970s, but these procedures were viewed as rather ineffective, having many exemptions and few effective guarantees for public participation.[183] In 1985, the European Economic Community defined harmonized rules for a European EIA regime through Directive 85/337/EEC.[184] The EIA Directive was amended several times, and was eventually recodified as Directive 2011/92/EU.[185] Meanwhile, Directive 2001/42/EC established an SEA framework for plans and programmes.[186] These Directives define general norms that Member States must transpose into national law, often with some discretion in doing so.[187] Non-member European states adopted EA instruments from the 1980s onwards (eg Switzerland,[188] Turkey,[189] pre-accession

[176] Environmental Protection and Enhancement Procedures 1973.

[177] Resource Management Act 1991, Pt 6. See Burrell E Montz and Jennifer E Dixon, 'From Law to Practice: EIA in New Zealand' (1993) 13 Environmental Impact Assessment Review 89.

[178] Cabinet decision, 'On the Environmental Conservation Measures Relating to Public Works' (June 1972), cited in Hidefumi Kurasaka, 'Japanese Environmental Impact Assessment Law: Before and after' (2001) 27 Built Environment 16, 16.

[179] Cabinet Decision, 'On the Implementation of Environmental Impact Assessment' (August 1984), cited in Kurasaka (n 178) 17.

[180] Environmental Impact Assessment Act, Law No 81 of 1997, consolidated as of 2014, translation at <https://perma.cc/246Q-NHTJ>. See Kurasaka (n 178) 19; Ministry of the Environment (Japan), 'Environmental Impact Assessment in Japan' <https://perma.cc/2YHQ-KCYZ>.

[181] Décret 77-1141 du 12 octobre 1977 pris pour l'application de l'article 2 de la loi 76-629 du 10 juillet 1976 relative à la protection de la nature, JORF 13 October 1977.

[182] Grundsätze für die Prüfung der Umweltverträglichkeit öffentlicher Maßnahmen des Bundes (Principles for the environmental impact assessment of public federal measures), Cabinet decision of 12 September 1975 (Ger.).

[183] Eg William V Kennedy, 'Environmental Impact Assessment in the Federal Republic of Germany' (1980) 1 Environmental Impact Assessment Review 92; Michel Prieur, 'Le respect de l'environnement et les études d'impact' (1981) 6 Revue juridique de l'environnement 103; Norman Lee and Christopher Wood, 'Environmental Impact Assessment in the European Economic Community' (1980) 1 Environmental Impact Assessment Review 287.

[184] Council Directive 85/337/EEC of 27 June 1985 on the Assessment of the Effects of Certain Public and Private Projects on the Environment [1985] OJ L 175. See generally Elizabeth Fisher, Bettina Lange, and Eloise Scotford, *Environmental Law: Text, Cases, and Materials* (2nd edn, OUP 2019) 704–5.

[185] Directive 2011/92/EU (n 101).

[186] Directive 2001/42/EC (n 101).

[187] Directive 2011/92/EU (n 110) consolidated as of 15 May 2014, art 14; Directive 2001/42/EC (n 101) art 13. See David Langlet and others, *EU Environmental Law and Policy* (OUP 2016) 158.

[188] Message relatif à une loi fédérale sur la protection de l'environnement, FF 1979 III 741; Loi fédérale sur la protection de l'environnement, RO 1984 1122, art 9; Ordonnance relative à l'étude de l'impact sur l'environnement, RO 1988 1931.

[189] Environmental Code No 2872, Turkey Official Gazette 18132, 11 August 1983, art 10; Environmental Impact Assessment Regulation, Turkey Official Gazette 21489, 7 February 1993. See Deniz Tekayak, 'An Overview of Environmental Impact Assessment in Turkey: Issues and Recommendations' (2014) 13 Ankara Avrupa Çalışmaları Dergisi 133.

Poland,[190] and Ukraine);[191] in the following years, their procedures evolved, often in parallel with the evolution of EU law.[192]

Some developing countries established EA frameworks from the late 1970s (eg the Philippines,[193] South Korea,[194] and Venezuela)[195] or 1980s (eg Algeria,[196] Brazil,[197] Mexico),[198] but most of them followed in the 1990s (eg India,[199] Indonesia,[200] Nigeria,[201] Pakistan,[202] South Africa,[203] and Thailand)[204] or 2000s (eg China[205] and North Korea),[206] and almost all had a formal, mandatory procedure in place by the 2010s. Similarly, all post-Soviet states had an EA framework in place by the early 2000s.[207] Non-state actors such as international and private

[190] Order of the Minister of Environmental Protection, Natural Resources and Forestry of 23 April 1990 on investments particularly harmful to the environment and human health and the conditions to be met by an assessment of the environmental impact of investments and buildings prepared by an appraiser, Monitor Polski 16(126). See also Annie Donnelly, Barry Dalal-Clayton, and Ross Huges, *A Directory of Impact Assessment Guidelines* (2nd edn, Russel 1998).

[191] Law on Environmental Expertise, No 45/95-BP, Bulletin of the Supreme Council of Ukraine 1995/8.

[192] Tekayak (n 189) 141; Ayla Bilgin, 'Analysis of the Environmental Impact Assessment (EIA) Directive and the EIA Decision in Turkey' (2015) 53 Environmental Impact Assessment Review 40; Witold Woloszyn, 'Evolution of Environmental Impact Assessment in Poland: Problems and Prospects' (2004) 22 Impact Assessment & Project Appraisal 109; Mariya Krasnova, 'Legal Problems in the Implementation of the Environmental Impact Assessment in Ukraine: A Critical Review' (2021) 4 Grassroots Journal of Natural Resources 91.

[193] Office of the President, Philippine Environmental Policy, Exec Ord No 1151 (6 June 1977) <www.officialgazette.gov.ph/1977/06/06/presidential-decree-no-1151-s-1977/>.

[194] Environmental Preservation Law 1977 (South Korea). See Karl Kim and Duk Hee Lee Murabayashi, 'Recent Developments in the Use of Environmental Impact Statements in Korea' (1992) 12 Environmental Impact Assessment Review 295; Hong Sik Cho, 'An Overview of Korean Environmental Law' (1999) 29 Environmental Law 501.

[195] Organic Law on the Environment, No 5.833 (22 December 2006).

[196] Loi No 83-03 relative à la protection de l'environnement (5 February 1983).

[197] Resolution of the National Environmental Council No 1/1986 of 23 January 1986.

[198] Ley general del equilibrio ecológico y la protección al ambiente (General Law of Ecological Balance and Environmental Protection), DOF 28-01-1988, last revised by DOF 24-01-2024.

[199] Ministry of Environment and Forests, 'Environmental Impact Assessment Notification' (27 January 1994).

[200] Government Regulation No 27/1999, Analysis of Environmental Impacts (7 May 1999) <www.ecolex.org/details/legislation/government-regulation-no271999-re-analysis-of-environmental-impacts-lex-faoc036671/>.

[201] Environmental Impact Assessment Decree, No 86 of 1992, 73:79 SOG A979.

[202] Pakistan Environmental Protection Act No 34 of 1997 <https://faolex.fao.org/docs/pdf/pak115821.pdf>.

[203] Environmental Conservation Act 73 of 1989; Environmental Impact Assessment Regulations, GN R.1182 of GG 18261 (5 September 1997).

[204] Enhancement and Conservation of the National Environmental Quality Act, BE 2535 (1992).

[205] Environmental Assessment Act (China) (n 145).

[206] Law on the Environmental Impact Assessment, adopted by Decree No 1367 of the Presidium of the Supreme People's Assembly on 9 November 2005, as amended by Decree No 2195 of the Presidium of the Supreme People's Assembly on 27 March 2007. See Robert Ward and Dae Un Hong, 'Environmental Impact Assessment in North Korean Environmental Law: Origins, Evolution, and a Comparative Analysis' (2021) 48 Ecology Law Quarterly 38.

[207] See eg Law on Environmental Impact Assessment (18 March 1997) No 85-1 (Kazakhstan); Law No 73-II on Environmental Assessment (25 May 2000) (Uzbekistan); Regulations on the assessment of the impact of planned economic and other activities on the environment in the Russian Federation (Положение об оценке воздействия намечаемой хозяйственной и иной деятельности на окружающую среду в Российской Федерации), 16 May 2000 (No 372) (Russia). See Urszula Rzeszot,

financial institutions have also developed EA procedures, including the World Bank's operation directive of 1989[208] and the Equator Principles adopted by a group of private banks in 2003.[209]

While most jurisdictions have some sort of EA in place,[210] there is no denial that these frameworks differ in many important ways. Most obviously, the terminology used to refer to EA vary between jurisdictions.[211] Further, while most jurisdictions have a mandatory EIA procedure applicable to projects liable to cause significant environmental impacts, SEA is not as commonly or systematically required for more general plans, programmes, and policies, especially in developing countries.[212] The content of EA procedures, in particular the subjective component, differs significantly between liberal democracies—where public participation is often viewed as the 'soul' of EA procedures[213]—and illiberal regimes—where participation is not infrequently limited to a few, hand-picked 'experts'.[214] The implementation of EA faces various practical issues, in particular in developing countries, where it can be undermined by a lack of data availability, financial resources, technical expertise, and enforcement capacity, as well as corruption.[215] And even when these procedures are formally complied with, there is always a risk that they become a mere formality, thus failing to influence decision-making.[216] Nonetheless, these procedures seek to fulfil the same functions—to inform decision-making and to minimize environmental impacts—and, in doing so, to the extent that they are implemented, they tend to face similar issues.

'Environmental Impact Assessment in Central and Eastern Europe', in Judith Petts (ed), *Handbook of Environmental Impact Assessment* (Wiley-Blackwell 1999) vol 2, 123. See also Order on approval of requirements for environmental impact assessment materials (Приказ об утверждении требований к материалам оценки воздействия на окружающую среду), 1 December 2020 (No 999) (Russia).

[208] World Bank, Operational Directive 4.01, 'Environmental Assessment' (1989). See Jose O Castaneda, 'The World Bank Adopts Environmental Impact Assessments' (1992) 4 Pace Yearbook of International Law 241; Jean-Roger Mercier, 'The World Bank and Environmental Impact Assessment', in CJ Bastmeijer and Timo Koivurova (eds), *Theory and Practice of Transboundary Environmental Impact Assessment* (Brill 2007) 291.

[209] See Leonie Schreve, 'The Equator Principles: A Voluntary Approach by Bankers', in Bastmeijer and Koivurova (eds) (n 208) 327 at 327.

[210] See summary table in Glasson and Therivel (n 71) 44–45.

[211] Eg Wood (n 80) 5.

[212] See eg Glasson and Therivel (n 71) 316–29; Thomas B Fischer and Ainhoa González (eds), *Handbook on Strategic Environmental Assessment* (Edward Elgar 2021).

[213] Craik (n 125) 31. See also Wood (n 80) 1.

[214] See generally Chiara Armeni and Maria Lee, 'Participation in a Time of Climate Crisis' (2021) 48 Journal of Law & Society 549.

[215] Omolola Fasina, 'Environmental Impact Assessment for Oil and Gas Projects: A Comparative Evaluation of Canadian and Nigerian Laws' (Master's dissertation, University of Western Ontario, 2017); Alison Clausen, 'An Evaluation of the Environmental Impact Assessment System in Vietnam: The Gap between Theory and Practice' (2011) 31 Environmental Impact Assessment Review 136; John O Kakonge, 'Environmental Impact Assessment in Sub-Saharan Africa: The Gambian Experience' (2006) 24 Impact Assessment & Project Appraisal 57, 58.

[216] See Werner (n 90) 19 (reporting that, out of the first thousands of projects assessed in Thailand and the Philippines, 'not a single project was denied clearance due to environmental reasons'); Bartlett (n 79) 54.

3. The argument for CA

This subsection presents a preliminary argument for the application of EA as a tool for climate change mitigation. It observes that EA has often been applied to assess the transboundary and global effects of proposed activities, and that large quantities of GHG emissions can be associated with certain activities, before outlining the possible analogy between climate impacts and the environmental impacts that have typically been subjected to EA requirements.

a) The assessment of extraterritorial impacts

EA has often been approached as a tool not only to protect the national environment, but also to prevent extraterritorial environmental harm. This trend followed in part from the recognition of EA under international law, for instance in the World Charter for Nature of 1982[217] and the Rio Declaration on Environment and Development of 1992.[218] There is no obvious reason why EA, a quasi-universal practice promoted by international law instruments, would only consider harms unfolding within a state's own territory. Since the late 1970s, the UN Environment Programme (UNEP) advocated for the application of EA to activities likely to cause transboundary environmental harm.[219] Several treaties require the implementation of EA as a tool to protect shared resources such as the marine environment,[220] the Antarctic,[221] and biological diversity.[222] The Member States of the UN Economic Commission for Europe (UNECE)[223] adopted two complementary treaties promoting the implementation of EA procedures in

---

[217] UN General Assembly Resolution 37/7, 'World Charter for Nature' (28 October 1982) para 11(c). EAs were also included in early drafts of the Stockholm Declaration on the Human Environment but withdrawn from the final version due to objections by developing States. See Wade Rowland, *The Plot to Save the World: The Life and Times of the Stockholm Conference on the Human Environment* (Clarke 1973) 54.

[218] Rio Declaration on Environment and Development, A/CONF 151/26 (Vol I) (12 August 1992), principle 17.

[219] UNEP, 'Draft Principles of Conduct in the Field of the Environment for the Guidance of States in the Conservation and Harmonious Utilization of Natural Resources Shared by Two or More States' (1978) 17 ILM 1094, principle 4; UNEP, Governing Council resolution 14/25 (n 76) principle 11.

[220] UNCLOS (n 110) arts 204–06; Agreement under the United Nations Convention on the Law of the Sea on the Conservation and Sustainable Use of Marine Biological Diversity of Areas Beyond National Jurisdiction (adopted 19 June 2023, not yet in force) C.N.202.2023.TREATIES-XXI.10 (BBNJ Agreement), pt IV. See generally Robin Warner, 'Environmental Assessment in Marine Areas beyond National Jurisdiction', in Rosemary Rayfuse (ed) *Research Handbook on International Marine Environmental Law* (Edward Elgar 2015) 291.

[221] Protocol on Environmental Protection to the Antarctic Treaty (adopted 4 October 1991, entered into force 14 January 1998) 30 ILM 1455 (1991) (Madrid Protocol) art 8.

[222] Convention on Biological Diversity (adopted 5 June 1992, entered into force 29 December 1993) 1760 UNTS 79, art 14(1)(a).

[223] The UNECE comprises 56 states located in Europe, Northern America, and Central Asia. See UNECE, 'Member States and Member States Representatives' (22 November 2023) <https://unece.org/member-states-and-member-states-representatives>.

relation to transboundary environmental harm: the Espoo Convention on EIA in a Transboundary Context in 1991[224] and its Kiev Protocol on SEA in 2003.[225]

At times, consideration of transboundary environmental impacts has been limited on the basis of reciprocity. In particular, both the Espoo Convention and its Kiev Protocol focus on cases where an activity implemented in a state causes environmental harm in an area within the jurisdiction of an 'affected Party',[226] to the exclusion of transboundary harm affecting states not parties to these treaties. Yet, national practice has sometimes evolved beyond this condition of reciprocity. For instance, while the EU's EIA and SEA Directives require notification and consultations only when the affected states are EU Member States,[227] they recognize the transboundary nature of the impact as a factor that EA procedures should consider.[228]

States not parties to the Espoo Convention have also applied EA procedures to extraterritorial environmental impacts. For instance, NEPA directed agencies to 'recognize the worldwide and long-range character of environmental problems',[229] and courts have sometimes applied NEPA as requiring consideration of transboundary environmental harm, at least when it affects global commons.[230] In addition, Executive Order 12114 of 1979 establishes a distinct EA procedure applicable to federal action with 'environmental effects abroad', including effects on 'global commons outside the jurisdiction of any nation (eg the oceans or Antarctica)'.[231] Likewise, Canada's successive statutory EA frameworks have required consideration of any environmental impact 'occur[ing] within or outside Canada';[232] Australia's federal EA framework prohibits, in principle, any

---

[224] Convention on Environmental Impact Assessment in a Transboundary Context (adopted on 1 March 1991, entered into force 10 September 1997) 1989 UNTS 309 (Espoo Convention).

[225] Protocol on Strategic Environmental Assessment to the Convention on Environmental Impact Assessment in a Transboundary Context (adopted on 21 May 2003, entered into force 11 July 2010) 2685 UNTS 140 (Kiev Protocol).

[226] See Espoo Convention (n 224) art 1(viii); Kiev Protocol (n 225) arts 2(4), 10.

[227] Directive 2011/92/EU (n 101), consolidated as of 15 May 2014, art 7; Directive 2001/42 (n 101) art 7.

[228] Directive 2011/92/EU (n 101), consolidated as of 15 May 2014, Annex III para 3(c) and Annex IV para 5; Directive 2001/42 (n 101), Annex II para 2.

[229] 42 USC (2024) § 4332(I). See also CEQ, 'Guidance on NEPA Analyses for Transboundary Impacts' (1 July 1997) <https://ceq.doe.gov/docs/ceq-regulations-and-guidance/memorandum-transboundary-impacts-070197.pdf>.

[230] See *Environmental Defense Fund Inc v Massey* (DC Cir 1993) 986 F 2d 528; *Natural Resources Defense Council Inc v US Department of Navy* (CD California 17 September 2002) CV-01-07781 CAS(RZX), 2002 WL 32095131. But see *Natural Resources Defense Council Inc v Nuclear Regulatory Commission* (DC Cir 1981) 647 F 2d 1345, 1365–66; *Nepa Coalition of Japan v Aspin* (D DC 1993) 837 F Supp 466, 467 (both invoking foreign policy consideration to justify that an assessment of transboundary impacts in foreign countries is not required); *Basel Action Network v Maritime Administration* (D DC 2005) 370 F Supp 2d 57, 71–72 (holding that NEPA does not require an assessment of impacts occurring in the high sea).

[231] Executive Order 12114 (4 January 1979) 44 Federal Register 1957, s 2(3)(a). The main limitation of this Executive Order, by contrast to NEPA, is that it does not 'create a cause of action' and, thus, cannot readily be enforced by federal courts. See ibid, s 3(1).

[232] Canadian Environmental Assessment Act 1992 (n 170) s 2(1) (definition of 'environmental effect'). See also Canadian Environmental Assessment Act 2012 (n 171), ss 4(1)(g), 5(b)(iii); Impact

'significant impact on the environment inside or outside the Australian jurisdiction';[233] and New Zealand's Resource Management Act requires consideration for both 'the national [and] global environment'.[234] Neil Craik and Kristine Gu noted that EIA 'is broadly accepted as a legal requirement for managing the marine environment in areas beyond national jurisdiction'.[235]

As part of this trend, the decision of the International Court of Justice (ICJ) in *Pulp Mills* identified 'a requirement under general international law to undertake an environmental impact assessment where there is a risk that the proposed industrial activity may have a significant adverse impact in a transboundary context, in particular, on a shared resource'.[236] The same rule was identified in subsequent judicial pronouncements, including in *Activities in the Area*, *Certain Activities*, *South China Sea*, and *Commission of Small Island States on Climate Change and International Law (COSIS)*.[237] This requirement was not only applied when harm occurred within the territory of another state, but also when harm affected shared resources[238] and, as discussed below,[239] in the context of climate change.

The existence of this extraterritorial EA requirement can be justified in two ways. International courts have generally presented it as an implication of the obligation of states to prevent transboundary environmental harm, noting that this implication had been widely recognized in state practice.[240] An alternative

---

Assessment Act 2019 (n 108), ss 2(b), 6(1)(l); Environmental Assessment Act, No. 10 of 2011, gazette of 30 June 2011 (Botswana), § 68.

[233] Environment Protection and Biodiversity Conservation Act 1999 (Cth), consolidated as of 15 December 2023, s 28. EA frameworks established by Australian states and territories exclude extraterritorial environmental impacts, seemingly for reasons of constitutional competence. See eg Environment Protection Act 2019 (NT), consolidated as of 1 March 2024, s 3 (defining as an object of the act 'to protect the environment of the Territory'); Environmental Protection Act 1986 (WA), consolidated as of as of 24 October 2023, s 4A ('The object of this Act is to protect the environment of the State').

[234] Resource Management Act (New Zealand), consolidated as of 24 August 2023, s 45(2)(b). See also ibid, s 142(3)(a)(i), (v).

[235] Neil Craik and Kristine Gu, 'Strategic Environmental Assessment in Marine Areas beyond National Jurisdiction: Implementing Integration' (2022) 37 International Journal of Marine & Coastal Law 189.

[236] *Pulp Mills on the River Uruguay (Argentina v Uruguay)*, Judgment [2010] ICJ Rep 14 [204].

[237] *Responsibilities and Obligations of States with Respect to Activities in the Area*, Advisory Opinion [2011] ITLOS Reports 10 [141]–[150]; *Certain Activities Carried Out by Nicaragua in the Border Area (Costa Rica v Nicaragua) and Construction of a Road in Costa Rica along the San Juan River (Nicaragua v Costa Rica)*, Judgment [2015] ICJ Rep 665 [101]–[105], [146]–[162]; *South China Sea (Philippines v China)* (2016) 33 RIAA 153 [988]–[991]; *Request for an Advisory Opinion submitted by the Commission of Small Island States on Climate Change and International Law*, Advisory Opinion, 12 May 2024, <www.itlos.org/fileadmin/itlos/documents/cases/31/Advisory_Opinion/C31_Adv_Op_21.05.2024_orig.pdf> [355].

[238] Thus, the ICJ in *Pulp Mills* approached the Uruguay River as a shared resource under joint management in application of a bilateral treaty framework. See *Pulp Mills* (n 236) [81], [173]. See also *South China Sea* (n 237).

[239] See subsection B.1(a).

[240] See *Pulp Mills* (n 236) [204]. See also ILC, Draft Articles on Prevention of Transboundary Harm from Hazardous Activities, in (2001) II(2) ILC Yearbook 148; *Certain Activities* (n 237) [104]; *Activities in the Area* (n 237) [147]–[149]; *South China Sea* (n 237) [944], [964]; *COSIS* (n 237) [354].

theory, presented in particular by John Knox, is that an extraterritorial EA requirement originates from a principle of non-discrimination in international environmental law, according to which 'countries should apply the same environmental protections to potential harm in other countries that they apply to such harm in their own'.[241] While the *Pulp Mills* decision appears to suggest that every state must conduct EA procedures that comply with the same minimal standards, the non-discrimination approach would reflect states' different capacity to carry out effective EA procedures. The arbitral tribunal in *South China Sea* appeared to consider the non-discrimination approach when it found that China's environmental impact studies were insufficient based not only on the practice of other states, but also on China's domestic statutory requirements.[242]

b) The climate impact of proposed activities

Many activities cause large amounts of GHG emissions. Some of the most obvious examples are fossil-fuel fired power plant projects. For instance, the operation of the James H Miller coal-fired power plant, in Alabama, caused 24 Mt CO_2 in 2022.[243] Adding other sources of GHG emissions (eg methane emissions from the transportation of the coal and one-off emissions related to the construction and demolition of the plant) and assuming a lifespan of several decades,[244] the project's total lifetime emissions can be expected to exceed a gigatonne of carbon dioxide equivalent. Land-use projects can also result in large quantities of GHG emissions. For instance, the conversion of a single square kilometre of tropical rainforest to human settlement results on average in the emission of about 15 kt CO_2 from above-ground biomass.[245] As such, large infrastructure projects—for instance the development of Nusantara, the planned new capital city of Indonesia—could

---

[241] John H Knox, 'The Myth and Reality of Transboundary Environmental Impact Assessment' (2002) 96 American Journal of International Law 291, 292. See also Alan Boyle and others, *Birnie, Boyle, and Redgwell's International Law and the Environment* (4th edn, OUP 2021) 187–88; Henri Smets, 'Le principe de non-discrimination en matière de protection de l'environnement' (2000) 4 Revue européenne de droit de l'environnement 3; Craik (n 125) 55–59; OECD Council, 'Implementation of a Regime of Equal Right of Access and Non-Discrimination in Relation to Transfrontier Pollution', Recommendation C(77)28(Final) (17 May 1977), annex, principle 3(a).

[242] See *South China Sea* (n 237) [990].

[243] Environmental Protection Agency, 'eGRID' (database) (30 January 2024) <www.epa.gov/egrid/download-data>, sheet PLNT22, cell AT190. See also Ian Tiseo, 'Biggest polluters in the European Union in 2022' (Statista, 12 June 2023) <www.statista.com/statistics/1130785/biggest-polluters-europ ean-union/> (reporting the existence of several power plants in the European Union with comparable emissions, including the Bełchatów power station, in Poland, with about 38 Gt CO_2 per year).

[244] The plant's four units were put online between 1978 and 1991. See EPA (n 243), sheet UNT22, cells AF691–AF694.

[245] Calculated based on default above-ground biomass in natural forests for tropical rainforest (300 tonnes dry matter per hectare) and a carbon fraction of 0.5, as per Harald Aalde and others, 'Forest Land', in Simon Eggleston and others (eds), *2006 IPCC Guidelines for National Greenhouse Gas Inventories* (IGES 2006) vol 4, ch 4, 63; Jennifer C Jenkin and others, 'Settlements', in Eggleston and others (eds) (n 245) vol 4, ch 8, 19.

cause tens or hundreds of millions of tonnes of carbon dioxide emissions from land conversion alone.[246]

Other activities facilitate large amounts of GHG emissions that occur elsewhere. These include projects aimed at supplying fossil fuels, whether through mining coal and extracting oil and gas or by transporting these fuels towards their end-users, as the combustion of these fuels causes GHG emissions. For instance, the case of *Mid States Coalition for Progress v Surface Transportation Board* related to a railway project that would facilitate the transportation of 100 million tonnes of low-sulphur coal per year from coal mines in Wyoming to Minnesota, where it could be further shipped on the Mississippi river.[247] The combustion of this coal would result in 182 Mt CO_2e per year.[248] In *Waratah Coal Pty Ltd v Youth Verdict Ltd*, the Land Court of Queensland considered that the coal from the proposed Galilee mine would result in 1.58 Gt CO_2e of combustion emissions during its lifespan.[249] But indirect climate impacts may also occur in projects without any significant local environmental impact. For instance, heavy electricity consumption by data centres, in particular crypto-mining facilities, often prompt additional GHG emissions from power plants.[250]

New infrastructure can lock societies in GHG-intensive development pathways or, to the contrary, facilitate the development of cleaner technologies.[251] For instance, the extension of an airport enables more flights, with important implications for GHG emissions.[252] Performance standards applicable on new buildings or vehicles or urban-planning policies applicable to entire neighbourhoods or cities can also have long-term effects on GHG emissions.[253]

Broader plans, programmes, and policies may have wider implications for GHG emissions than individual projects. For instance, the electricity sectors in China and the United States are estimated to cause, respectively, 4.5 and 1.5 Gt CO_2e in annual emissions.[254] In either of these countries, an SEA could conceivably have

---

[246] See discussion in Dennis Normile, 'Indonesia's Utopian New Capital May Not Be as Green as It Looks' (2022) 375 Science 479.

[247] *Mid States Coalition for Progress v Surface Transportation Board* (8th Cir 2003) 345 F 3d 520, 548.

[248] Calculated based on default net calorific value (19 terajoules per gigagram) and emission factors (eg 96,100 kilogrammes of CO_2 per terajoule) for subbituminous coal, as per Amit Garg and others, 'Introduction', in Eggleston and others (eds) (n 245) vol 2, ch 1, 18; Darío R Gómez and others, 'Stationary Combustion', in Eggleston and others (eds) (n 245) vol 2, ch 2, 16.

[249] *Waratah Coal Pty Ltd v Youth Verdict Ltd (No 6)* [2022] QLC 21 [649].

[250] See Camilo Mora and others, 'Bitcoin Emissions Alone Could Push Global Warming above 2°C' (2018) 8 Nature Climate Change 931.

[251] See Gregory Unruh, 'Understanding Carbon Lock-In' (2000) 28 Energy Policy 817; Gregory Unruh, 'Escaping Carbon Lock-In' (2002) 30 Energy Policy 317.

[252] See Benoit Mayer and Wu Lan, 'The Environmental Impact Assessment Ordinance: Two Decades, no Change' *Hong Kong Lawyer* (June 2019); Hacan, 'The Environmental Impact of Heathrow' (January 2021) <https://hacan.org.uk/wp-content/uploads/2021/04/150121-Environmental-Impact-of-Heathrow.pdf> 2.

[253] See eg *Center for Biological Diversity v Department of Fish and Wildlife* (2015) 62 Cal 4th 204.

[254] China, Third Biennial Update Report under the UNFCCC (20 December 2023) <https://unfccc.int/documents/636696>, 10; US, Fifth Biennial Report under the UNFCCC (29 December 2022) <https://unfccc.int/documents/624756>, 53 ('[e]lectric [p]ower [s]ector', 2020).

to deal with a policy concerning the development of the energy sector over several years or decades, with implications for tens of billions of tonnes of caron dioxide emissions. Likewise, decisions concerning the development of transport infrastructure (eg China's high-speed railway system[255] or India's regional airport development programme),[256] industrial development, waste management, and land use can also have massive implications for GHG emissions.

Admittedly, even a large amount of GHG emissions does not directly cause any concrete harm, and its contribution to causing a change in the climate system is marginal. For instance, a back-of-the-envelope calculation shows that a billion tonnes of carbon dioxide emissions would increase the global average temperature by only about half-a-thousandth of a degree Celsius.[257] A large amount of GHG emissions may nonetheless be significant because it contributes to an extremely large problem, with ecological, social and economic risks unfolding at the global scale and over centuries. In other words, one can reasonably be concerned by a very small contribution to a very large problem. Indeed, many governments have shown considerable interest in incremental action to reduce and avoid GHG emissions, and EA appears as one of the tools they could use to pursue this goal.

B. The Adoption of CA Requirements

This section retraces the adoption of EA as a tool for climate change mitigation in multiple jurisdictions across the world. In particular, it documents three mutually reinforcing developments: an evolving international legal context, the reinterpretation of existing EA instruments, and the reform of these instruments.

1. International legal context

Section A has shown that international legal developments have contributed to the diffusion of EA as a national instrument.[258] The present subsection considers more specifically how these developments have contributed to the recognition of EA as a tool for climate change mitigation.[259] It shows that several international

---

[255] Qiong Shen, Yuxi Pan, and Yanchao Feng, 'The Impacts of High-Speed Railway on Environmental Sustainability: Quasi-Experimental Evidence from China' (2023) 10 Humanities and Social Sciences Communications (article #719) 1–19.

[256] K Chandrashekhar Iyer and Nivea Thomas, 'A Critical Review on Regional Connectivity Scheme of India' (2020) 48 Transportation Research Procedia 47.

[257] See Allan and others (n 2) 28 (noting that '[e]ach 1000 Gt CO2 of cumulative CO2 emissions … cause [about] 0.45°C' of global warming).

[258] See in particular subsection A.2(c).

[259] EU developments, albeit international in nature, are addressed in Section C as statutory or regulatory reform, given their greater similarities with national reform.

legal instruments requiring EA in a transboundary context could be interpreted as requiring an assessment of global environmental harm, including climate impacts.

a) Multilateral treaty requirements

No multilateral environmental treaty contains a clear and specific requirement for states to implement EA as a tool for climate change mitigation. In particular, climate treaties do not specifically recommend the use of EA as a tool for climate change mitigation. While Article 4(1)(f) of the UNFCCC requires Parties to '[t]ake climate change considerations into account, to the extent feasible, in their relevant social, economic and environmental policies and actions',[260] this does not necessarily imply a formal EA procedure. Article 4(1)(f) further recommends 'appropriate methods, for example impact assessments', but only with the view of minimizing the adverse impacts of climate action, and not as a way to mitigate climate change.[261] Consistently, the COP has promoted the use of EA to promote adaptation to climate change[262] and to reduce the unintended impacts of climate action,[263] but not to mitigate climate change. This silence of climate treaties and COP decisions on CA is in line with negotiators' unwillingness to prescribe specific measures and policies in application of what they identified as a 'principle of sovereignty of States in international cooperation to address climate change'.[264]

By contrast to climate treaties, several multilateral environmental agreements prescribe the use of EA procedures as a tool to prevent harm to other aspects of the global commons.[265] In particular, the UN Convention on the Law of the Sea (UNCLOS) requires States to assess the 'potential effects' of planned activities under their jurisdiction or control which they have 'reasonable grounds for believing ... may cause substantial pollution of or significant and harmful changes to the marine environment'.[266] Similarly, the Convention on Biological Diversity

---

[260] UNFCCC (n 17) art 4(1)(f).

[261] Ibid. On the distinction between EA as a tool for climate change mitigation and EA as a tool to minimize the adverse effects of climate action, see Chapter I, Section B.

[262] Decision 5/CP.7, 'Implementation of Article 4, Paragraphs 8 and 9, of the Convention', FCCC/CP/2001/13/Add 1 (21 January 2002) 32, para 7(a)(iii).

[263] Decision 3/CMP.1, 'Modalities and Procedures for a Clean Development Mechanism as Defined in Article 12 of the Kyoto Protocol', FCCC/KP/CMP/2005/8/Add 1 (28 November 2005) 6, Annex para 37(c); Decision 5/CMP.1, 'Modalities and Procedures for Afforestation and Reforestation Project Activities under the Clean Development Mechanism in the First Commitment Period of the Kyoto Protocol', FCCC/KP/CMP/2005/8/Add 1 (30 March 2006) 61, Annex para 12(c); Decision 6/CMP.1, 'Simplified Modalities and Procedures for Small-Scale Afforestation and Reforestation Project Activities under the Clean Development Mechanism in the First Commitment Period of the Kyoto Protocol and Measures to Facilitate their Implementation', FCCC/KP/CMP/2005/8/Add 1 (30 March 2006) 81, Annex para 14(c); Decision 9/CMP.1, 'Guidelines for the Implementation of Article 6 of the Kyoto Protocol', FCCC/KP/CMP/2005/8/Add 2 (30 March 2006) 2, Annex para 33(d); Decision 23/CMA.4, 'Report of the Forum on the Impact of the Implementation of Response Measures', FCCC/PA/CMA/2022/10/Add 3 (17 March 2023) 39, paras 20(c), 27.

[264] UNFCCC (n 17) preamble para 10.

[265] Craik (n 125) 96–105.

[266] UNCLOS (n 110) art 206.

calls, 'as far as possible and as appropriate', for the use of EIA procedures of 'proposed projects that are likely to have significant adverse effects on biological diversity'[267] The Agreement on Marine Biological Diversity of Areas beyond National Jurisdiction (BBNJ Agreement) further obliges its Parties to ensure that an EIA is conducted for planned activities under their jurisdiction or control when 'the Party has reasonable grounds for believing that the activity may cause substantial pollution of or significant and harmful changes to the marine environment'[268] Nothing in these provisions, or in the definition that UNCLOS provides of 'pollution of the marine environment',[269] suggests that the harm to the marine environment must be local rather than global in nature. Further, it is well known that climate change has major adverse effects on both the marine environment and biological diversity.[270]

Thus, the International Tribunal for the Law of the Sea found in *COSIS* that the EIA requirement under Article 206 of UNCLOS could be applied to climate change. According to the Tribunal, '[a]ny planned activity, either public or private, which may cause substantial pollution to the marine environment or significant and harmful changes thereto through anthropogenic GHG emissions, including cumulative effects, shall be subjected to an environmental impact assessment'[271] On the other hand, the advisory opinion says little about the content of this requirement or the conditions under which it applies.

b) Espoo Convention and its Kiev Protocol

The UNECE has played an important role in promoting EA procedures, including in relation to global environmental harm.[272] In 1991, negotiations convened by the UNECE led to the adoption of the Espoo Convention, which requires its Parties to adopt EIA procedures applicable to activities likely to cause significant adverse transboundary impact.[273] The Kiev Protocol, adopted in 2003, further calls for the adoption of SEA procedures for 'plans and programmes which are likely to have significant environmental, including health, effects'[274] Participation to the Espoo Convention and its Kiev Protocol is opened to any UN Member State[275] (although any non-UNECE state is yet to grasp this opportunity), and the UNECE Member

---

[267] Convention on Biological Diversity (n 222) art 14(1).
[268] BBNJ Agreement (n 220) art 30(1)(b).
[269] UNCLOS (n 110) art 1.1(4).
[270] Pörtner and others, 'Summary for Policymakers' (n 5) 14.
[271] *COSIS* (n 237) [367].
[272] See Benoit Mayer, 'Environmental Assessments in the Context of Climate Change: The Role of the UN Economic Commission for Europe' (2019) 28 Review of European, Comparative & International Environmental Law 82.
[273] Espoo Convention (n 224) art 2(2).
[274] Kiev Protocol (n 225) art 4(1).
[275] Amendment to the Convention on Environmental Impact Assessment in a Transboundary Context (adopted 27 February 2001, entered into force 26 August 2014) 2999 UNTS 351, art 17(3); Kiev Protocol (n 225) art 23(3).

States have sought to present these treaties as a model that non-UNECE states could follow.[276]

The application of the Espoo Convention and its Kiev Protocol to climate impacts is not clear, as neither of these treaties expressly mentions climate change. The Espoo Convention applies only in relation to transboundary impacts that are 'not exclusively of a global nature',[277] a caveat that could be read as excluding climate impacts from the scope of the Convention. By contrast, the Kiev Protocol applies to any environmental impact, notwithstanding whether it is of a transboundary, global, or purely national nature. However, both treaties have occasionally been invoked as a legal ground for a CA requirement.[278]

Indeed, the Parties to the Espoo Convention and its Kiev Protocol have repeatedly suggested that the two treaties may apply to climate impacts. In 2004, the Meeting of the Parties to the Convention adopted a guidance document on the practical application of the Convention which mentioned 'activities with linkages to climate change' among long-range transboundary impacts within the scope of the Convention.[279] In the following years, the Parties organized several events to discuss the relevance of EAs to climate change mitigation, in which they often agreed on the potential role of, in particular, SEAs.[280] Thus, in 2011, UNECE Member States declared that SEA 'can be an appropriate mechanism to introduce the consideration of climate change impacts in plans and programmes that are prepared for regional development planning'.[281]

Following the adoption of the Paris Agreement, the UNECE Secretariat took a more assertive approach to the potential role of the Convention and its Protocol in addressing climate change. In 2016, the Secretariat affirmed that the Kiev Protocol 'provides a mechanism for integrating climate change considerations into sectoral

---

[276] Minsk Declaration, adopted by the representatives of the UNECE Member States and the European Union in June 2017, ECE/MP.EIA/23/Add 1–ECE/MP.EIA/SEA/7/Add 1 (19 September 2017) 35 paras 13, 16. See also CM Kersten, 'Rethinking Transboundary Environmental Impact Assessment' (2009) 34 Yale Journal of International Law 173, 178.

[277] Espoo Convention (n 224) art 1(viii).

[278] See eg Peter Splinter, 'Slow Down Mr. Sunak: The UK Has an Obligation to Consult its Neighbours before You Can Authorise Millions of Tonnes of CO_2 Emissions' *OpinioJuris* (29 March 2024) <https://opiniojuris.org/2024/03/29/slow-down-mr-sunak-the-uk-has-an-obligation-to-consult-its-neighbours-before-you-can-authorise-millions-of-tonnes-of-CO2-emissions/>.

[279] Meeting of the Parties to the Convention on Environmental Impact Assessment in a Transboundary Context, Decision III/4, 'Guidelines on Good Practice and on Bilateral and Multilateral Agreements', ECE/MP.EIA/6 (13 September 2004) 56, app para 26.

[280] See Report of the meeting of the Parties to the Convention on Environmental Impact Assessment in a Transboundary Context on its Fourth Meeting, held in Bucharest from 19 to 21 May 2008, ECE/MP.EIA/10 (28 July 2008) 10 paras 32–33; Swedish Ministry of the Environment, Cooperation on the EIA Convention in the Baltic Sea subregion: Report of a Seminar in Vilnius 22–23 October 2009 (March 2010) <https://unece.org/fileadmin/DAM/env/eia/documents/Events/VilniusOct09/VilniusReport.pdf>, 7. See also Meeting of the Parties to the Convention on Environmental Impact Assessment in a Transboundary Context, Report of the Working Group on Environmental Impact Assessment on its Fourteenth Meeting, UN Doc ECE/MP.EIA/WG.1/2010/5 (18 January 2011), 11.

[281] Declaration adopted by the high-level representatives of the UNECE Member States and of the European Union in June 2011, ECE/MP.EIA/SEA/2 (16 August 2011) 31 para 8.

development plans and programmes'.[282] In 2017, the Secretariat cautioned states that 'the provisions of the Protocol or the Convention are not yet consistently and fully used for addressing climate change'.[283] That year, a high-level panel discussion was organized during the seventh session of the Meeting of the Parties to the Convention,[284] and the Parties affirmed the role of SEA as 'a key tool for the development of national climate change action and planning, and for the incorporation of specific climate change mitigation and adaptation measures into regional development and sectoral plans, programmes and policies'.[285]

The 2023 meetings of the Parties to the Convention and its Protocol reflected an emerging consensus on an interpretation of the two treaties as requiring the application of EA as a tool for climate change mitigation. UNECE Executive Secretary Tatiana Molcean observed that '[o]ver the past decade', the treaties had become 'a more widely applied tool for climate change ... mitigation' by 'promoting carbon neutrality and accelerating the energy transition', in particular 'through incorporating climate related considerations and alternatives into planning'.[286] The EU Commission stated that the '[c]orrect implementation' of the two treaties 'is an important building block in our fight against climate change'.[287] The UNECE Member States agreed to call for a strengthening of the implementation of the Convention and its Protocol 'with the aim of accelerating the energy transition, and to promote carbon neutrality ... through the consideration of ... climate impacts of projects, plans and programmes'.[288] A programme of informal consultations funded by Italy led to a recognition of the need for further guidance on the way EAs can best be deployed as a tool for climate change mitigation.[289]

---

[282] UNECE, 'UNECE and Climate Change' (2016) <www.unece.org/fileadmin/DAM/information/ 1529385_UNECE_climate_change_interactive.pdf>, 9.

[283] UNECE, Information on Panel Discussion on the Role of the Protocol and the Convention in Addressing Climate Change, ECE/MP.EIA/2017/INF.10, 23 May 2017) para 7.

[284] Report of the Meeting of the Parties to the Convention on its Seventh session and of the Meeting of the Parties to the Convention Serving as the Meeting of the Parties to the Protocol on its Third Session, ECE/MP.EIA/23–ECE/MP.EIA/SEA/7 (19 September 2017) paras 53–62.

[285] Meeting of the Parties to the Convention on Environmental Impact Assessment in a Transboundary Context and Meeting of the Parties to the Protocol on Strategic Environmental Assessment, Decision VII/7–III/6, 'Development of a Strategy and an Action Plan for the Future Application of the Convention and the Protocol', ECE/MP.EIA/23/Add 1–ECE/MP.EIA/SEA/7/Add 1 (19 September 2017) 34 para 9.

[286] Tatiana Molcean, Statement at the 9th Session of the Meetings of the Parties to the Espoo Convention and the 5th Session of the Meeting of the Parties to the Protocol on SEA (14 December 2023) <https://unece.org/sites/default/files/2023-12/UNECE_ES_statement_Espoo_MOPs_14D ec2023_pm_0.pdf>.

[287] European Commission, Statement at the 9th Meeting of the Parties to the Espoo Convention and 5th Meeting of the Parties to the Protocol on SEA (14 December 2023) <https://unece.org/sites/defa ult/files/2023-12/EuroeanCommission_HLS_EU_Statement_2023-12-14_final.pdf>.

[288] Geneva Declaration, draft for adoption by the high-level representatives of the UNECE member States and the European Union in December 2023, ECE/MP.EIA/2023/11–ECE/MP.EIA/SEA/2023/ 11 (15 September 2023) para 2.

[289] See eg Guiseppe Magro, statement at a High-Level Event on Green Financing (14 December 2023) <https://unece.org/sites/default/files/2023-12/Italy_Speech_HLE_2023_ProfMagro_ Green%20financing_ENG.pdf>; UNECE, Note by the Bureau prepared with support from two

c) Customary law

Besides its conventional legal bases, EIA can also be approached as a legal obligation under customary international law.[290] In *Pulp Mills*, the ICJ identified 'a requirement under general international law to undertake an environmental impact
assessment where there is a risk that the proposed industrial activity may have a
significant adverse impact in a transboundary context, in particular, on a shared
resource.'[291] The existence of this customary norm was confirmed by subsequent
decisions.[292] The International Tribunal for the Law of the Sea, in particular, noted
that this requirement 'may also apply to activities with an impact on the environment in an area beyond the limits of national jurisdiction.'[293]

It is unclear whether this customary EIA requirement extends to climate impacts. The ICJ suggested that this requirement could be identified from an inductive approach, based on state practice accepted as law, but also by deduction
from other norms of general international law: 'due diligence, and the duty of vigi-
lance and prevention which it implies, would not be considered to have been exer-
cised' if a state was to implement a project liable to cause significant environmental
harm without undertaking an EIA.[294] From an inductive approach, CA appears to
be a widespread practice, but not necessarily one reflective of acceptance as law.[295]
By contrast, a deductive approach provides stronger support to the idea that CA
may be a customary international law requirement. It is generally accepted that
each state has a general international law obligation to exercise due diligence with
the view of respecting the rights of other states, as a corollary of the existence of
such sovereign rights under general international law.[296] This due diligence obligation has frequently been applied when the activity would cause transboundary
environmental harm with direct consequences for another state,[297] as fewer disputes were concerned with global environmental harm. However, this obligation

consultants, Energy transition, circular economy and green financing: Role of the Espoo Convention
and its Protocol, ECE/MP.EIA/2023/14–ECE/MP.EIA/SEA/2023/14 (2 October 2023) para 27
and box 2.

[290] See ILC, Prevention of Transboundary Harm (n 240) 148.
[291] *Pulp Mills* (n 236) [204].
[292] See eg *Certain Activities* (n 237) [104].
[293] *Activities in the Area* (n 237) [147]–[149], [148]. See also *South China Sea* (n 237) [987]–[993].
[294] See *Pulp Mills* (n 236). On the deductive and inductive approaches to the identification of customary international law, see generally Anthea Elizabeth Roberts, 'Traditional and Modern Approaches
to Customary International Law: A Reconciliation' (2001) 95 American Journal of International
Law 757.
[295] See discussion in Benoit Mayer, 'Climate Change Mitigation as an Obligation under Customary
International Law' (2023) 48 Yale Journal of International Law 105.
[296] This obligation is the corollary of the existence of sovereign rights. See eg *Island of Palmas
(Netherlands v United States)* (1928) 2 RIAA 829, 839; Jutta Brunnée, 'Sic utere tuo ut alienum non
laedas', in Anne Peters (ed), *Max Planck Encyclopedias of International Law* (online edn, OUP 2024).
[297] See *Trail Smelter Arbitration (United States v Canada)* (1938 and 1941) 3 RIAA 1905; *Legality of
the Threat or Use of Nuclear Weapons*, Advisory Opinion [1996] ICJ Rep 226 [29]; *Pulp Mills* (n 236)
[204]; *Certain Activities* (n 237) [104].

appears to apply a fortiori when the harm is suffered not by one or a few states in particular, but by most or all states.[298] Thus, most scholars agree that this due diligence obligation implies an obligation to mitigate climate change,[299] which, in turn, could imply an obligation to conduct CA before approving projects that may cause a significant climate impact.

However, one could object that while the general obligation of due diligence implies an EIA requirement in a transboundary context, it may not have the same implication in a global context. Such would be the case, in particular, if EIA was not as effective a tool to prevent global environmental harm as it is in relation to transboundary harm, or if its effectiveness was not as well recognized by states. Admittedly, treaties that require the implementation of EIAs in a transboundary context do not necessarily impose the same obligation in a global context. Yet, in *COSIS*, ITLOS noted that 'most of the participants in the ... proceedings were of the view that there is an obligation to conduct an environmental impact assessment' under both UNCLOS and customary international law.[300] Indeed, as will appear in the following subsections, there is growing state practice consistent with the recognition by states of an obligation to conduct EIA in a global context, in particular as a tool for climate change mitigation, which provides some evidence of the emergence of a customary norm.[301]

2. Reinterpretations of EA instruments

This subsection shows that courts, national agencies, and project proponents have interpreted existing EA instruments as requiring an assessment of climate impacts even when these instruments did not contain any specific provision in this regard. These interpretations were generally based, at least implicitly, on an analogy between climate and environmental impacts.

a) Judicial decisions
Many national EA instruments require the assessment of 'environmental impacts' in broad terms, which can be interpreted as including climate impacts.[302] Building

---

[298] See ILC, Draft Guidelines on the Protection of the Atmosphere, in ILC Report, 76th Session, A/76/10 (2021) 10, Guideline 3.

[299] See eg International Law Association, Resolution 2/2014, 'Declaration of Legal Principles Relating to Climate Change', reproduced in (2014) 76 International Law Association Reports of Conferences 21, art 7A; Benoit Mayer, *International Law Obligations on Climate Change Mitigation* (OUP 2022) 88.

[300] *COSIS* (n 237) [353].

[301] See also Benoit Mayer, 'Climate Assessment as an Emerging Obligation under Customary International Law' (2019) 68 International & Comparative Law Quarterly 271.

[302] See Jacqueline Peel, 'Environmental Impact Assessment and Climate Change', in Michael Faure (ed), *Elgar Encyclopedia of Environmental Law* (Edward Elgar 2016) vol 1, 348, 351.

on this, courts have identified implicit CA requirements in existing EA instruments since at least the early 2000s.

US federal courts were among the first to interpret existing EA instruments as requiring an assessment of climate impacts.[303] NEPA, a succinct piece of legislation, only requires consideration for 'the environmental impact of the proposed action',[304] without specifying what constitutes such an 'impact'. In 2003, the District Court for the Southern District of California and the Court of Appeal for the Eighth Circuit reached independently the same conclusion: the carbon dioxide emissions of coal power plants could fall within the scope of NEPA reviews due to their 'potential environmental impacts'.[305] Five years later, the Court of Appeal for the Ninth Circuit highlighted the relevance of NEPA reviews as a tool for climate change mitigation in *Center for Biological Diversity v National Highway Traffic Safety Administration*, a case concerning the NEPA review of a new fuel economy standard applicable to light trucks.[306] The Court found that the EA report failed to assess the effect of the new policy on carbon dioxide emissions, in particular the '*actual* environmental effects resulting from those emissions'.[307] While the respondents emphasized the global natural of climate change,[308] the Court asserted that '[t]he impact of greenhouse gas emissions on climate change is precisely the kind of cumulative impacts analysis that NEPA requires agencies to conduct'.[309] State courts have made similar interpretations of 'mini-NEPA' instruments, for instance in California[310] and Montana.[311]

Other early decisions were made in Australian states.[312] For instance, the 2004 decision by the Victoria Civil and Administrative Tribunal in *Australian Conservation Foundation v Latrobe City Council* concerned the EA relating to the proposed amendment of the planning scheme of the Latrobe municipality that would allow an extension of the Hazelwood coal mine to supply fuel to the Hazelwood power plant.[313] The Tribunal had to decide whether a panel tasked with conducting the assessment had to consider submissions concerning GHG

---

[303] See Michael Gerrard, 'Climate Change and the Environmental Impact Review Process' (2008) 22 Natural Resources & Environment 20.

[304] 42 USC (2024) § 4332(C)(i).

[305] *Border Power Plant Working Group v Department of Energy* (SD California 2003) 260 F Supp 2d 997, 1029. See also *Mid States Coalition for Progress* (n 247) 550.

[306] *Center for Biological Diversity v National Highway Traffic Safety Administration* (9th Cir 2008) 538 F 3d 1172, 1181.

[307] Ibid, 1216 (emphasis in the original).

[308] Ibid, 1217.

[309] Ibid.

[310] See *Center for Biological Diversity v Department of Fish and Wildlife* (n 253) 218.

[311] See *Held v State of Montana* (Montana District Court 14 August 2023) CDV-2020-307, 2023 WL 5229257; *Montana Environmental Information Center v Montana Department of Environmental Quality* (Montana District Court 6 April 2023) DV21-01307 <https://climatecasechart.com/wp-content/uploads/case-documents/2023/20230406_docket-DV21-01307_order.pdf>.

[312] See eg *Gray v Minister for Planning* [2006] 152 LGERA 258 (NSWLEC); *Hunter Environment Lobby Inc v Minister for Planning* [2011] NSWLEC 221 [100].

[313] *Australian Conservation Foundation v Latrobe City Council* (2004) 140 LGERA 100 [3].

emissions. The Tribunal opined that consideration for GHG emissions fell within the scope of the EA procedure,[314] in particular because these emissions affected the statutory objectives concerned with 'the maintenance of ecological processes' and the balancing of 'the present and future interests of all Victorians'.[315]

More recent cases have discussed the need to implement a CA at different stages of complex decision-making processes. For instance, the Norwegian Supreme Court's 2020 decision in *Greenpeace Nordic Association v Ministry of Petroleum and Energy* was satisfied that a CA was not necessary at the stage of granting oil production licences on the ground that the climate impact of oil production were to be considered either at the earlier stage (when the area was 'opened' for petroleum activities) or at a later stage (when the actual exploitation projects would be proposed).[316] The same year, the UK Supreme Court held that the climate impact of expanding Heathrow Airport would primarily be assessed during the EIA of the project itself, rather than during an SEA of the national policy statement establishing a framework for this project to be proposed.[317] French administrative courts, meanwhile, clarified that CA had to be conducted for the exploitation permit of power plants, but not in relation to other administrative procedures that power plants had to follow.[318] These developments reflect an improving integration of CA with other decision-making processes.

Relevant decisions were also made in developing countries, for instance in relation to new coal-fired power plant. In 2017, in the case of *Earthlife Africa Johannesburg v Minister of Environmental Affairs*, the South Gauteng High Court in Johannesburg had to consider whether competent national authorities were required to consider the climate impact of a proposed coal-fired power station before authorizing it under the National Environmental Management Act.[319] The Act required consideration 'of all relevant factors, which may include ... any pollution, environmental impacts or environmental degradation likely to be caused if the application is approved'.[320] The Court held that this provision 'logically expects consideration of climate change'.[321] Two years later, Kenya's National Environmental

---

[314] Ibid [49].

[315] Planning and Environment Act 1987 (Vic), s 4(1)(b), (g). See also *Australian Conservation Foundation v Latrobe City Council* (n 313) [43].

[316] *Greenpeace Nordic Association v Ministry of Petroleum and Energy* (2020) Case No 20-051052SIV-HRET (Supreme Court) [185]–[192], [217]–[223] (Høgetveit Berg J). But see also ibid [259]–[267], [272]–[275], [280]–[287] (dissenting opinion of Webster J).

[317] *R (Friends of the Earth Ltd) v Heathrow Airport Ltd* [2020] UKSC 52, [2021] 2 All ER 967 [98], [113], [132], [157], [166].

[318] See eg Conseil d'État (State Council) (6ème–5ème chambres réunies), 10 February 2022, 455465, ECLI:FR:CECHR:2022:455465.20220210 [4]–[5].

[319] *Earthlife Africa Johannesburg v Minister of Environmental Affairs* [2017] 2 All SA 519 (GP).

[320] National Environmental Management Act 107 of 1998 (South Africa) (amended by National Environmental Management Laws Amendment Act 2 of 2022) s 24O(1)(b).

[321] See *Earthlife Africa Johannesburg* (n319) [78]. But see *Trustees of the Groundwork Trust v Acting Director-General: Department of Water and Sanitation*, WT02/18/MP, [2020] ZAWT 1 [20] (finding that a CA is not a necessary component of the water use licence application, given that 'an application

Tribunal reached a similar decision in *Save Lamu v National Environmental Management Authority*.[322] Interpreting a regulatory provision according to which an EIA report must state 'the potential environmental impacts of the project',[323] the Tribunal found that '[c]limate change issues are pertinent' in a coal plant project[324] and, therefore, had to be included in the assessment of this project.[325] Similarly, in 2022, the Supreme Court of Chile interpreted the national EIA requirement as requiring the assessment of the climate impact of a power plant.[326] The same year, the Administrative Court of Bandung, in Indonesia, revoked the environmental permit of a coal plant on the ground that the EIA had not considered the plant's climate impact.[327]

Another remarkable decision was reached by the National Green Tribunal of India in 2019 in the case of *Pandey v Union of India*.[328] The petition contended that the Indian government had failed to take adequate action to mitigate climate change, among other things because it had not adequately implemented the national EIA framework.[329] The 2006 EIA Notification requires the assessment of 'all relevant environmental concerns'.[330] The petitioner argued that this had to be interpreted as implying a requirement for a project proponent to 'divulge information as to how the proposed project would impact the climate',[331] and that national authorities had failed to enforce this requirement.[332] The National Green Tribunal agreed with the petitioner's interpretation of the EA instrument—'[t]he issue of climate change is certainly a matter covered in the process of impact assessment'[333]— but asserted that the government had properly applied this requirement.[334]

for environmental authorisation must invariably precede, and be submitted with [a water use licence] application').

[322] *Save Lamu v National Environmental Management Authority*, NET 196/2016, judgment (26 June 2019) <https://climatecasechart.com/wp-content/uploads/non-us-case-documents/2019/20190626_Tribunal-Appeal-No.-Net-196-of-2016_decision.pdf>.
[323] Environmental (Impact Assessment and Audit) Regulations 2003, Cap 387, s 7(1)(f).
[324] See *Save Lamu* (n322) [138].
[325] Ibid [155].
[326] Corte Suprema de Justicia (Supreme Court), 19 April 2022, *Asociación de Prestadores de Servicios Turísticos de Mejillones v Director Regional del Servicio de Evaluación Ambiental*, Rol 71628-2021, civil. See also *Fundación Greenpeace Argentina v Estado Nacional*, Cámara Federal de Apelaciones de Mar Del Plata (Federal Court of Appeal of Mar del Plata), 3 June 2022 <https://climatecasechart.com/non-us-case/greenpeace-argentina-et-al-v-argentina-et-al/> (Argentina) (ordering an assessment of the climate impact of a fossil-fuel project).
[327] Administrative Court of West Java in Bandung, *WALHI v Provincial Government of West Java*, 13 October 2022, Decision No 52/G7LH/2022/PTUN.Bdg <https://elaw.org/resource/id_tanjungjatia_13oct22> [20].
[328] *Pandey v Union of India*, 187/2017, [2019] NGT 843 https://perma.cc/32JF-JQAB.
[329] Ibid [36].
[330] Ministry of Environment and Forests, 2006 EIA Notification (n 109) s 7(i)II(i).
[331] See *Pandey v Union of India*, 187/2017, petition (25 March 2017) https://perma.cc/3XSR-SCXG [39].
[332] Ibid [40].
[333] See *Pandey* (2019) (n 328) [2].
[334] Ibid [3]. An appeal is pending before the Supreme Court. See *Pandey v Union of India* (2021) Supreme Court of India (CA No 388/2021), diary no 14040/2019.

These judicial decisions interpreting EA instruments have raised many questions about the modalities of these implicit CA requirements. In a number of cases where the existence of a CA requirement was not necessarily in question, courts were called upon to specify some of these modalities.

Thus, multiple cases have sought to determine whether and to what extent a CA should account for the activity's indirect effects on GHG emissions, in particular with regard to projects on the extraction of fossil fuel, where downstream emissions from the combustion of the fuel tend to be far more significant than on-site emissions from the extraction of the fuel. At times, the downstream use occurs on the same site and under the control of the same company. For instance, the proposed extension of the Hazelwood coal mining project at issue in *Australian Conservation Foundation v Latrobe City Council* was directly related to the continued operation of the Hazelwood power plant.[335] As such, the Victorian Civil and Administrative Tribunal had no difficulty holding that the EA had to account for the GHG emission from both the extraction and the combustion of the coal.[336]

However, the inclusion of downstream emissions in CAs is more problematic when these emissions occur under the control of another company and in a different jurisdiction. In *Gray v Minister for Planning*, the applicant challenged the validity of the EA of a proposed coal mining project on the ground that the EA did not consider the impact of the GHG emissions from the downstream combustion of the coal.[337] The competent authority had requested a 'detailed greenhouse gas assessment' to be included in the EA report,[338] but it had accepted a report that only assessed the direct emissions generated by the mine itself.[339] In its 2006 decision, the Land and Environment Court of New South Wales held that the failure to take downstream emissions into account constituted a breach of the principles of intergenerational equity and precaution due to the significance of these emission and their likely impact on Australia and New South Wales.[340] Several subsequent cases on Australian coal mines were decided, consistently with *Gray*, as requiring the assessment of downstream emissions from the combustion of exported coal.[341]

---

[335] See *Australian Conservation Foundation v Latrobe City Council* (n 313) [3].

[336] See ibid [49].

[337] See *Gray* (n 312). See generally Anna Rose, '*Gray v Minister for Planning*: The Rising Tide of Climate Change Litigation in Australia' (2007) 29 Sydney Law Review 725.

[338] See *Gray* (n 312) [36].

[339] Ibid [20]–[24].

[340] Ibid [97]–[98], [126], [135].

[341] See *Gloucester Resources Ltd v Minister for Planning* [2019] NSWLEC 7; *Waratah Coal Pty Ltd v Youth Verdict Ltd (No 6)* [2022] QLC 21; *Wollar Property Progress Association v Wilpinjong Coal Pty Ltd* [2018] NSWLEC 92 (confirming the requirement for CA in EIA, including an assessment of downstream emissions); *Australian Coal Alliance Inc v Wyong Coal Pty Ltd* [2019] NSWLEC 31 (para 84: 'the PAC [Planning Assessment Commission] has had regard, as it was obliged to by cl 14(1) and (2) of the Mining SEPP, [to] the question of downstream emissions that will arise from the burning of the coal proposed to be produced from this mine'); *KEPCO Bylong Australia Pty Ltd v Independent Planning Commission* [2021] NSWCA 216. But see also, by contrast, *Coast and Country Association of Queensland Inc v Smith* [2016] QCA 242 (Queensland).

US courts have encountered comparable questions with regard to projects aimed at transporting fossil fuels or at transmitting the electricity produced from it. In particular, the case of *Border Power Plant Working Group v Department of Energy* concerned the building of electricity transmission lines that would connect Mexican coal-fired power plants to the Southern Californian power grid.[342] A key question was whether, beyond the environmental impacts from the construction and operation of the transmission lines, the NEPA review should have assessed the environmental impacts of the operation of the power plants.[343] The District Court for the Southern District of California found that there was a sufficient causal nexus between the construction and operation of the transmission lines on the one hand and the operation of the power plant on the other hand for the latter to fall within the scope of the NEPA review of the former.[344] The same year, the Court of Appeal for the Eighth Circuit held in *Mid States Coalition for Progress* that the NEPA review for a vast railways development project aimed at transporting coal from mines to power plants should consider the emissions resulting from the effect on coal consumption.[345]

Yet, the question remains far from being settled, as recent cases illustrate, in particular in Europe. Until recently, the Scottish Court of Session (in 2021) and the Court of Appeal of England and Wales (in 2022) found that the EIAs of oil extraction projects did not need to include an assessment of the GHG emissions from the use of the oil to conform with national or EU law.[346] For Lord Carloway of the Court of Session, '[t]he ultimate use of a finished product is not a direct or indirect significant effect of the project' and, at any rate, the assessment of the GHG emissions from the use of the fuel would 'not be practicable' as part of the EIA of a single project.[347] The UK Supreme Court's 2024 decision in *R (Finch) v Surrey County Council* rejected this view and clarified the law applicable in the United Kingdom by holding that the EIA for an oil and gas extraction project had to consider the effect of the project on downstream emissions.[348] By contrast, the Supreme Court of Norway found in 2020 that an EA for oil extraction activities did not need to consider the climate effects of oil exports on the ground that those effects were assessed on a regular basis through political processes.[349]

Beside the scope of assessment, courts have had to determine the minimum content requirements of CAs, absent any specific provisions in EA instruments. At the very least, courts have verified that decisions were properly justified and internally

---

[342] See *Border Power Plant Working Group* (n 305) 1006.
[343] Ibid 1012.
[344] Ibid 1017.
[345] See *Mid States Coalition for Progress v Surface Transportation Board* (n 247), 550.
[346] See *Greenpeace Ltd v Advocate General* [2021] CSIH 53, 2021 Scot (D) 9/10 [63]; *R (Finch) v Surrey County Council* [2022] EWCA Civ 187, [2022] All ER (D) 93 (Feb).
[347] *Greenpeace Ltd v Advocate General* (n 346) [63], [68].
[348] *Finch* (SC) (n 89). See also *R (Friends of the Earth) v South Lakeland Action on Climate Change* [2024] EWHC 2349 (Admin).
[349] See *Greenpeace Nordic* (2020) (n 316) [241].

consistent. Thus, in 2008, the Canadian Federal Court in *Pembina Institute for Appropriate Development v Canada* directed a review panel to provide 'a rationale' for its conclusion that recommendations would ensure that the project at issue—an oil sands mine—would have no significant climate impact.[350] Other courts have gone further in controlling the methodological choices that informed EA reports. US courts refrained from 'prescribing a specific metric for the agency to use'[351] when assessing the significance of climate impacts, but they also observed that, 'if an accurate method exists', the agency had to 'perform that analysis or explain why it ha[d] not'.[352]

b) The practice of agencies

In the absence of provisions in EA instruments or judicial developments, CA procedures have sometimes been implemented at the initiative of governmental agencies. For instance, while Mauritius' 2002 Environment Protection Act did not originally mention climate change,[353] it was reported that the Department of Environment adopted guidelines requiring the incorporation of climate change in EIA procedures[354] before a 2020 statutory reform integrated this administrative practice in new statutory provision.[355] In Minnesota, courts found no implicit requirement to consider GHG emissions under the state's EA framework,[356] and lawmakers decided not to add such a requirement in the relevant legislation,[357] but, after years of debate, the Minnesota Environmental Quality Board decided to require information on GHG emissions within a form that proponents must submit when proposing new activities.[358]

China is another jurisdiction where national agencies have led the charge with regard to integrating climate change in EA procedures. China's EA laws and

---

[350] *Pembina Institute for Appropriate Development v Canada (Attorney General)* [2008] FCJ 324, 2008 FC 302 (Federal Court).

[351] *350 Montana v Haaland* (9th Cir 2022) 50 F 4th 1254, 1271.

[352] *Diné Citizens against Ruining Our Environment v Haaland* (10th Cir 2023) 59 F 4th 1016, 1042.

[353] Environment Protection Act No 19 of 2002 (5 September 2002).

[354] See C Elrick-Barr, N Aldum, A Travers, and R Kay, *Integrating Climate Change into Environmental Impact Assessment in the Republic of Mauritius: recommendations for mainstreaming climate change into the EIA framework* (UNCP African Adaptation Programme 2012) 15.

[355] See Climate Change Act No 11 of 2020 (27 November 2020), s 30(2).

[356] *Minnesota Center for Environmental Advocacy v Holsten* (Minn Ct App 22 September 2009) A08-2171, 2009 WL 2998037, at *3–*4 (finding that the agency had lawfully held that a CA is not 'within the current state of the art').

[357] See Thaddeus R Lightfoot, 'Climate Change and Environmental Review: Addressing the Impact of Greenhouse Gas Emissions under the Minnesota Environmental Policy Act' (2010) 36 William Mitchell Law Review 1068, 1099–1103.

[358] 'Climate Assessments' (Minnesota Environmental Quality Board, nd) <www.eqb.state.mn.us/environmental-review/climate-assessments>; Kirsti Marohn, 'Minnesota Changes Environmental Review to Measure Climate Impacts', *MPR News* (26 December 2022) <www.mprnews.org/story/2022/12/26/minnesota-changes-environmental-review-to-measure-climate-impacts>. See Minnesota Environmental Quality Board, 'Environmental Assessment Worksheet' (December 2022) <www.eqb.state.mn.us/environmental-review/overview/environmental-assessment-worksheet-eaw-process>, s 18.

regulations do not contain any reference to climate change,[359] and no Chinese court appears to have ruled on the applicability of the national EA framework to climate impacts. Nonetheless, two of the documents issued by the Ministry of Environmental Protection to guide the implementation of particular aspects of the national EA framework suggest that China's EIA and SEA should include a CA. One such guidance document indicates that technical reviews of EIA reports should consider the feasibility and effectiveness of any measure described in the project which seeks to reduce GHG emissions.[360] The other document includes carbon dioxide (but none of the other major GHGs) among the air pollutants to be documented in SEAs.[361] In practice, it has been reported that about a fifth of SEA reports involve some sort of consideration for GHG emissions, although this rarely amounted to a systematic appraisal.[362] In this context, scholars have advocated for a reform of China's EA framework towards a more effective and systematic approach of CA, including—as in Mauritius—through the integration of consideration of climate impacts in the national EA instruments.[363]

c) The approach of project proponents

At times, proponents include information on GHG emissions in the absence of any requirement to do so. Unsurprisingly, such spontaneous CA tends to be carried out in relation to proposed activities likely to achieve net emission reductions, such as renewable energy projects. In Hong Kong, for instance, GHG emission reductions were spontaneously documented, in spite of their small scale, in the EIA reports relating to a small windfarm project[364] and to road improvement works in the development of a quarry,[365] with project proponents being keen to point out

---

[359] See Environmental Assessment Act (China) (n 145); Regulation on Environmental Impact Assessment of Planning (规划环境影响评价条例) of 17 August 2009.

[360] Ministry of Environmental Protection, Guideline for Technical Review of Environment Impact Assessment (建设项目环境影响技术评估导), HJ616-2011, 1 September 2011, para 6.3.2.8.

[361] Ministry of Environmental Protection, Technical Guidelines for Strategic Environmental Assessment: General Principles (规划环境影响评价技术导则: 总纲), HJ-130-2014, 1 September 2014, A6.

[362] See Wu Yanan and Ren Jingming, 规划环评中纳入气候变化因素的现状调查与分析 ('Survey and Analysis of the Status Quo of the Climate Change Factors in Strategic Environmental Assessments') [2014] 中国环境科学学会学术年会 (Annual Meeting of the Chinese Society of Environmental Sciences) 1192; Wu Hao and Zhang Yixin, 关于中国将气候变化因素融入环境影响评价的探讨 ('Discussion of China's Integration of Climate Change Factors into Environmental Impact Assessment') (2011) 33 Environmental Pollution and Control 91.

[363] See eg Chen Ying, Wang Yanan, and Zhang Zhansheng, 'Suggestions to Response to Climate Change by Environmental Impact Assessment Mechanisms Innovation' (2016) 41 Environment & Sustainable Development 17 (in Chinese); He Xiangbai, 'Integrating Climate Change Factors within China's Environmental Impact Assessment Legislation: New Challenges and Developments' (2013) 9 Law, Environment & Development Journal 50.

[364] Environmental Resources Management on behalf of HK Electric, 'Development of a 100MW Offshore Wind Farm in Hong Kong', EIA-177/2009, AEIAR-152/2010 (January 2010), executive summary, §§2.1, 3.14.

[365] Ove Arup & Partners Hong Kong Ltd on behalf of Civil Engineering and Development Department, 'Agreement No CE 18/2012 (CE) Development of Anderson Road Quarry – Investigation:

that these projects contribute to the government's decarbonization initiatives. By contrast, the EIA report for the conversion of a Hong Kong coal-fired power plant into one fuelled by natural gas simply declared that the project aimed at 'reducing [the] carbon intensity of local electricity generation',[366] without assessing whether the project would actually achieve this goal or rather lock the territory into a GHG-intensive energy system. Likewise, the EIA for the extension of the Hong Kong airport did not document the project's climate impact.[367]

3. Reforming EA instruments

In some other jurisdictions, lawmakers and governments have reformed existing EA procedures to apply them as a tool for climate change mitigation. The following briefly documents several examples of such reforms in the United States, Canada, the European Union, South Africa, and several international financial institutions. Other prominent examples where EA instruments were amended to require an assessment of climate impacts include US states (eg California,[368] Hawaii,[369] and Massachusetts),[370] Canadian provinces (eg Quebec),[371] EU Member States[372] and other European countries,[373] Australian states and territories,[374] New Zealand,[375] Japan,[376] and Mexico.[377]

Environmental Impact Assessment Report', EIA-043/2000, 227724-REP-037-03 (June 2014), §3.3.1.4 (Table 3.1).

[366] Environmental Resources Management on behalf of Castle Peak Power Co Ltd, 'Additional Gas-Fired Generation Units Project', EIA-237/2016, AEIAR-197/2016 (April 2016), executive summary, § 1.1.
[367] Mott MacDonald on behalf of the Hong Kong Airport Authority, 'Expansion of Hong Kong International Airport into a Three-Runway System', EIA-223/2014, AEIAR-185/2014 (June 2014).
[368] See eg California Pub Res Code (2024) § 21183.6; California Code Regs (2024) title 14, § 15064.4.
[369] Hawaii Code R (2024) § 11-200.1-13.
[370] Massachusetts Gen Laws (2024) ch 30, § 61.
[371] Regulation on Environmental Impact Assessment and Review Procedure of Certain Projects, D 287-2018, GOQ II, 1719A (23 March 2018).
[372] See eg Code de l'environnement (France) arts L122-1(III.3), R122-5 II.4, II.5(f); Gesetz über die Umweltverträglichkeitsprüfung (n 108) §2(1)(3), Sch 4 para 4(b), (c)(gg) (Ger); Town and Country Planning (Environmental Impact Assessment) Regulations 2017 (n 116) s 4(2)(c) and Sch 4 paras 4, 5(f) (UK). See also Town Planning (Environmental Impact Assessment) Regulations 2019, Sch 4 para 4 (Gibraltar).
[373] See eg Forskrift om konsekvensutredninger (Regulations on impact assessment), FOR-2017-06-21-854, consolidated as of 1 January 2024, ss 10, 21 (Norway).
[374] See eg Environmental Planning and Assessment Regulation 2021 (NSW), consolidated as of 14 December 2023, ss 29(3), 35BA, 35C, 35D, 79A(3)(c), 102(3)(a); Environmental Protection Act 2017 (Vic), consolidated as of 20 December 2023, Sch 1 para 8.5–8.7.
[375] Resource Management Act 1991, ss 61, 66, 74.
[376] Environmental Agency Public Notice No 87, Basic Matters Relating to the Guidelines to be Established by the Competent Minister in Accordance with the Provisions of the Environmental Impact Assessment Act (環境影響評価法の規定による主務大臣が定めるべき指針等に関する基本的事項), 12 December 1997, s 1(2)(4) (Japan).
[377] Ley general del equilibrio ecológico y la protección al ambiente (n 198) art 41.

a) United States

In 1970, NEPA created the CEQ, as part of the Executive Office of the President, as the agency responsible for the implementation of NEPA.[378] Along with the courts, the CEQ has played an important role in clarifying the application of NEPA as a tool for climate change mitigation, in particular through the adoption of guidance documents. While CEQ guidance documents are not legally binding on agencies,[379] they are persuasive authorities that courts often follow when interpreting NEPA.[380]

Thus, in 1997, CEQ mentioned the 'release of greenhouse gases resulting in climate modification' as an example of the 'cumulative effects' to be assessed in NEPA reviews.[381] The same year, it issued an initial draft guidance on consideration of 'global climatic change', directing agencies to 'determine whether and to what extent their actions affect greenhouse gases'.[382] The public was consulted on new draft guidance memorandums issued in 2010 and 2014,[383] and a 'Final Guidance' document was adopted in 2016.[384] This 2016 Guidance affirmed that the effects of climate change 'fall squarely within NEPA's purview'[385] and aimed 'to provide for greater clarity and more consistency in how agencies address climate change in the environmental impact assessment process'.[386]

The following year, the Trump administration ordered the CEQ to rescind its 2016 Guidance.[387] Nonetheless, courts continued to rely on the 2014 Draft Guidance[388] or on the 2016 Guidance[389] when interpreting NEPA 'to the

---

[378]   42 USC (2024) § 4342.

[379]   Ibid, fn 3.

[380]   See eg 'Legal Status of CEQ's Final Guidance on Climate Change in Environmental Reviews under NEPA', CRS Reports & Analysis (17 August 2016) <https://sgp.fas.org/crs/misc/ceq-nepa.pdf>.

[381]   CEQ, 'Considering Cumulative Effects under the National Environmental Policy Act' (January 1997) <https://ceq.doe.gov/docs/ceq-publications/ccenepa/exec.pdf>, 13.

[382]   CEQ, 'Draft Guidance Regarding Consideration of Global Climate Change in Environmental Documents Prepared Pursuant to the National Environmental Policy Act' (8 October 1997), cited in JB Ruhl, 'Climate Change and the Endangered Species Act: Building Bridges to the No-Analog Future' (2008) 88 Boston University Law Review 1, 44 (fn 178).

[383]   CEQ, 'Draft NEPA Guidance on Consideration of the Effects of Climate Change and Greenhouse Gas Emissions' (18 February 2010) <https://perma.cc/4P7U-RFYW>; CEQ, 'Revised Draft Guidance for Federal Departments and Agencies on Consideration of Greenhouse Gas Emissions and the Effects of Climate Change in NEPA Reviews' (24 December 2014) 69 Federal Register 77802. See also Jessica Anne Wentz, 'Draft NEPA Guidance Requires Agencies to Consider both GHG Emissions and the Impact of Climate Change on Proposed Actions' (2015) 26 Environmental Law in New York 57.

[384]   CEQ, 'Final Guidance for Federal Departments and Agencies on Consideration of Greenhouse Gas Emissions and the Effects of Climate Change in National Environmental Policy Act Reviews' (1 August 2016) <https://perma.cc/2A9H-GPNH>.

[385]   Ibid, 2.

[386]   Ibid.

[387]   Executive Order 13783 (28 March 2017) 82 Federal Register 16093, s 3(c). See also CEQ, 'Withdrawal of Final Guidance for Federal Departments and Agencies on Consideration of Greenhouse Gas Emissions and the Effects of Climate Change in National Environmental Policy Act Reviews' (5 April 2017) 82 Federal Register 16576.

[388]   *AquAlliance v US Bureau of Reclamation* (ED California 2018) 287 F Supp 3d 969, 1028.

[389]   *San Juan Citizens Alliance v US Bureau of Land Management* (D New Mexico 2018) 326 F Supp 3d 1227, 1243 and fn 5; *WildEarth Guardians v Zinke* (D Montana 11 February 2019) CV-17-80-BLG-SPW-TJC, 2019 WL 2404860, at *10.

extent [that] the reasoning is logically sound and consistent with case law'[390] and 'provide[s] helpful context'.[391] In 2019, the CEQ adopted a new Draft Guidance,[392] a shorter document that set less stringent standards than the 2016 Guidance.[393] For instance, the 2019 Draft Guidance suggested that agencies were only required to quantify an activity's GHG emissions when this could readily be done by relying on 'high quality' information.[394]

In turn, in 2021, the Biden administration directed the CEQ to rescind its 2019 Draft Guidance and to 'review, revise, and update' its 2016 Guidance.[395] The CEQ immediately suggested that federal agencies follow the 2016 Guidance pending completion of the review process.[396] In 2023, the CEQ adopted a new interim guidance document, which largely builds upon the 2016 Guidance.[397] In particular, the 2023 Guidance recommends that Agencies 'quantify the reasonably foreseeable direct and indirect GHG emissions of their proposed actions and reasonable alternatives'.[398] Agencies should assess these emissions by using the Social Cost of Greenhouse Gas—a monetization tool developed by an interagency working group[399]—and, when relevant, by explaining 'how the proposed action and alternatives would help meet or detract from achieving relevant climate action goals and commitments'.[400] The assessment should consider both the direct and the indirect effects, including 'reasonably foreseeable emissions related to a proposed action that are upstream or downstream of the activity resulting from the proposed action'.[401] To avoid or reduce climate impacts, agencies are directed to contemplate 'a range of reasonable alternatives, as well as reasonable mitigation measures if not already included in the proposed action or alternatives, consistent with the level of NEPA review'.[402] In turn, CEQ guidance has prompted some federal agencies to develop their own guidelines on the implementation of CA requirements to NEPA reviews within their purview.[403]

[390] *San Juan Citizens Alliance* (n 389) (referring to the 2016 Guidance).

[391] See *350 Montana* (n 351) 1284–85.

[392] CEQ, 'Draft National Environmental Policy Act Guidance on Consideration of Greenhouse Gas Emissions' (26 June 2019) 84 Federal Register 30097.

[393] For example, Sabin Center for Climate Change Law and Environmental Defense Fund, Letter to the CEQ, 'Proposed Amendments to the Regulations Implementing the Procedural Provisions of the National Environmental Policy Act' (10 March 2020) <https://perma.cc/BZQ6-9MGW>.

[394] CEQ, 2019 Draft Guidance (n 392) 30098.

[395] Executive Order 13990 (20 January 2021) 86 Federal Register 7037, § 7(e).

[396] CEQ, 'National Environmental Policy Act Guidance on Consideration of Greenhouse Gas Emissions' (19 February 2021) 86 Federal Register 10252.

[397] CEQ, 2023 Guidance (n 112).

[398] Ibid, 1201.

[399] Ibid, 1202.

[400] Ibid, 1203.

[401] Ibid, 1204.

[402] Ibid.

[403] See eg Federal Aviation Administration, Order 1050.1F, 'Environmental Impacts: Policies and Procedures' (16 July 2015) 80 Federal Register 44208, ch 4 at 5; Office of Surface Mining Reclamation and Enforcement, 'Handbook on Procedures for Implementing the National Environmental Policy Act' (July 2019) <www.osmre.gov/sites/default/files/pdfs/directive995_NEPAHandbook.pdf>, ch 7 at

In 2024, the NEPA regulations were revised, among other things, to include an explicit statement that the 'impacts' to be assessed under NEPA include 'the contribution for a proposed action and its alternatives to climate change'.[404] One of the questions CEQ considered when revising these regulations was whether to incorporate the 2023 Guidance into the NEPA regulations.[405] Among the comments the CEQ received, some pointed out that it 'would be premature to codify the guidance and that retaining it as guidance would provide flexibility to continue to update the manner in which agencies address climate change in NEPA reviews'.[406] CEQ decided not to integrate the guidance in the NEPA regulations for the time being, but noted that it might consider doing so 'in a future rulemaking'.[407]

b) Canada

Canada and some of its provinces and territories have established EA frameworks from the 1970s forward.[408] These procedures did not initially mention GHG emissions among the 'environmental effects' that were to be assessed.[409] In 2003, the Federal-Provincial-Territorial Committee on Climate Change and Environmental Assessment published a 'general guidance for practitioners' on 'incorporating climate change considerations in environmental assessment'.[410] This document was aimed at exploiting EA's 'potential to link project planning to the broader management of climate change issues'.[411] It included, at the time, relatively advanced technical guidance on estimating GHG emissions.

At the federal level, the assessment of climate impacts was formalized with the adoption of the 2019 Impact Assessment Act.[412] This overhauled EA instrument observes that EA can 'contribute ... to Canada's ability to meet its ... commitments

---

4; Missile Defense Agency, 'National Environmental Policy Act Implementation Procedures' (8 August 2014) 79 Federal Register 46410, 46411; National Oceanic and Atmospheric Administration, 'Policy and Procedures for Compliance with the National Environmental Policy Act and Related Authorities' (13 January 2017) <https://perma.cc/TLR8-K29B>, 11.

[404] CEQ, 'National Environmental Policy Act Implementing Regulations Revisions Phase 2' (1 May 2024) 89 Federal Register 35442, 35575 (to be codified as 42 USC §1508.1(i)(4)).

[405] Ibid at 35494.

[406] Ibid.

[407] Ibid.

[408] See Benidickson (n 167) 257–65; Jeff Nishima-Miller, 'Environmental Assessment Reform in Canada', in Kevin Hanna (ed), *Routledge Handbook of Environmental Impact Assessment* (Routledge 2022) 332, 332–33.

[409] Eg Environmental Assessment and Review Process Guidelines Order, SOR/84-467; Canadian Environmental Assessment Act 1992 (n 170).

[410] Federal-Provincial-Territorial Committee on Climate Change and Environmental Assessment, 'Incorporating Climate Change Considerations in Environmental Assessment: General Guidance for Practitioners' (2003) <https://publications.gc.ca/collections/collection_2012/acee-ceaa/En106-50-2003-eng.pdf> 1.

[411] Ibid.

[412] Impact Assessment Act 2019 (n 108).

in respect of climate change'.[413] Among other things, it requires the competent authorities to consider how 'the effects of the ... project hinder or contribute to the Government of Canada's ability to meet ... its commitments in respect of climate change'.[414]

In complement to the 2019 Impact Assessment Act, the Government of Canada has published several documents aimed at 'enabl[ing] consistent, predictable, efficient and transparent consideration of climate change throughout the impact assessment process'.[415] These documents outline innovative approaches to assessing the GHG emissions of projects, such as the 'best-in-class' standard of assessment for oil-and-gas projects—namely, the requirement that a project should achieve the lowest emission intensity achieved by any comparable projects.[416] These federal guidance documents often influenced, and were influenced by, similar developments at the provincial level.[417]

c) The European Union

The original version of Directive 85/337/EEC required EIAs to assess the effects of projects on the 'climate',[418] though not specifically on climate change. The Directive was amended in 1997 with the view of ensuring compliance with the UNECE Espoo Convention on EIA in a transboundary context,[419] and it was recodified in 2011 as Directive 2011/92/EU,[420] still with no clear reference to climate change. Likewise, the SEA procedure under Directive 2001/42/EC included 'climatic factors' among the information to be provided in environmental reports,[421] without

---

[413] Ibid, preamble para 9.

[414] Ibid, ss 22(1)(i), 63(e).

[415] Government of Canada, 'Strategic Assessment of Climate Change' (October 2020) <www.canada.ca/en/services/environment/conservation/assessments/strategic-assessments/climate-change.html> 1. See also Government of Canada, 'Draft Technical Guide Related to the Strategic Assessment of Climate Change: Guidance on Quantification of Net GHG Emissions, Impact on Carbon Sinks, Mitigation Measures, Net-Zero Plan and Upstream GHG Assessment' (August 2021) <www.canada.ca/en/environment-climate-change/corporate/transparency/consultations/draft-technical-guide-strategic-assessment-climate-change.html>; Government of Canada, 'Policy Context: Considering Environmental Obligations and Commitments in Respect of Climate Change under the Impact Assessment Act' (17 January 2020) <https://perma.cc/JC7N-D6D6>.

[416] Government of Canada, 'Draft Guidance for the Submission of Information Demonstrating Best-in-Class GHG Emissions Performance by Oil and Gas Projects Undergoing a Federal Impact Assessment' (2022).

[417] See eg Environmental Assessment Office, 'Guideline for the Selection of Valued Components and Assessment of Potential Effects' (2013) <www2.gov.bc.ca/assets/gov/environment/natural-resource-stewardship/environmental-assessments/guidance-documents/eao-guidance-selection-of-valued-components.pdf> (BC); Ministry of the Environment and Climate Change, 'Considering Climate Change in the Environmental Assessment Process' (2017) <https://perma.cc/P5C9-X46Z>; Nova Scotia Environment, 'Guide to Considering Climate Change in Environmental Assessments in Nova Scotia' (February 2011) <https://perma.cc/F4QX-GS75>.

[418] Council Directive 85/337/EEC (n 184) art 3. See also ibid, Annex III para 3.

[419] Council Directive 97/11/EC of 3 March 1997 amending Directive 85/337/EEC on the assessment of the effects of certain public and private projects on the environment, [1997] OJ L 73/5.

[420] Directive 2011/92/EU (n 101).

[421] Directive 2001/42/EC (n 101) Annex I para (f).

specifying whether this implied an assessment of the activity's impact on climate change or only an assessment of the impacts of climate change on the activity.

The EU Commission's 2009 reviews on the implementation of EIA Directive 85/337/EEC and SEA Directive 2001/42/EC observed that the effects of climate change were not adequately considered in EIA and SEA procedures. In particular, the Commission found that EIA reports were 'often not go[ing] beyond evaluating existing emissions and ensuring that ambient air quality standards are met',[422] while consideration of climate change issues in SEA was 'on a case-by-case basis'.[423] The Committee of the Regions, an advisory body of the European Union, received these reports and noted the need for 'a well-established methodology to determine the impacts of climate change' in EIAs and SEAs[424]—though it did not mention the need for an assessment of the impacts *on* climate change.

In the following years, the EU directed its Member States to include consideration for climate change in EA in two ways: by considering the impacts of climate change on activities and also by considering the impacts of activities on climate change. Thus, in 2013, the Commission adopted guidance documents to advise practitioners and authorities on how to integrate climate change mitigation and climate change adaptation (along with the protection of biological diversity) at various stages of EIA and SEA procedures.[425] The following year, the European Parliament and the Council amended EIA Directive 2011/92/EU to require an assessment of 'the impact of projects on climate (for example greenhouse gas emissions) and their vulnerability to climate change'.[426] In particular, EIA reports were explicitly required to assess 'the nature and magnitude of greenhouse gas emissions'.[427]

The EU has thus played an important role in promoting the use of EA procedures as a tool for climate change mitigation in Member States, with probably some influence on other jurisdictions. However, the Commission's guidance documents do not define an assessment methodology as clearly and specifically as the guidelines that were adopted in the United States or Canada. For instance, the EU guidance documents do not specify whether and under which circumstances a

---

[422] Commission of the European Communities, 'Report on the Application and Effectiveness of the EIA Directive' (23 July 2009) COM(2009) 378 final <https://perma.cc/3HHZ-RQ7N>, s 3.5.4.

[423] Commission of the European Communities, 'Report on the Application and Effectiveness of the Directive on Strategic Environmental Assessment (Directive 2001/42/EC)' (14 September 2009) COM(2009) 496 final <https://perma.cc/2DVH-7ZCQ> s 4.4.

[424] Committee of the Regions, Opinion, 'Improving the EIA and SEA Directives', [2010] OJ C 232/41 para 6.

[425] EU Commission, 'Guidance on Integrating Climate Change and Biodiversity into Environmental Impact Assessment' (2013); EU Commission, 'Guidance on Integrating Climate Change and Biodiversity into Strategic Environmental Assessment' (2013).

[426] Directive 2014/52/EU of the European Parliament and of the Council of 16 April 2014 amending Directive 2011/92/EU on the assessment of the effects of certain public and private projects on the environment, [2014] OJ L 124/1, preamble para 13.

[427] Ibid, Annex IV para 5(f). See also *Greenpeace Nordic Association v Ministry of Energy* (2024) Case No 23-099330TVI-TOSL/05 (District Court of Oslo), s 3.5.4.

quantitative assessment of GHG emissions is required or how the significance of these emissions is to be evaluated. Lastly, SEA Directive 2001/42/EC is yet to be amended in the same way as the EIA Directive 2011/92/EU to include a clear and explicit requirement for consideration of GHG emissions.

d) South Africa

Four years after the High Court decision in *Earthlife Africa Johannesburg*,[428] the Department of Forestry, Fisheries, and the Environment circulated a draft national guideline for consideration of climate change implications in EA procedures, which it explicitly presented as a response to the Court's decision, and called for input on this draft.[429] The Department noted that, '[w]hile some effort has already been made to incorporate climate change considerations in EIAs, this Guideline puts forward a consistent approach in providing interested and affected parties ... with the minimum requirements to consider when undertaking a climate change assessment.'[430] The Guideline relates to three different statutory procedures relating to applications for environmental authorizations, atmospheric emission licences, and waste management licences. If adopted, the Guideline would not be prescriptive, but it would 'provide best practice guidance to improve the quality of specialist input related to climate change.'[431] Nonetheless, the Guideline would provide a benchmark on which courts are likely to fall back in judicial reviews building on the precedent of *Earthlife Africa Johannesburg*.

e) Financial institutions

International financial institutions have also adopted EA to assess the environmental and climate impacts of the project they consider financing. Thus, the World Bank's 1999 Safeguard Policies mandated an EA extending to 'global environmental aspects' of projects proposed for the Bank's financing, including 'global environmental issues' such as 'climate change.'[432] In 2016, the Bank replaced these policies with an Environmental and Social Framework, in which GHG emissions are considered as a type of air pollution that must be assessed when relevant and 'technically and financially feasible.'[433] This development reflected the Bank's understanding that '[e]stimation of project GHGs is part of good international industry practice.'[434]

---

[428] *Earthlife Africa Johannesburg* (n 319).

[429] Department of Forestry, Fisheries and the Environment, 'Consultation on Intention to Publish the National Guideline for Consideration of Climate Change Implications in Applications for Environmental Authorisations, Atmospheric Emission Licences and Waste Management Licences', GN 559 in GG 44761 (25 June 2021).

[430] Ibid, 7.

[431] Ibid, 11.

[432] World Bank, Operational Policy 4.01, 'Environmental Assessment' (1999) para 3 and fn 4.

[433] World Bank, 'Environmental and Social Framework' (2016) 19 (para 27). See also ibid 20 (para 35).

[434] World Bank, 'Environmental and Social Framework (Proposed Third Draft)' (4 August 2016) <https://perma.cc/T6JN-FLFS> para 59.

Other international financial institutions have followed suit. For instance, the Asian Development Bank's 2009 Safeguard Policy Statement requires the borrower to monitor GHG emissions throughout the project and to seek effective options to reduce or offset these emissions.[435] Similarly, the Asian Infrastructure Investment Bank requires its clients to design and implement their projects in ways that 'minimize emissions' and promote the realization of 'the aims of the Paris Agreement' and the achievement of NDCs.[436] The African Development Bank conducts 'environmental and social assessments' to promote 'climate change considerations' into all lending operations and project activities.[437] And the UN Development Programme requires an assessment of the GHG emissions resulting from its projects and consideration for options to reduce these emissions.[438]

Private financial institutions have also adopted EA procedures, in particular for activities they finance in countries with less stringent environmental law or ineffective implementation. Over a hundred institutions that have adopted the Equator Principles in 2003 pledge to implement certain procedures to assess the environmental and social risk arising from the projects they finance.[439] The fourth update of the Equator Principles, in 2020, inserts requirements on climate change, including consideration for the impacts of climate change on projects as well as the impact of projects on climate change.[440]

C. Opposition to CA

This section identifies and discusses the main lines of arguments against the use of EA as a tool for climate change mitigation. The first subsection identifies instances where EA frameworks were not implemented as a tool for climate change mitigation. The second subsection engages with concerns about potential redundancies between CA and other national policies on climate change mitigation, showing that such concerns do not withstand scrutiny. The third subsection acknowledges more difficult questions relating to the analogy between environmental and

---

[435] Asian Development Bank, 'Safeguard Policy Statement' (2009), 38. See also ibid 16.

[436] Asian Infrastructure Investment Bank, 'Environmental and Social Framework' (2019) 33.

[437] African Development Bank Group, 'Environmental and Social Assessment Procedures (ESAP)' [2015] 1(4) *Safeguards and Sustainability Series* 9.

[438] UN Development Programme, 'Social and Environmental Standards' (2021) 24.

[439] See Demetri Sevastopulo, 'Banks Commit to Socially Responsible Lending', *Financial Times* (3 June 2003); Suellen Lazarus, 'The Equator Principles at Ten Years' (2014) 5 Transnational Legal Theory 417; 'The "Equator Principles": An Industry Approach for Financial Institutions in Determining, Assessing and Managing Environmental and Social Risk in Project Financing' (June 2013). See also Equator Principles Association, 'Members & Reporting' (October 2022) <https://equator-principles.com/members-reporting/>.

[440] Equator Principles, version 4 (July 2020), principle 2 and Annex A. See also Equator Principles, 'Guidance Note on Climate Change Risk Assessment' (September 2020) 4; Equator Principles, 'Guidance Note on Climate Change Risk Assessment' (May 2023) 4.

climate impacts. The following chapters provide a systematic treatment of these questions.

As a caveat, this section focuses on arguments informed by the premises that every jurisdiction should take ambitious action to mitigate climate change and that EA is a generally relevant tool to minimize or avoid environmental impacts. As such, it does not engage with broader political or economic arguments against EA as a whole or against climate action, even though such arguments may at times have had an influence on the development of CA law and practice.[441]

1. Jurisdictions without CA

Some EA frameworks do not require an assessment of the climate impacts of proposed activities. This subsection documents developments in five jurisdictions where climate impacts are not (or were not until recently) included within the scope of their EA framework. Administrative inertia contributes to explaining the absence of CA requirement in some of these jurisdictions, in particular India and Hong Kong. In other jurisdictions, the exclusion of CA was the outcome of a deliberate decision. These jurisdictions include Kazakhstan, where this exclusion persists to date, as well as New Zealand and Montana, where decisions were reversed in 2022 and 2023.

a) India

The absence of CA in India appears to have more to do with political inertia, hesitancy, or reluctance, than with a deliberate decision. The Indian EA regulatory framework—the Environmental Protection Act 1986 supplemented by a 'notification' issued by the Ministry of Environment, Forest and Climate Change—requires project proponents to prepare an EIA Report based on terms of reference that 'address ... all relevant environmental concerns'.[442] In practice, terms of reference generally do not include considerations for GHG emissions, even in the projects most likely to cause large amounts of such emissions.[443] In 2019, the National Green Tribunal found that climate change considerations were already covered by the national EA framework,[444] a decision that might prompt the Ministry to revise standard terms of references.

---

[441] The clearest example might be the decisions of the Trump administration reforming the CA under NEPA. See Executive Order 13783 (n 387).

[442] Ministry of Environment and Forests, 2006 EIA Notification (n 109) s 7(i)(II).

[443] See eg Ministry of Environment and Forests, 'Terms of Reference [TOR] for EIA Report for Activities/Projects Requiring Environmental Clearance' (August 2009); Ministry of Environment and Forests, 'Standard Terms of Reference [TOR] for EIA/EMP Report for Projects/Activities Requiring Environment Clearance under EIA Notification 2006' (April 2015).

[444] See *Pandey* (2019) (n 328) [2].

b) Hong Kong

Until recently, Hong Kong was one of the jurisdictions where the debate on the assessment of climate impacts just did not seem to have sparked off. In the EIA legislation adopted in 1997, the environmental impacts that may have to be assessed are listed in a 'technical memorandum',[445] which include issues such as air quality, water pollution and waste,[446] but not climate change or GHG emissions. The potential use of this framework as a tool for climate change mitigation was rarely considered in local legal or political debates.[447]

In 2022, the government of Hong Kong initiated the first review of this legal framework in a quarter century. During preliminary consultations, members of the public proposed the inclusion of 'carbon emissions', 'carbon neutrality', and 'climate change' in the EIA framework.[448] The Environmental Protection Department opposed any such suggestion, on two grounds. First, the Department observed that 'internationally established acceptance criteria have not yet been formulated for greenhouse gas emission', thus suggesting that 'topics on climate change' could not 'be included in the scope of EIA studies'.[449] Second, the Department asserted that the government considers 'climate change and carbon emissions' as part of the 'engineering feasibility studies' of 'some large-scale development projects'.[450] By contrast to EIA reports, such feasibility studies do not aim at assessing or reducing environmental impacts; further, they are not the object of a legal requirement and they are may not be readily available to the public. On these grounds, the government decided not to include any provision requiring an assessment of climate impacts as part of the territory's EIA framework. The reform of the framework was confined to a few minor changes to the list of projects subjected to EIA, such as the addition of wind power plants to the list of designated projects.[451]

---

[445] Environmental Impact Assessment Ordinance (1997) cap 499, s 5(6).

[446] Environmental Protection Department, 'Technical Memorandum on Environmental Impact Assessment Process' (16 May 1997).

[447] But see David Gallacher, 'Climate Change and Environmental Impact Assessment in Hong Kong' (Newsletter of the Hong Kong Institute of Environmental Impact Assessment, June 2017) <https://hkieia.org.hk/wp-content/uploads/2021/11/newsletter_jun2017.pdf> 1; Benoit Mayer, 'Hong Kong's Outdated Environmental Impact Law Needs to Move with the Times', South China Morning Post (30 March 2018) <www.scmp.com/comment/insight-opinion/article/2139332/hong-kongs-outdated-environmental-impact-law-needs-move>.

[448] Legislative Council Panel on Environmental Affairs, 'Optimising the Environmental Impact Assessment Ordinance Process', LC Paper No CB(1)883/2022(01) (12 December 2022) para 22 and Annex 1, 1.

[449] Ibid Annex 1, 2.

[450] Ibid.

[451] Environmental Impact Assessment Ordinance (1997) cap 499, as modified by Environmental Impact Assessment Ordinance (Amendment of Schedules 2 and 3) Order (2023) LN 77/2023, Sch 2 s D.

c) New Zealand (2002–20)

New Zealand decided early on to exclude GHG emissions from the scope of its national EA framework. In 2002, the Environment Court expressed 'considerable disquiet about the efficacy, appropriateness and reasonableness' of a climate change mitigation condition that a regional authority had imposed on a gas-fired power plant project.[452] The Court considered that climate change mitigation should be carried out exclusively through national policies, as only those could 'guarantee an efficiency compatible with achieving best social, environmental and economic outcomes'.[453]

Two years later, the Parliament endorsed this reasoning through an amendment to the Resource Management Act that excluded considerations for 'the effects on climate change of discharges into air of greenhouse gases'.[454] Member of Parliament Pete Hodgson explained that this amendment would avoid duplication of efforts and reduce administrative costs, as a national carbon pricing mechanism would ensure that New Zealand complies with its quantified emissions limitation and reduction commitment under the Kyoto Protocol.[455] On the other hand, the amendment allowed consideration for the GHG-emission reductions achieved by renewable energy projects.[456] This selective inclusion of GHG emissions in EAs may appear inconsistent with EAs' intended purpose of providing complete and objective information. If decision-makers are to take into account the benefits of reductions in GHG emission in the assessment of some proposed activities, it is unclear why they should not also recognize the costs of additional GHG emissions in the assessment of other activities.

The exclusion of climate impacts from the scope of national EA framework came under increasing scrutiny. In 2020, a report by the Ministry for the Environment acknowledged suggestions by members of the public that EA procedures 'should be fully available and used to support the transition to a low emission economy'.[457]

---

[452] *Environmental Defence Society Inc v Auckland Regional Council*, A183/2002 [2002] NZEnvC 315, [2002] NZRMA 492, (2003) 9 ELRNZ 1 [88]. See also *Environmental Defence Society Inc v Taranaki Regional Council*, A184/2002 [2002] NZEnvC 441 [24].

[453] *Environmental Defence Society Inc v Auckland Regional Council* (n 452) [88].

[454] Resource Management (Energy and Climate Change) Amendment Act 2004, s 3(b)(ii). See also *Greenpeace New Zealand Inc v Genesis Power Ltd* [2008] NZSC 112, [2009] 1 NZLR 730; *West Coast ENT Inc v Buller Coal Ltd* [2013] NZSC 87, [2014] 1 NZLR 32 [177]; and, generally, Ceri Warnock, 'Global Atmospheric Pollution: Climate Change and Ozone', in Peter Salmon and David Grinlinton (eds), *Environmental Law in New Zealand* (Thomson Reuters 2015) 789, 813–17; Kierra Parker, 'Litigating at the Source: Attributing Climate change Impacts to Coal Mines' (2020) 37 Environmental & Planning Law Journal 67, 78–79.

[455] See (5 August 2024) 610 NZPD (first treading of the 'Resource Management (Energy and Climate Change) Amendment Bill 2003'). See also *Greenpeace New Zealand Inc v Genesis Power Ltd* (n 454) [40]. An emissions trading scheme was established in 2009. See Climate Change Response (Emissions Trading) Amendment Act 2008.

[456] Resource Management Act 1991, s 70A, as amended by Resource Management (Energy and Climate Change) Amendment Act 2004, s 6.

[457] Ministry for the Environment, 'Impact summary: Linking the Zero Carbon Act 2019 with the Resource Management Act 1991' (13 February 2020) <https://environment.govt.nz/publications/imp act-summary-linking-the-zero-carbon-act-2019-with-the-resource-management-act-1991/> para 26.

The Ministry acknowledged that the national EA framework was 'increasingly misaligned with the more urgent and far-reaching aims of climate change policy to achieve a low emissions climate-resilient economy'.[458] A few months later, the Resource Management Act was amended to repel the provisions that had been inserted in 2004 and explicitly require consideration for GHG emissions in relevant EA procedures.[459]

d) Montana (2011–23)

As the Montana Environmental Policy Act (MEPA) 'was patterned almost word for word after NEPA',[460] the interpretation of NEPA as requiring an assessment of climate impacts could readily be applied to the Montana EA framework. Thus, the Montana Department of Environmental Quality conducted several CA procedures in the 2000s, either on its own[461] or along with Federal agencies.[462] However, in 2011, the Montana State Legislature decided to limit the scope of MEPA by precluding the review of 'actual or potential impacts beyond Montana's borders' and, more specifically, 'actual or potential impacts that are regional, national, or global in nature'.[463] This limitation was interpreted as, in effect, a 'climate change exception'.[464] Yet, in 2023, the Montana District Court held in *Montana Environmental Information Center v Montana Department of Environmental Quality* that this provision did not 'absolve [the Montana Department of Environmental Quality] of its MEPA obligation to evaluate a project's environmental impacts within Montana', including 'the greenhouse gas effects of this project as it relates to impacts within the Montana borders'.[465] This argument seems to ignore that, while an activity's GHG emissions may have significant impact on a global scale, only a tiny fraction of this impact would unfold within the territory

---

[458] Ibid para 21.

[459] Resource Management Amendment Act 2020 ss 17–21, 35, 36, amending Resource Management Act 1991 (repealing ss 70A, 70B, 104E, and 104F and amending ss 61, 66, and 74) with effect in November 2022.

[460] Legislative Environmental Policy Office, 'A Guide to the Montana Environmental Policy Act' (revised edn, Hope Stockwell 2021) <https://leg.mt.gov/content/Publications/Environmental/2021-mepa-handbook.pdf> 55.

[461] See eg *Roundup Power Project*, Montana Department of Environmental Quality, Draft Environmental Impact Statement (November 2002) <https://leg.mt.gov/content/Publications/MEPA/2002/deq1120_2002001.pdf> ch 4, 20–22.

[462] See eg *Highwood Generating Station*, US Department of Agriculture and Montana Department of Environmental Quality, Final Environmental Impact Statement (January 2007) vol 1 <https://leg.mt.gov/content/Publications/MEPA/2007/deq0202_2007004.pdf> ch 4 at 53–56. See generally *Held v State of Montana* (Montana Dist Ct 16 October 2023) CDV-2020-307, Declaration of Sonja Nowakowski in Support of Defendants' Motion for Clarification and for Stay of Judgment Pending Appeal <https://climatecasechart.com/wp-content/uploads/case-documents/2023/20231016_docket-CDV-2020-307_declaration.pdf> paras 7–14.

[463] Montana Environmental Policy Act, Montana Code Annotated (2011) § 75-1-201(2)(a), as amended by Montana Laws 2011, ch 396, § 1, effective 12 May 2011.

[464] Patrick Parenteau, 'The Atmosphere as A Global Public Good' (2023) 16 Journal of Law and Public Policy, St Thomas U 217, 229.

[465] See *Montana Environmental Information Center* (n 311) *29.

of Montana, and the impacts unfolding in Montana are unlikely to be significant when considered on their own.[466]

Montana's legislature almost immediately reacted to this 'overreach by the court'[467] by passing a new law to amend MEPA—this time, directly proscribing the 'evaluation of greenhouse gas emissions and corresponding impacts to the climate in the state or beyond the state's borders'.[468] To drive the point home, the Legislature further modified MEPA, a week later, to prevent judicial action against the decision of an agency not to assess the climate impact of a proposed activity.[469] A few months later, however, the Montana District Court found in *Held v State of Montana* that these two amendments were unconstitutional and permanently enjoined.[470] While the Supreme Court accepted to hear the state's appeal,[471] the courts rejected the state's application for a stay of execution of the order pending appeal.[472]

e) Kazakhstan

Kazakhstan's Environmental Code was amended in 2011 to provide for the creation of an emission trading scheme (ETS) applicable to GHG emissions.[473] This amendment also inserted in the Environmental Code a provision that excluded 'the impact of greenhouse gas emissions' from the scope of EAs.[474] The 2021 overhaul of the Environmental Code reaffirmed that '[t]he impacts caused by greenhouse gas emissions are not taken into account in the environmental impact assessment process'.[475] The circumstances suggest that the exclusion of climate impacts from the scope of the national EA framework aimed at avoiding a potential overlap between the ETS and the EA framework.

This decision has been criticized by international observers. The UNECE, in particular, lamented that the Kazakh government continued to approach climate change as 'a separate topic that must be managed by a specific authority designated

---

[466] See discussion below, this chapter, subsection C.3(b).

[467] Representative Joshua Kassmier, presentation of House Bill 971, House Natural Resources committee (17 April 2023).

[468] Montana Environmental Policy Act, Montana Code Annotated (2023) § 75-1-201(2)(a), as amended by Laws 2023, ch 450, § 1, effective 10 May 2023.

[469] Montana Environmental Policy Act, Montana Code Annotated (2023) § 75-1-201(6)(a)(ii), as amended by Montana Laws 2023, ch 703, § 1, effective 19 May 2023.

[470] See *Held* (Montana District Court 14 August 2023) (n 311) [8]–[9].

[471] *Held v State of Montana* (Montana Sup Ct 17 October 2023) DA 23-0575 <https://climatecasech art.com/case/11091/>.

[472] *Held v State of Montana* (Montana District Court 21 November 2023) CDV-2020-307 <https:// climatecasechart.com/case/11091/>; *Held v State of Montana* (Montana Sup Ct 16 January 2024) DA 23-0575 <https://climatecasechart.com/case/11091/>.

[473] Law amending the Environmental Code (3 December 2011) No 505-IV, art 94-7.

[474] Ibid, art 31-1.2(13)-(14), modifying Environmental Code (9 January 2007) No 212-III-ZRK, art 38.2(1).

[475] Environmental Code (2 January 2021) No 400-VI-LRK, art 66(6), unofficial translation by European Union—Central Asia Water, Environment and Climate Change Cooperation (WECOOP) https://wecoop.eu/wp-content/uploads/2021/04/2021-KZ-ENV-Code_full-text_en.pdf.

as being in charge of climate change issues', in spite of the fact that 'climate change is of a cross-sectoral nature'.[476] Indeed, the 2021 overhaul of the Environmental Code led to provisions on climate change being contained in a 'special part' of the Code.[477] According to article 284, GHG emissions are to be regulated through a market instrument for carbon units.[478]

Furthermore, observers have also suggested that the Kazakh ETS 'is not fully functional yet to deliver tangible emissions reductions'.[479] Issues with the monitoring, reporting, and verification system led the government to suspend temporarily the ETS's application in 2016 and 2017.[480] At best, this scheme only covers emissions of carbon dioxide (to the exclusion of other GHGs such as methane and nitrous oxide)[481] from some sectors (eg electricity and mining, but not transportation and waste management).[482] Overall, the ETS relies entirely on grandfathering of emission allowances rather than on their auctioning, which limits the incentive it creates for emission reduction.[483]

In these circumstances, it is perhaps unsurprising that Kazakhstan's climate change mitigation strategy has proven rather ineffective. Most electricity production continues to rely on bituminous coal, despite a strong economic case for a switch to natural gas.[484] National statistics suggest that Kazakhstan did not reach its 2020 emission-reduction goal[485] and cast serious doubts on its ability to achieve the unconditional mitigation objectives of its NDC.[486]

2. Concerns about policy redundancy

Some objections to the implementation of EA frameworks as a tool for climate change mitigation relate to concerns for policy redundancies. Most jurisdictions

[476] UNECE, 3rd Environmental Performance Review: Kazakhstan (June 2019) 156.

[477] Environmental Code 2021 (n 475) pt 14. See ibid, Pt 18.

[478] Ibid, art 284(2)(2).

[479] Takeshi Kuramochi and others, 'Greenhouse Gas Emission Scenarios in Nine Key Non-G20 Countries: An Assessment of Progress toward 2030 Climate Targets' (2021) 123 Environmental Science & Policy 67, 73.

[480] UNECE, Kazakhstan (n 476) 74.

[481] Environmental Code 2021 (n 475) art 289(1).

[482] UNECE, Kazakhstan (n 476) 73.

[483] See ibid 74; International Carbon Action Partnership, 'Kazakhstan Emissions Trading System' (2022).

[484] Peter Howie and Zauresh Atakhanova, 'Assessing Initial Conditions and ETS Outcomes in a Fossil-Fuel Dependent Economy' (2022) 40 Energy Resources Review 1, 7.

[485] UNFCCC, Report on the Technical Review of the Fourth Biennial Report of Kazakhstan, FCCC/TRR.4/KAZ (4 June 2021) para 60.

[486] See Kazakhstan First NDC (6 December 2016); Kazakhstan, NDC update (27 June 2023) (both committing to an economy-wide 15 percent reduction in GHG emissions by 2030 compared with 1990 levels, including land-sector emissions); UNFCCC (n 485) table 7 (indicating a predicted 7.0 percent increase, excluding land-sector emissions); table 2 (showing a significant increase in land-sector emissions from 1990 to 2018). See also Kuramochi and others (n 479) 73 (assessing that Kazakhstan will 'likely miss its NDC target with existing policies').

already have policies and measures in place that aim at mitigating climate change. One could question the added value of EA in this already crowded landscape—all the more, given the cost and burden EA inevitably imposes on proponents and, more generally, on society.

These objections take multiple forms depending on the type of policies and measures that are in place in a particular jurisdiction. For one, one could question whether EA should impose requirements extending beyond those already defined by existing regulatory standards. Another issue concerns the relation between EA and cap-and-trade mechanisms: when emissions are 'capped' and trading in emission allowances is permitted, reducing the GHG emissions from one project would only free emission allowances that other actors could purchase at a discounted price, resulting in no net mitigation outcome. In yet other circumstances, one may be concerned that two or several EA frameworks may overlap, in particular in federal countries where such frameworks can be established at the same time by the federal and state government.

The counter-arguments are generally that, to mitigate climate change, governments need to rely on more than one policy. Most governments have taken multiple concurrent measures and policies, some of which are applicable to several sectors, others only to particular activities.[487] These coexisting measures, which often overlap, can be mutually reinforcing as part of a multipronged approach aimed at addressing emissions from all sectors on various time scales through complementary standards and incentives. CA is certainly not the only tool for climate change mitigation,[488] but it could nonetheless be a useful addition to national toolboxes—not only as a way to ensure that climate objectives are achieved, but also to make sure that this is done in a fair and efficient manner.

a) Policy strategy

A first objection to CA relies on the premise that climate change should be addressed through comprehensive policy strategies adopted at a general and abstract level, rather than through activity-based EA. For instance, the Environment Court of New Zealand noted in 2002 that the government had expressed a clear preference for addressing GHG emissions at the national level: this, in the Court's view, excluded consideration for climate impacts in EA procedures that, under New Zealand's law, are generally carried out by regional councils.[489] The Supreme Court

---

[487] See generally Dubash and others (n 68) 1396. See also UNFCCC Subsidiary Body for Implementation, Report by the Secretariat, 'Compilation and synthesis of fifth biennial reports of Parties included in Annex I to the Convention', FCCC/SBI/2023/INF 7 (17 October 2023) para 28 (noting that the 39 Annex I Parties that had submitted their fifth biennial report had reported 3,255 policies and measures); UN Climate Change, International Consultation and Analysis: Facilitating Climate Action through Transparency (UNFCCC 2020) <https://unfccc.int/ICA2020> s 3.2.

[488] See R (Finch) v Surrey County Council [2020] EWHC 3566 (Admin), [2021] PTSR 1160 [105] (objecting to 'the proposition that there are no other measures in place' to address certain GHG emissions).

[489] Environmental Defence Society Inc v Auckland Regional Council (n 452) [88].

of Norway made a comparable argument in *Greenpeace Nordic* when discussing the need for an EA to consider the downstream emissions resulting from the exportation of oil and gas. The Court suggested that the Government may decide to deal with these emissions 'on a superior level—i.e. as a part of the Norwegian climate policy—rather than addressing them in the individual environmental assessment'.[490] Similarly, the Hong Kong government asserted that 'the issue of carbon emissions involves a territory-wide strategy, and it can be tackled more effectively and properly at the policy level than by assessing under individual projects'.[491]

This objection relies on a false dichotomy: the suggestion that policy strategy and CA are mutually exclusive, so that a choice must be made between the two. Instead, one could approach EA as one of the tools to implement a policy strategy on climate change mitigation. EA procedures often assess the significance of proposed activities based on relevant policy objectives.[492] In Hong Kong, for instance, the Director for Environmental Protection is required to 'have regard', among other things, to 'the attainment and maintenance of an acceptable environmental quality' when deciding whether to permit a project as part of an EIA procedure.[493] It is unclear why the Director could not also have regard to the achievement of the territory's climate change mitigation objectives, except for the fact (problematic in itself) that these climate objectives are not incorporated in territorial legislation.

By contrast, in the cases at issue, Norway had no policy to limit or reduce downstream emissions occurring overseas, and New Zealand had yet to decide on how to implement the Kyoto Protocol, which had not yet entered into force.[494] The absence of a relevant policy strategy does not mean that climate impacts should have no weight in decisions weighing the merits of proposed activities. It may be difficult to conduct a CA without relying on an underlying policy strategy on climate change mitigation, but this would certainly be preferable to a complete exclusion of any consideration of climate impacts. At the very least, conducting a CA could then prompt the governments to develop relevant policy strategies.

b) Measures on climate change mitigation

Other objections to CA invoke the existence of more specific measures aimed at limiting and reducing GHG emissions. Governments have frequently imposed technical standards to mitigate climate change, consisting for instance in performance standards for road vehicles,[495] biofuel-content requirements on fuel supply,[496]

---

[490] See *Greenpeace Nordic* (2020) (n 316) (unofficial translation by the Court) [234].
[491] Legislative Council Panel on Environmental Affairs (n 448) Annex 1 at 1.
[492] See Chapter III, subsection C.2(b).
[493] Environmental Impact Assessment Ordinance (1997) cap 499, s 10(2)(b).
[494] See *Environmental Defence Society Inc v Auckland Regional Council* (n 452) [84]–[85].
[495] See eg Regulation (EU) 2019/631 (n 63); Environmental Protection Agency, 'Vehicle Greenhouse Gas Emissions Standards' (n 63).
[496] Eg 42 USC (2024) § 7545; Directive 2018/2001 (n 64); Clean Fuel Regulations (n 64).

or even the banning of certain technologies.[497] They have implemented industrial policies in favour of new technologies[498] and directed companies towards lower-emissions activities.[499] Some other prominent measures that government have taken seek to impose an economic incentive against GHG emissions.[500] Carbon taxes impose a fixed price on GHG emissions or related activities (eg the purchase of fuel),[501] whereas cap-and-trade mechanisms such as the EU Emissions Trading Scheme require some companies to acquire and surrender tradable allowances in proportion to their GHG emissions.[502] Subsidies can also be established to support sustainable activities, for instance by guaranteeing feed-in tariffs to help renewable energy producers to recoup their initial investment.[503]

When such technical regulations or economic incentives are in place, an argument can be made that a CA would be redundant: existing measures are enforced through other means (eg criminal prosecution), and—the argument goes—compliance with these measures should be sufficient to justify the implementation of the proposed activity. Thus, the Environment Court of New Zealand justified the exclusion of CA based on the assumption that the government's policies, in particular the economic incentives that the government was preparing to impose, would address GHG emissions 'in a more consistent and efficient manner than controls under the [Resource Management Act]'.[504] As noted above, Kazakhstan's decision to exclude climate impacts from the scope of its EA procedure also coincided with the establishment of a cap-and-trade mechanism.[505]

It is common for EA procedures to consider environmental impacts that are already subjected to detailed regulatory standards. In such circumstances, EA is an opportunity not only for an early check of the compatibility of a proposed activity with existing standards (and a clarification of these standards, if needed),[506] but also for a more thorough consideration of the activity's unique characteristics.[507] Technical regulation cannot fully replace a case-by-case assessment of the merit of a proposed activity. For instance, an environmental impact that could be acceptable if it is necessary for an activity that benefits society may not otherwise

---

[497] On a future ban of internal combustion engines, see Gitlin (n 65); 'EU ban on the sale of new petrol and diesel cars from 2035 explained', News EU Parliament (3 November 2022).
[498] See generally Rissman and others (n 66).
[499] See eg Benoit (n 67); Haonan Qu and others, 'South Africa Carbon Pricing and Climate Mitigation Policy' (International Monetary Fund 2023) para 14; Loi fédérale sur les objectifs en matière de protection du climat, sur l'innovation et sur le renforcement de la sécurité énergétique, FF 2022 2403 (Switzerland) art 5(1).
[500] See generally World Bank, 'State and Trends of Carbon Pricing 2022' (n 58).
[501] See eg Carbon Tax Regulation (n 60), s 12.
[502] Directive 2003/87/EC (n 59).
[503] See eg European Commission, Report on the Performance of Support for Electricity from Renewable Sources (n 62).
[504] See Environmental Defence Society Inc v Auckland Regional Council (n 452) [37(ii)].
[505] See text above at n 476.
[506] Glasson and Therivel (n 71) 128–29.
[507] Gibson and others (n 106) 167.

be acceptable. Further, contemplating the project's unique characteristics may shed light on the potential for minor alterations that would significantly reduce its impacts at a little cost. In other words, CAs (and EAs generally) do more than assessing compliance with existing measures[508] (which at any rate may be varied during the implementation of the project).[509]

Beside technical regulation, however, objectors have often invoked potential redundancies with economic incentives, as for instance in New Zealand and Kazakhstan. Here again, however, the objection relies on shaky grounds: economic incentives to emission reduction cannot replace a case-by-case assessment of the merits of proposed activities. This is because the ability of the project proponent to pay does not necessarily mean that the implementation of the proposed activity is in the public interest. While economic incentives are theorized to be the most cost-effective way for society to decrease its GHG emissions,[510] cost-effectiveness is not the only relevant criterion to evaluate policies and measures on the mitigation of climate change—among other things, these policies and measures may also be expected to be just and equitable.[511]

Overall, economic incentives have well-known practical limitations.[512] For one, economic incentives often incentivize short-term, incremental emission reductions, rather than long-term structural changes.[513] Further, these instruments only apply to emissions from specified sources (eg carbon dioxide emissions from the power sector), whereas CAs can contemplate emissions in other sectors and gases (eg nitrous oxide emissions from land use). Thus, New Zealand came to the realization that, even while its emissions trading scheme 'remains a critical lever to drive emission reduction', it remains that 'there are also many emissions reductions options available that are not responsive to [this] price signal, but can be cost-effectively captured through other measures'.[514]

c) The waterbed effect
A more specific objection to CA regards its potential interaction with cap-and-trade mechanisms. As such mechanisms impose a cap on certain emissions—the objection goes—anything a CA could do to reduce emissions under this cap would only allow other actors to emit more. In practical terms, reducing emissions from

---

[508] California Code Regs (2024) title 14, § 15064.4(b)(3).

[509] See *Gray* (n 312) [138].

[510] See eg Dubash and others (n 68) 1385.

[511] See ibid, 1383.

[512] See eg Charles F Sabel and David G Victor, *Fixing the Climate: Strategies for an Uncertain World* (Princeton University Press 2022); Signe Krogstrup and William Oman, 'Macroeconomic and Financial Policies for Climate Change Mitigation: A Review of the Literature' (2019) IMF Working Paper WP/19/185; Jessica F Green, 'Does Carbon Pricing Reduce Emissions? A Review of ex-post Analyses' (2021) 16 Environmental Research Letters (article #43004) 1–17. See also this chapter, subsection C.1(e), on the ineffective implementation of the Kazakhstan emission-trading scheme.

[513] Ministry for the Environment (n 457) para 49(a)(i).

[514] Ibid para 25.

one project would free emission allowances that other actors could purchase at a discount.[515] As a result of this 'waterbed effect', a CA would appear ineffective, being unable to achieve any genuine net emission reduction.

The waterbed effect has been invoked not just against CA, but against various other measures aimed at limiting GHG emissions in jurisdictions with a cap-and-trade mechanism. For instance, such arguments were made, in the, UK against a carbon price aimed at reducing emissions from fuel combustion which were already partly subjected to the EU ETS,[516] and, in Germany, against a draft policy to reduce reliance on low-grade coal for power generation.[517] With regard to CA, such concerns may have played an instrumental role in the decisions of Kazakhstan and New Zealand to exclude climate impacts from the scope of their EAs.[518]

On the other hand, it is noteworthy that some jurisdictions that have established cap-and-trade mechanisms, such as the EU and California, do nonetheless apply EA frameworks as a tool for climate change mitigation. Dutch courts in *Urgenda v the Netherlands* and *Milieudefensie v Royal Dutch Shell* have rejected the argument that the waterbed effect precludes the imposition of additional emission-reduction obligations on activities that are subjected to the EU ETS.[519] And, in *Climate Resolve v County of Los Angeles*, the Superior Court of California confirmed the applicability of a CA requirement despite the respondents' insistence that most of the project's GHG emissions would be regulated by the state's cap-and-trade mechanism.[520]

Indeed, further analysis shows that, in relation to CA, concerns with the waterbed effect are largely groundless, for two reasons. First, CA has often a much wider scope than cap-and-trade mechanisms. Even the most advanced

---

[515] Knut Einar Rosendahl, 'EU ETS and the Waterbed Effect' (2019) 9 Nature Climate Change 734; Johannes Jarke-Neuert and Grischa Perino, 'Energy Efficiency Promotion Backfires under Cap-and-Trade' (2020) 62 Resource & Energy Economics (article #101189) 1–21.

[516] See eg Steven Sorrell, 'Brexit: An Opportunity to Rethink UK Carbon Pricing', *Energy Post EU* (28 September 2016)('Any additional abatement in the UK simply "frees up" EU allowances that can be either sold or banked, and hence used for compliance elsewhere within the EU ETS, with the result that the CPF [carbon price floor] achieves no additional reduction in carbon emissions').

[517] RWE Statement, 'Proposals of Federal Ministry for Economic Affairs and Energy Endanger the Future Survival of Lignite' (20 March 2015) <https://web.archive.org/web/20150524111127/http://www.rwe.com/web/cms/en/113648/rwe/press-news/press-release/?pmid=4012793> ('The proposals [to reduce lignite generation] would not lead to a CO2 reduction in absolute terms. [The number of] certificates in the ETS would remain unchanged and as a result emissions would simply be shifted abroad').

[518] See this chapter, subsections C.1(c) and (e).

[519] *Urgenda v Netherlands*, ECLI:NL:RBDHA:2015:7145 (District Court of The Hague, 24 June 2015), ILDC 2456 (NL 2015) s 4.81; *Milieudefensie v Royal Dutch Shell*, ECLI:NL:RBDHA:2021:5337 (District Court of The Hague, 26 May 2021) s 4.4.46.

[520] *Climate Resolve v County of Los Angeles* (Superior Court of California, County of Los Angeles 5 April 2021) Order, 19STCP01917 <https://climatecasechart.com/wp-content/uploads/case-documents/2021/20210405_docket-19STCP02100_order.pdf> *40–*44. See also *Center for Biological Diversity v County of Los Angeles* (Superior Court of California, County of Los Angeles 22 March 2023) Judgement Granting Peremptory Writ of Mandate, 19STCP02100 <https://climatecasechart.com/wp-content/uploads/case-documents/2023/20230322_docket-19STCP02100_judgment.pdf> *2.

cap-and-trade mechanisms tend to focus on carbon dioxide emissions to the exclusion of other GHGs, and to a few economic sectors (eg power generation and industrial facilities)[521] to the exclusion of others (eg transport, agriculture, land, and waste). The EU ETS, for instance, covered only 38 percent of the EU's total GHG emissions in 2021.[522] Thus, a CA can be useful as a complement to cap-and-trade mechanisms at least with regard to the emission that do not fall within the scope of these mechanisms. Further, while cap-and-trade mechanisms create an incentive for incremental, short-term emission reduction,[523] CAs may assess the long-term climate impact of a proposed activity—for instance, over the lifetime of a coal mine, power plant, or airport. And while cap-and-trade mechanisms are generally confined to emissions occurring within the state's territory, a CA may consider ways to reduce extraterritorial emissions such as the downstream emissions from fossil-fuel projects.

Second, the cap in a cap-and-trade mechanism can typically be adjusted to reflect variations in the demand for emission allowances. In the short-term, national agencies can often decide to distribute fewer emission allowances or even to buy existing emissions on the secondary market as a way to keep trading prices within an acceptable range.[524] Under the EU ETS, excess emission allowances in the market stability reserve can be invalidated,[525] which, in effect, allows for a lowering of the cap on emissions in response to low demand for emission allowances.[526] On the longer term, regulators determine caps on emissions applicable to future implementation periods based in part on the demand they anticipate for such emissions.[527] To the extent that CA achieves emission reductions that fall within the scope of a cap-and-trade mechanism, it will likely prompt the cap to be lowered.

d) Overlapping EA frameworks

Some other objections related to policy overlaps are not objections to CA as such, but to its implementation by one institution rather than by another. Policy overlap

---

[521] Dubash and others (n 68) 1384.

[522] International Carbon Action Partnership, 'EU Emissions Trading System (EU ETS)' (2024) <https://icapcarbonaction.com/en/ets/eu-emissions-trading-system-eu-ets>. See also EU, Eighth National Communication (n 69) 199 ('approximately 40% of the EU's greenhouse gas emissions').

[523] Sabel and Victor (n 512).

[524] Decision (EU) 2015/1814 of the European Parliament and of the Council of 6 October 2015 concerning the establishment and operation of a market stability reserve for the Union greenhouse gas emission trading scheme and amending Directive 2003/87/EC, [2015] OJ L 264/1. See Cameron Hepburn and others, 'The Economics of the EU ETS Market Stability Reserve' (2016) 80 Journal of Environmental Economics & Management 1. See also California Code Regs (2024) title 17, §§ 95910–15.

[525] Decision (EU) 2015/1814 (n 524) art 1(5a).

[526] See Communication from the Commission of 15 May 2023 of the total number of allowances in circulation in 2022 for the purposes of the Market Stability Reserve under the EU Emissions Trading System established by Directive 2003/87/EC, [2023] OJ C 172/1, 4 (noting that 2.5 billion emission allowances became invalid on 1 January 2023). See generally Rosendahl (n 515).

[527] See Adam Whitmore, 'Puncturing the Waterbed Myth', Sandbag (October 2016) 17.

may also occur among EA frameworks established at different levels of government, in particular in federal countries, where environmental protection is often a competence shared by federal and provincial authorities. For instance, the GHG emissions of a project implemented in New York City may fall within the scope of NEPA,[528] the New York State Environmental Quality Review Act,[529] and even New York City's Environmental Quality Review.[530] Such redundancies can make environmental policies more reliable and effective,[531] in particular 'in a complex policy domain such as the environment',[532] but they can also impose additional costs and delays affecting project proponents and ultimately society. As such, federal and provincial governments have often sought to avoid such redundancies, for instance by concluding bilateral agreements (eg between a federal and a provincial government),[533] by delegating parts of an EA procedure to another jurisdiction,[534] by allowing one EA procedure to substitute fully to another,[535] or by arranging a joint implementation of overlapping procedures.[536]

Whether GHG emissions should be addressed by federal or provincial government is an open legal and political debate in many federal countries. One view is that climate impacts fall more naturally within the competence of federal authorities because climate change, as a collective action problem, is best addressed through international cooperation, which generally falls within the competence of the federal government.[537] However, deferring EA decisions to central or federal authorities for any project with significant GHG emissions could shift power relations in unacceptable ways. Thus, the Canadian Supreme Court held that the Impact Assessment Act 2019 could not lawfully permit federal decision-making merely 'on the basis that a project emits [GHGs] that cross provincial and national borders'.[538] This decision will force the Canadian government to review the scope

---

[528] 42 USC (2024) § 4332.

[529] New York Env't Conserv Law (McKinney 2024) § 8-0109. See also New York State Department of Environmental Conservation, Commissioner Policy 49 (issued 22 October 2010, revised 14 December 2022) <https://extapps.dec.ny.gov/docs/administration_pdf/cp492022.pdf>, 1 (requiring an assessment of GHG emissions).

[530] New York City Executive Order 91 of 1977, ch 6 §6-12. See also Mayor's Office of Environmental Coordination, CEQR Technical Manual (December 2021) <www.nyc.gov/site/oec/environmental-quality-review/technical-manual.page>, ch 18, 1–2 (requiring an assessment of GHG emissions).

[531] Martin Landau, 'Redundancy, Rationality, and the Problem of Duplication and Overlap' (1969) 29 Public Administration Review 346. See also Daniel C Esty, 'Revitalizing Environmental Federalism' (1996) 95 Michigan Law Review 570.

[532] Robyn Hollander, 'Rethinking Overlap and Duplication: Federalism and Environmental Assessment in Australia' (2010) 40 Publius: The Journal of Federalism 136, 137.

[533] Eg Impact Assessment Act 2019 (n 108), s 114(1)(f); Government of Canada, 'Agreements related to assessments' (6 July 2016) <www.canada.ca/en/impact-assessment-agency/corporate/acts-regulations/legislation-regulations/environmental-assessment-agreements.html>.

[534] Eg Impact Assessment Act 2019 (n 108), s 29.

[535] Eg ibid s 31; Environment Protection and Biodiversity Conservation Act 1999 (Cth), consolidated as of 15 December 2023, s 47.

[536] Eg Impact Assessment Act 2019 (n 108), s 39.

[537] See eg *Environmental Defence Society Inc v Auckland Regional Council* (n 452) [20] ('climate change is an international issue, and should therefore be dealt with at a national level').

[538] *Reference re Impact Assessment Act* [2023] SCC 23 [184].

of the federal CA framework, possibly by defining a clearer or higher threshold of significance for climate impacts.[539]

On the other hand, while climate treaties only create obligations for sovereign states (and the European Union), the COP has also pointed to the role of subnational governments in addressing climate change.[540] Federal courts, in the United States, have acknowledged that adopting policies on climate change mitigation was also within the competence of federated states,[541] and several states have implemented their EA framework as a tool for climate change mitigation.[542] Similarly, CAs have been widely implemented by federated states or provinces in Australia[543] and Canada.[544] When a proposed activity is not otherwise of national or federal importance, provincial or even local EA procedures may appear as a more attractive forum for meaningful political deliberations on the comprehensive assessment of the costs and benefits of the proposed activity.

3. Concerns about the nature of climate impact

More frequent objections to the inclusion of climate impacts in EA procedures raise more difficult questions with regard to the nature of climate impacts. Climate impacts differ from the environmental impacts that have typically been subjected to EA frameworks, in particular because climate impacts are diffuse and abstract, unfold on a global scale, and may be particularly difficult to predict. This subsection introduces some of the difficulties this rases, while the following chapters further consider whether EA can usefully be implemented as a tool for climate change mitigation in spite of the distinctive nature of climate impacts.

a) Diffuse and abstract nature
An important objection to CA builds on the tenuous causal link between the GHG emissions of a proposed activity and any concrete social or ecological harm. For instance, the Federal Court of Australia has questioned the possibility of considering that the GHG emissions from a particular project might 'cause an impact upon a protected matter'.[545] In particular, it noted that the applicants had not suggested

---

[539] See ibid [351] (Karakatsanis and Jamal JJ, dissenting) (arguing that the act is lawful because it limits federal EA decision-making powers to circumstances where 'an individual project's GHG emissions would cause a non-trivial change to the environment in another province or outside Canada').

[540] See eg Decision 1/CP.21, 'Adoption of the Paris Agreement', FCCC/CP/2015/10/Add 1 (29 January 2016) 2 paras 133–35.

[541] See United States v California (ED California 2020) 444 F Supp 3d 1181, 1196–97, and references cited.

[542] See eg California Environmental Quality Act, 1970 California Stat 2780.

[543] Eg State Environmental Planning Policy (Resources and Energy) 2021 (NSW).

[544] Eg Environmental Assessment Act, SBC 2018, c 51, s 25(2)(h).

[545] Wildlife Preservation Society of Queensland Proserpine v Minister for the Environment and Heritage (2006) 232 ALR 510 [72].

'that the mining, transportation or burning of coal from [a] proposed mine would *directly* affect' specific environmental resources,[546] or could be a 'substantial cause' of any concrete environmental damage.[547] Similarly, the Court of Appeal of New Zealand noted that, 'given ... the infinitesimal contribution which any particular project could make [to global GHG emissions], there could be no demonstrable linkage between GHG emissions associated with any particular project and climate change generally'.[548] More recently, a dissenting opinion by R Nelson CJ at the US Court of Appeal for the Ninth Circuit questioned whether one could 'really expect scientists to agree on how many forest fires or other environmental harms ... can be allocated to a 0.04% increase in annual global GHG emissions'.[549] Public comments on a revision of the NEPA regulations, in 2024, questioned the interpretation of NEPA as requiring an assessment of climate impacts,[550] insisted that climate impacts could not properly be predicted,[551] or contended that EA was only aimed at assessing the proximate effects of proposed activities.[552]

This objection is based on a sound factual observation: a GHG-emitting activity does not directly affect any particular population in any significant way. Any proposed activity contributes only an infinitesimal fraction to global annual GHG emissions that add to the stock of GHGs in the atmosphere that causes climate change. Further, while climate change impacts exacerbate physical hazards (eg the frequency and severity of some extreme weather events),[553] the harm resulting from these physical hazards depends largely on social and political conditions, such as the exposure, vulnerability, and resilience of populations.[554] When taking these observations into account, only a vanishingly tenuous causal link can be established between the GHG emissions of a proposed activity and the concrete harm that a population may suffer.[555] The economic harm resulting from a forest fire in Canada can certainly not be attributed, in any meaningful way, to the GHG emissions of a coal power plant in Australia.

These observations, however, do not imply that GHG emissions cause no impact. The IPCC observed that '[e]very ton of CO_2 adds to global warming'.[556]

---

[546] Ibid.

[547] *Environment Council of Central Queensland Inc v Minister for the Environment and Water* [2024] FCAFC 56 [89].

[548] *Genesis Power Ltd v Greenpeace New Zealand Inc* [2007] NZCA 569, [2008] 1 NZLR 803 [17], cited in *West Coast ENT* (n 454) [123].

[549] See *350 Montana* (n 351) 1281.

[550] Council on Environmental Quality, 'National Environmental Policy Act Implementing Regulations Revisions Phase 2' (n 404) 35494.

[551] Ibid 35527.

[552] Ibid 35508.

[553] Pörtner and others, 'Summary for Policymakers' (n 5) 9.

[554] Ibid 12–13, 20.

[555] See Jacqueline Peel, 'Environmental Impact Assessments and Climate Change', in Daniel A Farber and Marjan Peeters (eds), *Climate Change Law* (Edward Elgar 2016) 350, 352.

[556] IPCC Core Writing Team and others, *Climate Change 2023: Synthesis Report* (WMO & UNEP 2023), 83.

As noted above, a billion tonnes of carbon dioxide can be expected to increase the global average temperature by about half-a-thousandth of a degree Celsius.[557] This effect, while extremely small, appears far from negligible when one considers that it contributes to extremely broad ecological, social, and economic consequences at the global scale and on the long term. GHG-emitting activities can have a significant climate impact, even though this impact is abstract and diffuse.

The objection to CA, therefore, must rely on a tacit assumption that EA procedures should only consider impacts that are concrete in nature and would directly affect a population or an ecosystem—for instance by causing a forest fire—to the exclusion of diffuse and abstract impacts such as the impacts of GHG emissions. Yet it is unclear how this assumption could be justified. The mere observation that EA procedures ordinarily deal with concrete impacts does not imply that these frameworks should not also extend impacts that are diffuse and abstract in nature. And the diffuse nature of an impact does not make it any less real or significant— it is beyond doubt that GHG emissions contribute marginally to a wide range of ecological, social, and economic impacts unfolding in the long term.[558] Rather, one could argue that the purpose of informing decision-makers about the consequences of their decisions is best served by documenting all consequences that could influence the decision, *especially* those impacts that are less self-evident because of their diffuse and abstract nature.

By including climate impacts in the scope of EA procedures, lawmakers, national agencies, and courts have overwhelmingly rejected the conception of EA as being limited to the assessment of concrete harm that underlies this objection. For instance, the US Supreme Court in *Massachusetts v Environmental Protection Agency* highlighted the importance of considering even 'a small incremental step' towards addressing climate change.[559] As the Court noted, '[a] reduction in domestic emissions would slow the pace of global emissions, no matter what happens elsewhere.'[560]

There remain nonetheless some difficulties with the assessment of climate impacts due to their diffuse and abstract nature. From a procedural perspective, issues may arise with regard to public participation in EA procedures in the absence of affected populations and with regard to standing in judicial proceedings in the absence of victims. Further, the abstract nature of climate impacts makes it more difficult for stakeholders to agree on thresholds of significance, whether at the preliminary stages of screening and scoping or, subsequently, when deciding on the merits of the proposed activity.

---

[557] See this chapter, subsection A.1(a).
[558] See CEQ, 2023 Guidance (n 112) 1206 (noting that '[a]ll types of GHG emissions contribute to real-world physical changes').
[559] *Massachusetts v Environmental Protection Agency* (2007) 549 US 497, 524.
[560] Ibid 525.

Lawmakers, agencies, and courts have been struggling with this issue for years, and sensible principles and methods have started to emerge. A growing understanding is that a mere indication of the magnitude of the emissions is insufficient: an assessment of the significance of the activity's climate impact requires a contextualization of this impact.[561] This contextualization could be carried out by comparing a project's local benefits with its contribution to local GHG emissions[562] or by assessing how a project may impede the achievement of local mitigation objectives.[563] Alternatively, as the US Council for Environmental Quality suggested, the significance of climate impacts could be assessed by translating these impacts 'into the more accessible metric of dollars', as a way to 'allow decision makers and the public to make comparisons, help evaluate the significance of an action's climate change effects, and [provide for a] better understand[ing] [of] the tradeoffs associated with an action and its alternatives'.[564]

b) Global scale

Another objection to CA relates to the global nature of climate impacts. Some EA instruments do not require consideration for the extraterritorial impacts of proposed activities, and some explicitly exclude such consideration. This is sometimes the case of EA frameworks adopted by federated states, reflecting a view that transboundary and global impacts fall more naturally within the competence of the federal government.[565] Thus, the Queensland Environmental Protection Act aims at protecting 'Queensland's environment',[566] whereas the New South Wales Environmental Planning and Assessment Act focuses on 'the State's natural and other resources'.[567] In Australia, however, the extraterritorial impact of a proposed activity would not necessarily be considered as part of a Commonwealth EA procedure, which is narrowly confined to an assessment of impacts on specified 'matters of national environmental significance'.[568]

---

[561] See eg IEMA, *Assessing Greenhouse Gas Emissions and Evaluating their Significance* (2nd edn, 2022), 24. See also *Diné Citizens* (n 352) 1042–44.

[562] See *350 Montana* (n 351) 1269; *Gray* (n 312) [98].

[563] See eg CEQ, 2023 Guidance (n 112) 1197, 1201; European Commission, 'Environmental Impact Assessment of Projects: Guidance on the Preparation of the Environmental Impact Assessment Report' (2017) 39; Code de l'environnement (France) art L181-3(II)(8); Northern Territory Environment Protection Authority, 'Referring a Proposal to the NT EPA: Environmental Impact Assessment Guidance for Proponents' (2021), 26; *Friends of Oroville v City of Oroville* (2013) 219 Cal App 4th 832, 842.

[564] CEQ, 2023 Guidance (n 112) 1098.

[565] See eg New South Wales, *Parliamentary Debates*, Legislative Assembly, 24 October 2019, 1576 (Rob Stokes, Minister for Planning and Public Spaces) (asserting that considerations for climate impacts are 'matters for the Commonwealth Government').

[566] Environmental Protection Act 1994 (Qld) s 3.

[567] Environmental Planning and Assessment Act 1979 (NSW) s 1.3(a).

[568] Environment Protection and Biodiversity Conservation Act 1999 (Cth), consolidated as of 15 December 2023, ss 12–25F. See also Rosemary Lyster and others, *Environmental and Planning Law in New South Wales* (5th edn, Federal Press 2021) 218; Victoria McGinness and Murray Raff, 'Coal and Climate Change: A Study of Contemporary Climate Litigation in Australia' (2020) 37 Environmental & Planning Law Journal 78, 97; Jacqueline Peel, 'The *Living Wonders* Case: A Backwards Step in

Similarly, MEPA once expressly noted that an EA procedure 'may not include a review of actual or potential impacts beyond Montana's borders', and, in particular, that it may not consider 'impacts that are regional, national, or global in nature'.[569] This development results in a potential gap in assessment, given the limited applicability of NEPA to federated states' activities.[570] Further, NEPA's application to extraterritorial effects has itself 'been a matter of controversy'.[571] In particular, the Trump administration amended the NEPA regulations to exclude NEPA's application to activities 'with effects located entirely outside of the jurisdiction of the United States',[572] though this provision does not preclude the assessment of the extraterritorial impacts of activities that also have a territorial effect.[573]

Courts have sometimes circumvented provisions excluding the assessment of impacts unfolding extraterritorially on the ground that climate change also has local effects. For instance, the Land and Environment Court of New South Wales noted in *Gray* that GHG emissions have an 'impact … on the Australian and consequently NSW environment' to justify including these emissions within the scope of an EA procedure.[574] Similarly, the Montana District Court decided in *Montana Environmental Information Center v Montana Department of Environmental Quality* that the exclusion of impacts beyond Montana's borders from the scope of the assessment did not justify an agency's failure to 'take a hard look at the greenhouse gas effects of [the] project as it relates to impacts within the Montana borders'.[575]

This argument is generally unconvincing. Excluding the extraterritorial effects of climate change leaves very little impact to be considered, especially in relatively small jurisdictions like New South Wales or Montana. Yet, neither court justified its tacit assumption that this small fraction of the activity's global climate impacts could, by itself, be of sufficient significance to justify an assessment. By contrast, the

---

Australian Climate Litigation on Coal Mines' (2024) 36 Journal of Environmental Law 125, 128; *Australian Conservation Foundation Inc v Minister for the Environment* (2016) 251 FCR 308.

[569] Montana Environmental Policy Act, Montana Code Annotated (2011) § 75-1-201(2)(a), as amended by Montana Laws 2011, ch 396, § 1, effective 12 May 2011.

[570] See 40 CFR (2024) § 1508.1(q)(1)(vi) (providing a general exemption from NEPA review for any '[n]on-[f]ederal projects with minimal [f]ederal funding or minimal [f]ederal involvement where the agency does not exercise sufficient control and responsibility over the outcome of the project').

[571] James W Spensley, 'National Environmental Policy Act', in Thomas FP Sullivan (ed), *Environmental Law Handbook* (24th edn, Bernan Press 2019) 681, 706.

[572] CEQ, 'Update to the Regulations Implementing the Procedural Provisions of the National Environmental Policy Act' (16 July 2020) 85 Federal Register 43304, 43375, modifying 40 CFR (2024) § 1508.1(q)(1)(i).

[573] See 42 USC (2024) § 4332(I) (requiring agencies to consider 'the worldwide and long-range character of environmental problems'); Executive Order 13990 (n 395), § 5(a) (calling on agencies to 'capture the full cost of greenhouse gas emissions as accurately as possible … by taking global damages into account'); CEQ, 2023 Guidance (n 112) 1203.

[574] See *Gray* (n 312) [100].

[575] See *Montana Environmental Information Center* (n 311) *29.

Federal Court of Australia's decision in *Wildlife Preservation Society of Queensland Proserpine v Minister for the Environment* observed that a coal project would likely not have a significant impact as far as only 'protected matters in Australia' were concerned.[576]

On the other hand, the decision to exclude extraterritorial harm from the scope of EA frameworks is highly questionable. While EA procedures have emerged historically in a context where the main concern was with local environmental impacts,[577] it is a well-accepted principle that '[a]n assessment cannot be restricted to "site specific" environmental effects'.[578] This general practice is partly reflected in binding norms of international law, including a norm of customary international law applicable in a transboundary context and for the protection of shared resources.[579] Scholars have suggested that a state's obligations of due diligence and cooperation under international environmental law preclude a discriminatory application of EA frameworks, whereby the state would give less weight to transboundary and global harm than to domestic harm.[580]

c) Uncertainties

A last objection to the application of EA frameworks to climate impacts relates to the difficulty of predicting the climate impacts of certain decisions due to the risk of carbon leakage and other indirect effects. Traditional environmental impacts depend on the location of the activity. Thus, preventing the construction of a steel mill would generally avoid local environmental impacts even if the same quantity of steel ends up being produced elsewhere. By contrast, as climate impacts do not depend on the location of the activity, one cannot properly assess a decision's climate impact without considering indirect effects. Preventing the construction of a new steel mill would likely limit global steel production at least temporarily (especially if other mills are already a maximum capacity), thus limiting global GHG emissions; but this could also cause more steel to be produced in less efficient mills and in jurisdictions with less stringent environmental regulation, which could ultimately increase global GHG emissions. Predicting indirect climate impacts can be a perilous exercise as these impacts depend, among other things, on

---

[576] See *Wildlife Preservation Society of Queensland Proserpine* (n 545) [72]. See also *Finch* (SC) (n 89) [96] ('it is wrong … to treat the impact on climate of GHG emissions as local to the places where the combustion occurs').

[577] See Charles H Eccleston, *Environmental Impact Assessment: A Guide to Best Professional Practices* (CRC Press 2011) 108 (noting that 'NEPA was designed to address geographically bounded environmental concerns—not global problems such as climate change').

[578] Murray Raff, 'Ten Principles of Quality in Environmental Impact Assessment' (1997) 14 Environmental & Planning Law Journal 207, 209.

[579] See this chapter, subsection B.1.

[580] Knox (n 241); Smets (n 241); Alan Boyle, 'Human Rights and the Environment: Where Next?' (2012) 23 European Journal of International Law 613, 639–40.

the evolution of global market conditions, policy decisions by other governments, technological innovation and development, geopolitical circumstances, as well as evolving cultural and demographic factors.

These observations have led some to suggest that the climate impacts of an activity cannot be predicted in a sufficiently reliable manner to inform decisions.[581] As R Nelson CJ put it in his dissenting opinion in *350 Montana v Haaland*, '[n]o other environmental concern is so intertwined with assumptions of the behavior of 200 other sovereign nations, the supply and demand projections of global energy models, or the personal energy usage decisions of 7 billion people worldwide'.[582] He added: '[i]t strains credibility to assume that such targeted issues can be adequately analyzed under NEPA with any scientific consensus'.[583] In other cases, courts and agencies have found that stopping fossil-fuel projects would 'have no impact on climate change because it [would] have no impact on ... global demand ... and therefore no impact on global GHG emissions'.[584]

These uncertainties are real and significant, yet one may question whether they justify excluding climate impacts from the scope of EA frameworks. As the Land and Environment Court of New South Wales noted in *Gray*, the difficulty of measuring the impact of burning coal when assessing a coal mine project 'does not suggest that the link to causation of an environmental impact is insufficient'.[585] EA is perhaps most useful in documenting such impacts that are not obvious and whose amplitude may be uncertain. At times, an assessment of climate impacts can reduce uncertainties. For instance, a market analysis can help determine, albeit with a large range of uncertainty, the effect that a fossil-fuel project would have on fossil-fuel consumption. Overall, a CA may help structure public deliberations on the merits of a project in spite of irreducible uncertainties by relying on the various techniques that EA practitioners have developed to handle such uncertainties.[586] A clearer understanding of all climate impacts of proposed activities, including the uncertainties associated with them, could usefully inform decisions.

---

[581] See eg Alexander Zahar, 'Environmental Impact Assessment for Greenhouse Gas Emissions Is Pie in the Sky', in Benoit Mayer and Alexander Zahar (eds), *Debating Climate Law* (CUP 2021) 297, 306–307.

[582] See *350 Montana* (n 351) 1281.

[583] Ibid.

[584] *Xstrata Coal Queensland Pty Ltd v Friends of the Earth* [2012] QLC 13 [559], [563]. See also *Hancock Coal Pty Ltd v Kelly (No 4)* [2014] QLC 12 [232]; *Adani Mining Pty Ltd v Land Services of Coast* [2015] QLC 48 [456]; *Flat Canyon Federal Coal Lease Tract*, UTU-77114, Bureau of Land Management, Final Environmental Impact Statement (January 2002) [581]; *East Lynn Lake Coal Lease*, EIS-ES-030-2008-0004, Bureau of Land Management, Final Land Use Analysis and Final Environmental Impact Statement (2009), 266, cited in Michael Burger and Jessica Wentz, 'Downstream and Upstream' (2017) 41 Harvard Environmental Law Review 110, 134 fn 120.

[585] *Gray* (n 312) [98].

[586] See eg Glasson and Therivel (n 71) 123–26 (mentioning techniques such as the use of a range of confidence, sensitivity analyses, and the production of uncertainty reports).

Conclusion

This chapter has taken stock of the integration of climate change mitigation in EA frameworks throughout the world. It has shown that most jurisdictions have extended existing EA frameworks to climate impacts, either through a reinterpretation or a revision of existing instruments. Often, the extension of existing EA frameworks to climate impacts has been justified through an analogy between climate impacts and other environmental impacts. Looking beyond the widespread acceptance of EA as a tool for climate change mitigation, the following chapters will shed light on the different modalities of implementation.[587]

Further, the last section has reviewed the arguments against this practice. Some objections are based on an ill-founded concern for potential overlaps between CA and other policies and measures on climate change mitigation. Stronger objections point to the limitations of the analogy between climate impacts and environmental impacts. Whereas EA frameworks typically apply to local, concrete, and relatively predictable impacts, climate impacts are global, diffuse, and often more uncertain. The cogency of these arguments depends in part on how EA frameworks can be adapted to address the climate impacts. In this regard, the following chapters consider, first, how the significance of a climate impact can be determined at various stages of the EA procedure; second, how the scope of CA can be decided; and, third, how the assessment of climate impact can inform final agency decisions.

---

[587] See also James Watkins and Bridget Durning, 'Carbon Definitions and Typologies in Environmental Impact Assessment: Greenhouse Gas Confusion?' (2012) 30 Impact Assessment & Project Appraisal 296.

III

Appraising Significance

Introduction

Authors have characterized the determination of the significance of environmental impacts as 'the very heart' of environmental assessment (EA)[1] and 'the core of decision-making throughout' any such procedure.[2] At the initial phases, the determination of significance informs screening decisions (on whether a proposed activity is to undergo an EA procedure) as well as scoping decisions (on what the procedure should assess). But significance also plays a central role in subsequent phases, including public deliberations and the substantive decision-making stage, where the significance of the predicted impacts is compared with that of its benefits.

Despite this central role in EA, however, significance is difficult to define or assess. There often is no clear touchstone to determine the significance of an impact. While this issue is faced in any EA procedure, it is more acute in relation to climate assessments (CAs) due to the abstract nature of climate impacts. The greenhouse gas (GHG) emissions of a proposed activity are not ordinarily more than a 'drop in the ocean':[3] a very small addition to the cumulative anthropogenic GHG emissions that cause climate change. As such, some have argued that no activity would have any 'significant' climate impact, in the sense that, when taken in isolation, no activity directly changes the global climate system in any measurable way or directly harms anyone or anything in particular.[4] However, this argument can be countered by pointing out that the addition of GHGs in the atmosphere exacerbates climate change, thus increasing a risk of harm on a global scale and in the long-term.[5]

---

[1] Gordon E Beanlands and Peter N Duinker, *An Ecological Framework for Environmental Impact Assessment in Canada* (Institute for Resource and Environmental Studies 1983) 43.

[2] John Glasson and Riki Therivel, *Introduction to Environmental Impact Assessment* (5th edn, Routledge 2019) 126.

[3] Jacqueline Peel, 'Issues in Climate Change Litigation' (2011) 5 Carbon & Climate Law Review 15, 16–17.

[4] See eg Takafumi Ohsawa and Peter Duinker, 'Climate-Change Mitigation in Canadian Environmental Impact Assessments' (2014) 32 Impact Assessment & Project Appraisal 222, 222 (noting the argument according to which 'each individual project's contribution to climate change is insignificant and essentially impossible to estimate').

[5] Benoit Mayer, 'The Emergence of Climate Assessment as a Customary Law Obligation', in Benoit Mayer and Alexander Zahar (eds), *Debating Climate Law* (CUP 2021) 285, 293–94. See also *Diné Citizens against Ruining Our Environment v Haaland* (10th Cir 2023) 59 F 4th 1016, 1042.

Environmental Assessment as a Tool for Climate Change Mitigation. Benoit Mayer, Oxford University Press.
© Benoit Mayer 2024. DOI: 10.1093/oso/9780198939184.003.0003

The harm may not be 'detectable' with bare eyes,[6] but it can still be assessed as statistically relevant increase in a risk of harm.

It remains however that not every emission of GHG can be considered as a significant climate impact. While some have asserted that '[n]o tonne of CO_2 [carbon dioxide] is immaterial',[7] it remains that small levels of emissions cannot reasonably be subjected to demanding assessment procedures, as this would impose an exceedingly burdensome requirement on project proponents, let alone be prohibited. The significance of climate impacts needs to be assessed, first, to determine whether a thorough CA procedure is required in a relation to a particular activity and, then, to inform a decision on whether and how to approve this activity.

This chapter considers how EA procedures can and should approach significance in relation to climate impacts when determining, for instance, what level of GHG emissions warrants a thorough scientific assessment or justifies the imposition of mitigating measures or even the rejection of the proposed activity. Courts, national agencies, lawmakers, and project proponents across the world have come up with multiple methods to determine the significance of climate impacts, ranging from little more than a line in the sand to more sophisticated standards, benchmarks, and methodologies. Simple but rudimentary approaches, such as an absolute emission threshold, may be appropriate for the preliminary assessment of significance at the screening and scoping stages. For what concerns the substantive decision-making stages, however, more sophisticated methods should be relied upon to evaluate the climate impact of the proposed activity in light of its purported benefits, despite the vastly different geographies and timeframes in which these impacts and benefits unfold. This can be done either by using relevant GHG inventories and climate change mitigation policies as benchmarks to assess the impacts of the proposed activity, or by relying on an economic valuation of GHG emissions.

The following section further discusses the need to determine significance and the issues this raises in relation to cumulative environmental impacts in general and to climate impacts more specifically. Sections B to D identify and discuss the three most common approaches to assessing the significance of climate impacts: the magnitude-based, benchmark-based, and valuation-based methods. While magnitude-based tools are rudimentary, Section B shows that they can be a convenient way to screen projects and scope their assessment. Section C observes that national and local climate change mitigation goals have a growing potential to be used as benchmarks of significance. And Section D sheds light on valuation

---

[6] Alexander Zahar, 'Environmental Impact Assessment for Greenhouse Gas Emissions Is Pie in the Sky', in Mayer and Zahar (eds) (n 5) 297, 298–99.
[7] *Waratah Coal Pty Ltd v Youth Verdict Ltd (No 6)* [2022] QLC 21 [1304] (citing a joint expert opinion).

tools, such as the social cost of GHG emissions (SC-GHG), as another effective approach to assessing the significance of climate impacts.

As a whole, the chapter argues in favour of the combined use of magnitude-based, benchmark-based, and valuation-based approaches as methods to inform the determination of significance. Magnitude-based approaches are more relevant at the screening and scoping stages. Benchmark- and valuation-based approaches can be used alternatively or in combination to assess significance at all stages, but in particular at the substantive decision stage.

A. Assessing Significance

This section identifies the central role of the concept of significance in EA procedures and shows that the determination of significance faces particular issues in relation to climate impacts. The first subsection considers how significance has generally been approached in EA procedures. The second subsection reviews the ways significance has been assessed in relation to cumulative environmental impacts. The third subsection shows the difficulty of assessing the significance of climate impacts.

1. The significance of significance

A determination of the significance of environmental impacts is central at several stages of a typical EA procedure.[8] At the initial stages of the procedure, screening aims 'to focus on those projects with potentially significant adverse environmental impacts',[9] while scoping seeks to narrow down the assessment, among all potential 'impacts, issues and alternatives', on 'the significant ones'.[10] Under the US National Environmental Policy Act (NEPA), for instance, agencies decide on 'the appropriate level of … review' by considering the likelihood of 'significant effects',[11] and then on the scope of any environmental impact statement by identifying 'significant issues'.[12]

In turn, the scientific assessment, public participation, and decision-making stages seek to determine whether the impacts within the scope of the assessment are significant.[13] These stages involve not only the identification of any change

---

[8] For a description of a typical EA procedure, see Chapter II, subsection A.2(b).
[9] Glasson and Therivel (n 2) 86.
[10] Ibid 88.
[11] 40 CFR (2024) § 1501.3(a).
[12] Ibid § 1501.9(a).
[13] Kevin Hanna and Lauren Arnold, 'An Introduction to Environmental Impact Assessment', in Kevin Hanna (ed), *Routledge Handbook of Environmental Impact Assessment* (Routledge 2022) 11.

likely to occur as a result of the proposed activity, but also the determination of 'whether or not this change is important' and, when taking into account any potential mitigating measure, 'acceptable'.[14] An assessment of significance is instrumental to a determination of the merits of the proposed activity, ordinarily by an agency decision but also at times by courts, whether in a merit review[15] or a judicial review procedure.[16] The significance of environmental impacts may also be considered subsequently, at the monitoring stage, for instance when evaluating any discrepancy between predicted and observed impacts.[17]

John Glasson and Riki Therivel emphasize an 'important distinction ... between the prediction of the likely magnitude ... and the significance ... of ... impacts'.[18] The magnitude of an impact refers to factual information such as the quantity of pollutant released or the effect on the population of an endangered species. By contrast, significance is partly a subjective concept as it involves value-based judgements about what is 'worthy of attention'.[19] Whether an impact is deemed significant depends on both the characteristics of the impact (eg its magnitude) and on its perceived importance or value.[20] While a scientific study can provide a better understanding of potential impacts, it cannot substitute for a value-based judgement.[21] For instance, whether one deems the ecological consequences of marine pollution significant depends largely on how much one cares about the preservation of the affected ecosystems. As the Federal Court of Canada noted, '[r]easonable people can and do disagree ... about the significance' of an activity's predicted impacts.[22]

Significance is best determined through political processes involving the public, independent commission, national agencies, and governments, rather than by deference to experts or agencies. However, this does not mean that the determination of significance can be an arbitrary decision. Although some degree of discretion is generally recognized to decision-makers, it remains that these political processes are guided by legal principles,[23] compliance with which can be controlled by courts.[24] Courts have sometimes pointed out that, in spite of this subjective

---

[14] Angus Morrison-Saunders, *Advanced Introduction to Environmental Impact Assessment* (Edward Elgar 2018) 54.

[15] See eg *Gloucester Resources Ltd v Minister for Planning* [2019] NSWLEC 7.

[16] See eg *Diné Citizens* (n 5).

[17] Morrison-Saunders (n 14) 68. See also Glasson and Therivel (n 2) 172.

[18] Glasson and Therivel (n 2) 115 (emphasis removed).

[19] 'Significance, n' (OED Online, OUP 2024). See eg Chris Wood, *Environmental Impact Assessment: A Comparative Review* (2nd edn, Routledge 2013) 179; Morrison-Saunders (n 14) 58.

[20] See Megan Jones and Angus Morrison-Saunders, 'Making Sense of Significance in Environmental Impact Assessment' (2016) 34 Impact Assessment & Project Appraisal 87, 88.

[21] See eg Glasson and Therivel (n 2) 115.

[22] *Alberta Wilderness Association v Express Pipelines Ltd* [1996] FCJ 1016, 137 DLR(4th) 177 (Federal Court of Appeal) [10].

[23] See eg Glasson and Therivel (n 2) 126–36.

[24] See eg *Pembina Institute for Appropriate Development v Canada (Attorney General)* [2008] FCJ 324, 2008 FC 302 (Federal Court) [79].

component, a determination of significance must be informed by evidence and based on 'cogent reasoning'.[25]

The EA literature acknowledges that there is 'no single agreed method for determining significance'.[26] Alan Ehrlich and William Ross have argued persuasively that significance determinations should be informed by prevalent societal values, that is, the social acceptability of these impacts.[27] In practice, EA frameworks point to objective criteria such as 'the magnitude and likelihood of the impact', 'its spatial and temporal extent', and 'the likely degree of the affected environment's recovery', as well as more subjective considerations such as 'the value of the affected environment, the level of public concern, and political repercussions'.[28] At the decision-making stages, agencies sometimes turn to 'the professional judgement of competent experts' not only to characterize the impacts, but also to determine their significance.[29] Other times, decision-makers explicitly seek to identify 'a societal threshold of significance' that 'separates the realm of the acceptable from the realm of the unacceptable',[30] including by seeking the views of interested parties.[31]

Various types of evaluation methods can be used to account for these criteria, 'including simple or complex, formal or informal, quantitative or qualitative, aggregated or disaggregated'.[32] At the stages of screening and scoping, the determination of significance tends to be done through relatively simple and straightforward methods. Some EA procedures rely on a predetermined list of categories of activities that are to have a detailed assessment.[33] Others rely on thresholds and criteria, with some room for a case-by-case assessment.[34] Yet others involve a combination of these two approaches.[35]

More sophisticated evaluation methods are often used to determine the significance of the impacts at the decision-making stages. These methods may rely on

---

[25] Alan Ehrlich and William Ross, 'The Significance Spectrum and EIA Significance Determinations' (2015) 33 Impact Assessment & Project Appraisal 87, 89.

[26] Morrison-Saunders (n 14) 58.

[27] Ehrlich and Ross (n 25) 90–91.

[28] Glasson and Therivel (n 2) 127.

[29] Environmental Protection Agency, 'Guidelines on the Information to be Contained in Environmental Impact Assessment Reports' (May 2022) (Ireland) 49.

[30] Ehrlich and Ross (n 25) 91–93. See also Department of Environmental Affairs and tourism, 'Strategic Environmental Assessment' (Department of Environmental Affairs and Tourism' (Integrated Environmental Management Information Series No 10, 2004) (South Africa), 15 (noting that judgements of significance reflect the 'public acceptability of impacts').

[31] See eg David P Lawrence, 'Impact Significance Determination: Designing an Approach' (2007) 27 Environmental Impact Assessment Review 730, 736; Impact Assessment Act, SC 2019, c 28 (Canada), s 84(a).

[32] Glasson and Therivel (n 2) 130.

[33] Environmental Protection Department, 'Technical Memorandum on Environmental Impact Assessment Process' (30 June 2023) (Hong Kong) ss 4.2.1(a)–(k), 4.3.1(a)–(d).

[34] See eg California Code Regs (2024) title 14, § 15064.7(a); European Commission, 'Environmental Impact Assessment of Projects: Guidance on Screening' (2017) 40 (box 18).

[35] See eg Directive 2011/92/EU of the European Parliament and of the Council of 13 December 2011 on the assessment of the effects of certain public and private projects on the environment, [2012] OJ L 26/1, art 4.

technical assessments, collaboration among stakeholders, or reasoned argumentation.[36] John Glasson and Riki Therivel note that '[m]ost EIAs [environmental impact assessments] use simple and pragmatic methods ... drawing on experience and expert opinion',[37] but that some use more structured methods. In particular, some analyses 'seek to apply monetary values to costs and benefits'[38] to allow an assessment of the acceptability of the proposed activity. Yet the apparent simplicity of monetary cost-benefit analysis is achieved by embedding value-based judgements—such as assumptions about the value of the extinction of a species—in the economic valuation of costs and benefits.[39] Alternative, 'multicriteria' methods 'seek to allow for a pluralist view of society, composed of "stakeholders" with diverse goals and with differing values concerning environmental changes', in particular by using 'some kind of simple scoring and weighting system'.[40] EA scholars and practitioners tend to consider it desirable for the significance of an environmental impact to be considered in light of several or 'blended' approaches.[41]

A finding of significance has important consequences, first for the procedural requirements applicable to the proposed activity, and then possibly for the substantive decision itself. However, a finding of significance does not necessarily preclude the approval of the proposed activity: environmental impacts may be considered as acceptable, albeit significant, when they are the unavoidable consequence of an activity that would bring about considerable benefits. Thus, the Supreme Court of California explained that an agency, upon finding that an activity would have a significant impact, could 'adopt whatever feasible alternatives and mitigation measures exist beyond [those] already incorporated in the project design'; and, if these measures were still insufficient to avoid significant impacts, the agency could 'still approve the project with a statement of overriding considerations'.[42] As such, significance can be conceptualized as an intermediary step between the scientific characterization of the impact and the political decision on the acceptability of the proposed activity.

2. Assessing the significance of cumulative impacts

Any proposed activity would only cause a very small increment to the total amount of anthropogenic GHG emissions that cause climate change. As such, the GHG emissions of a proposed activity do not directly 'cause' any detectable, concrete

---

[36] Lawrence (n 31).
[37] Glasson and Therivel (n 2) 130–31.
[38] Ibid 131.
[39] Ibid 132.
[40] Ibid 134.
[41] Lawrence (n 31) 749–52.
[42] *Center for Biological Diversity v Department of Fish and Wildlife* (2015) 62 Cal 4th 204, 231 (references omitted).

harm to anything or anyone in particular. This observation has led some scholars to argue that 'no single development will have a significant impact on climate change', even while acknowledging that 'the sum of all projects and all other human activities is having a very significant impact on the climate'.[43] Yet this argument confuses the significance of an impact with its magnitude. That the effect of an activity on the climate system is of extremely small magnitude—seldom causing more than half-a-thousandth of a degree Celsius of global warming[44]—does not mean that this impact is insignificant. In a context where climate change is widely recognized as an issue of extremely great concern, the mitigation of which justifies drastic measures on a global scale and over decades, even a very small contribution to climate change could be reasonably considered as significant.

The concept of cumulative impact has developed in EA law and practice to address environmental concerns that, like climate change, result from the combination of several or multiple activities. NEPA regulations define cumulative impacts as the impacts 'that result from the incremental effects of the action when added to the effects of other past, present, and reasonably foreseeable actions', including when these cumulative impacts 'can result from individually minor but collectively significant actions taking place over a period of time'.[45] As conceptualized by Harry Spaling, cumulative impacts are those caused by 'the accumulation of changes in environmental systems over time and across space in an additive or interactive manner'.[46] Besides climate change, typical examples of cumulative environmental impacts include deforestation, habitat fragmentation, overfishing, and eutrophication.[47]

Many EA frameworks provide for the assessment of cumulative impacts.[48] For instance, the California Environmental Quality Act (CEQA) requires a cumulative impact assessment if 'the possible effects of a project are individually limited

---

[43] Ayodele Olangunju and others, 'Cumulative Effects Assessment Requirements in Selected Developed and Developing Countries', in Jill AE Blakey and Daniel M Franks (eds), *Handbook of Cumulative Impact Assessment* (Edward Elgar 2021) 259, at 268.

[44] See Chapter II, subsection A.1(a).

[45] 40 CFR (2024) § 1508.1(g)(3). NEPA uses the phrase 'cumulative effects', but it uses the words 'effects' and 'impacts' interchangeably. See also, eg California Code Regs (2024) title 14, § 15064(h)(1).

[46] Harry Spaling, 'Cumulative Effects Assessment: Concepts and Principles' (1994) 12 Impact Assessment 231, 232.

[47] See Glasson and Therivel (n 2) 278; Morrison-Saunders (n 14) 131; *BT Goldsmith Planning Services Pty Ltd v Blacktown City Council* [2005] NSWLEC 210 [90].

[48] See eg Directive 2011/92/EU (n 35) consolidated as of 15 May 2014, Annex IV para 5; Impact Assessment Act (Canada) (n 31) s 22(1)(a)(ii); Resource Management Act 1991 (New Zealand), s 3(d); International Finance Corporation, 'Good Practice Handbook Cumulative Impact Assessment and Management: Guidance for the Private Sector in Emerging Markets' (31 July 2013) 42; Town and Country Planning (Environmental Impact Assessment) Regulations 2017, Sch 3. See also UNEP, 'Goals and Principles of Environmental Impact Assessment' (16 January 1987) UNEP/GC.14/17 Annex III, Principle 4(d); Equator Principles, 'Guidance Note to Support Effective Consistent Application of the Equator Principles' (July 2022) 37; Protocol on Environmental Protection to the Antarctic Treaty (adopted 4 October 1991, entered into force 14 January 1998) 30 ILM 1455 (1991) (Madrid Protocol) art 2(c)(ii).

but cumulatively considerable'.[49] An impact is deemed 'cumulatively considerable' under CEQA if 'the incremental effects of an individual project are considerable when viewed in connection with the effects of past projects, the effects of other current projects, and the effects of probable future projects'.[50] CEQA Guidelines note that a project's contribution to a cumulative impact may be 'less than cumulatively considerable if the project is required to implement or fund its fair share of a mitigation measure or measures designed to alleviate the cumulative impact'.[51]

Climate impacts have frequently been characterized as cumulative impacts.[52] Thus, the US Court of Appeal for the Ninth Circuit observed that '[t]he impact of greenhouse gas emissions on climate change is precisely the kind of cumulative impacts analysis that NEPA requires agencies to conduct'.[53] Similarly, the New South Wales Land and Environment Court noted that 'climate change is caused by cumulative emissions from a myriad of individual sources, each proportionally small relative to the global total of GHG emissions, and will be solved by abatement of the GHG emissions from these myriad of individual sources'.[54] The International Tribunal for the Law of the Sea (ITLOS), likewise, justified the application of the EIA requirement under the UN Convention on the Law of the Sea to GHG emissions by observing that the UN Convention on the Law of the Sea 'does not preclude' an assessment of cumulative impacts.[55]

These various judicial pronouncements suggest that the significance of a proposed activity's climate impact is to be assessed by taking the proposed activity's broader context into account, that is, by approaching this proposed activity as one part of a set of human activities that, when combined, cause a major environmental impact. Consistently, the Council for Environmental Quality recommended that, '[i]n evaluating a proposed action's cumulative climate change effects, an agency should consider the proposed action in the context of the emissions from past, present, and reasonably foreseeable actions'.[56] As ITLOS put it: '[i]n the context of pollution of the marine environment from anthropogenic GHG emissions, planned activities may not be environmentally significant if taken in

---

[49] California Pub Res Code (2024) § 21083(a)(2).

[50] Ibid. See also California Code Regs (2024) title 14, § 15064(h)(1).

[51] California Code Regs (2024) title 4, § 15130(a)(3). See also *Save Our Peninsula Committee v Monterey County Board of Supervisors* (2001) 87 Cal App 4th 99, 140.

[52] See eg Robert M Sanford and Donald G Holtgrieve, *Environmental Impact Assessment in the United States* (Routledge 2023) 14 (characterizing climate change as 'the ultimate cumulative impact').

[53] *Center for Biological Diversity v National Highway Traffic Safety Administration* (9th Cir 2008) 538 F 3d 1172, 1217.

[54] *Gloucester* (n 16) [516].

[55] *Request for an Advisory Opinion submitted by the Commission of Small Island States on Climate Change and International Law*, Advisory Opinion, 12 May 2024, <www.itlos.org/fileadmin/itlos/documents/cases/31/Advisory_Opinion/C31_Adv_Op_21.05.2024_orig.pdf> [365].

[56] See also CEQ, 'National Environmental Policy Act Guidance on Consideration of Greenhouse Gas Emissions and Climate Change' (9 January 2023) 88 Federal Register 1196, 1206.

isolation, whereas they may produce significant effects if evaluated in interaction with other activities'.[57]

Yet, climate impacts are a somewhat atypical type of cumulative impact. On the one hand, the context that one needs to take into account is far broader than in most cumulative environmental issues.[58] Climate change results from a far larger pool of activities than any other environmental harm caused by cumulative impacts. The harm itself unfolds at a global scale and over a long timeframe.

On the other hand, the assessment of climate impacts is, in some respects, more straightforward than the assessment of some other cumulative impacts as the relation between GHG emissions and climate impacts is nearly linear. EA scholars have shown that the assessment of some cumulative impacts is particularly difficult to assess because 'the system under examination is complex and ... responds to disturbance in a non-linear fashion',[59] so that a proposed activity may cause 'amplifying' or 'discontinuous' changes, or various other 'structural surprises'.[60] For instance, 'the addition of nutrients into a lake ... triggers algae blooms once critical concentrations are attained',[61] that is, when a threshold of concentration is reached. Assuming an analogy between climate impacts and these complex cumulative impacts, some authors have asserted that 'it is not possible to predict quantitatively how much the global temperature would increase due to GHG emissions from a project'.[62] However, this assertion lacks support in climate science, which has generally found a near linear relation between GHG emissions and the risk of social and ecological harm.[63] As each additional tonne of carbon dioxide added in the air has about the same cumulative impact as the previous or the next one, predicting other sources of emissions is not necessary to assess a project's climate impact.[64]

However, characterizing climate impacts as cumulative impacts says little about how significance can or should be determined. Cumulative impact assessment is

[57] COSIS (n 55) [365].

[58] Charles H Eccleston, 'Assessing Cumulative Significance of Greenhouse Gas Emissions: Resolving the Paradox – the Sphinx Solution' (2010) 12 Environmental Practice 105, 111.

[59] George Hegmann and GA (Tony) Yarranton, 'Alchemy to Reason: Effective Use of Cumulative Effects Assessment in Resource Management' (2011) 31 Environmental Impact Assessment Review 484, 489.

[60] Spaling (n 46) 244. See also Martin Broderick, Bridget Durning, and Luis E Sánchez, 'Cumulative effects', in Riki Therivel and Graham Wood (eds), Methods of Environmental and Social Impact Assessment (4th edn, Taylor & Francis 2017) 649; Glasson and Therivel (n 2) 278.

[61] Spaling (n 46) 244.

[62] Ohsawa and Duinker (n 4) 222.

[63] See eg Marshall Burke, Solomon M Hsiang, and Edward Miguel, 'Global Non-Linear Effect of Temperature on Economic Production' (2015) 527 Nature 235, 239 ('our projected global losses are roughly linear—and slightly concave—in temperature, not quadratic or exponential'); Richard SJ Tol, 'The Economic Impacts of Climate Change' (2018) 12 Review of Environmental Economics & Policy 7 (noting that most studies agree on a 'piecewise linear model'). See also Richard P Allan and others, 'Summary for Policymakers', in Valérie Masson-Delmotte and others (eds), Climate Change 2021: The Physical Science Basis. Working Group I Contribution to the Sixth Assessment Report of the Intergovernmental Panel on Climate Change (CUP 2021) 3, 28 (noting a near-linear relationship between GHG emissions and the level of global warming).

[64] See eg Waratah Coal (n 7) [27]–[29].

often 'considered one of the most challenging and least successful components of impact assessment'.[65] EA scholars have repeatedly observed the poor conceptualization and lack of clear agreed-upon definition of cumulative impacts.[66] For instance, Melissa Foley and colleagues observed that '[t]here are many suggested definitions of cumulative effect in the literature ... but not all of them explicitly state what types of impacts contribute to cumulative effects'.[67]

Overall, EA law and practice is yet to develop any clear method to determine the significance of cumulative environmental impact.[68] Two extreme approaches have been presented, neither of which appears particularly promising. The first extreme approach is that 'causing minor harm to an already degraded environment should automatically be considered significant'.[69] Thus, some courts have appeared to suggest that the impact of a proposed activity should be deemed significant if it adds, even in the most minimal way, to an already significant cumulative impact.[70] This approach suggests that every GHG-emitting activity should be considered as causing a significant climate impact. Yet, asserting that every GHG-emitting activity is significant does not help when determining which activities should undertake a CA requirement and when assessing whether a proposed activity should be of particular concern to decision-makers.

The other approach, at the other extreme, focuses on the significance of the activity's incremental impact. Thus, Charles Eccleston suggested a 'significant departure principle' according to which an activity's cumulative impact would be characterized as significant if it 'depart[s] significantly from conditions that would exist if the action is not pursued'.[71] The CEQA regulations hint to this approach when suggesting that '[t]he mere existence of significant cumulative impacts caused by other projects alone shall not constitute substantial evidence that the proposed project's incremental effects are cumulatively considerable'.[72] Yet it is not

[65] Chris Joseph and others, 'Improving Cumulative Effects Assessment: Alternative Approaches Based upon an Expert Survey and Literature Review' (2023) 41 Impact Assessment & Project Appraisal 162, 162. See also Jill Blakley and Jessica Russell, 'International Progress in Cumulative Effects Assessment: A Review of Academic Literature 2008–2018' (2022) 65 Journal of Environmental Planning & Management 186, 186; Rebecca Nelson and LM Shirley, 'The Latent Potential of Cumulative Effects Concepts in National and International Environmental Impact Assessment Regimes' (2022) 12 Transnational Environmental Law 150, 169; Elizabeth A Masden and others, 'Cumulative Impact Assessments and Bird/Wind Farm Interactions: Developing a Conceptual Framework' (2010) 30 Environmental Impact Assessment Review 1, 1.

[66] Richard K Morgan, 'Environmental Impact Assessment: The State of the Art' (2012) 30 Impact Assessment & Project Appraisal 5, 10.

[67] Melissa M Foley and others, 'The Challenges and Opportunities in Cumulative Effects Assessment' (2017) 62 Environmental Impact Assessment Review 122, 128.

[68] Morgan (n 66).

[69] Nelson and Shirley (n 65) 165.

[70] BT Goldsmith Planning Services Pty Ltd v Blacktown City Council (n 47) [90] ('Arguing that a single site is a tiny percentage of what remains is really an argument which fails to acknowledge cumulative impacts.').

[71] Charles H Eccleston, 'Applying the Significant Departure Principle in Resolving the Cumulative Impact Paradox: Assessing Significance in Areas That Have Sustained Cumulatively Significant Impacts' (2006) 8 Environmental Practice 241, 244 (emphasis removed).

[72] California Code Regs (2024) title 14, § 15064(h)(4).

clear how this approach reflects the cumulative nature of the environmental impacts it assesses. The comparison it suggests between two hypothetical scenarios, with and without the proposed activity, is precisely the type of analysis that would be conducted to assess any non-cumulative environmental impact. This approach is vulnerable to the argument that, when taken in isolation from its context, 'no single event can be said to have such a significant impact that it will irretrievably harm a particular environment'.[73] In other words, this approach does not allow cumulative impact assessment to play a role in preventing 'death by a thousand cuts'.[74]

As a result of this conceptual confusion, cumulative EA is often inadequate, formalistic, ineffective, and unhelpful.[75] Based on a survey of Canadian EIA practice, Wanda Baxter and colleagues noted that '[i]t is not clear to us, and to practitioners with whom we have discussed this matter, what [cumulative EA] legislative requirements mean in practical terms'.[76] Likewise, Lourdes Cooper and William Sheate reported that EA studies have been impeded by 'uncertainty in regulatory requirements' and 'lack of guidance' on the assessment of cumulative impacts.[77] Angus Morrison-Saunders notes that '[a]t present ... there is no clear guidance or expectation as to what a single proponent might reasonably accomplish' to limit or reduce cumulative environmental harm.[78] For lack of more specific guidance, decisionmakers have often resorted to what they contemplated to be 'appropriate and proportionate' to reach 'a sensible decision' when determining the significance of cumulative impacts.[79]

3. Assessing the significance of climate impacts

The difficulty of determining significance is not unique to climate change or to cumulative impacts. Yet, when an environmental impact has direct and concrete effects, in particular on society, reasonable observers are more likely to agree about what should count as significant and what should not. By contrast, the abstract and

---

[73] *Gray v Minister for Planning* (2006) 152 LGERA 258 (NSWLEC) [122] (denouncing this argument).

[74] See eg Jessica T Dales, 'Death by a Thousand Cuts: Incorporating Cumulative Effects in Australia's Environment Protection and Biodiversity Conservation Act' (2011) 20 Pacific Rim Law & Policy Journal 149; Rebecca Nelson, 'Breaking Backs and Boiling Frogs: Warnings from a Dialogue between Federal Water Law and Environmental Law' (2019) 42 University of New South Wales Law Journal 1179.

[75] See eg Broderick, Durning, and Sánchez (n 60) 649; Morrison-Saunders (n 14) 136.

[76] Wanda Baxter, William A Ross, and Harry Spaling, 'Improving the Practice of Cumulative Effects Assessment in Canada' (2001) 19 Impact Assessment & Project Appraisal 253, 261.

[77] Lourdes M Cooper and William R Sheate, 'Cumulative Effects Assessment: A Review of UK Environmental Impact Statements' (2002) 22 Environmental Impact Assessment Review 415, 435.

[78] Morrison-Saunders (n 14) 136. See also Glasson and Therivel (n 2) 279.

[79] Broderick, Durning, and Sánchez (n 60) 664.

diffuse nature of climate impacts makes it more difficult for observers to agree on what should be deemed a significant impact. Without further information or analysis, observers are unlikely to agree as to whether an extremely small increase in global average temperature should be considered 'significant'.

As for other cumulative impacts, two radical approaches to the assessment of climate impacts have been formulated, suggesting either that every climate impact is significant (the 'radical inclusion' thesis), or that no climate impact is significant (the 'radical exclusion' thesis). The radical exclusion thesis relies on the observation that 'even the largest GHG-emitting projects generally produce no discernible change' in GHG concentrations in the atmosphere 'because the incremental contribution of almost any imaginable proposal is dwarfed by the effect that hundreds of millions of other combined emitters have on the global concentration level'.[80] In other words, 'any one project's contribution is unlikely to be significant by itself',[81] and 'all agency actions causing an increase in GHG emissions will appear de minimis when compared to the regional, national, and global numbers'.[82] Relying on this thesis, the defendants in *350 Montana v Haaland* submitted to the US Court of Appeal for the Ninth Circuit that 'virtually *every* domestic source of GHGs may be deemed to have no significant impact as long as it is measured against total global emissions'.[83]

Yet, the proponents of the radical exclusion thesis fail to justify why every climate impact would necessarily be insignificant.[84] The US Court of Appeal for the Fifth Circuit found that a pollutant that forms 'a "very small portion" of a gargantuan source of water pollution' could still constitute a significant problem on its own terms.[85] Given the major consequences of climate change, it is perfectly plausible that exacerbating global warming by even a very tiny fraction of a degree Celsius could be deemed a significant impact, in the sense that this could result in a very slight increase in a very vast set of risks across the world and over time. In this sense, the Council for Environmental Quality rightly affirmed that 'a statement that emissions from a proposed Federal action ... represent only a small fraction of global or domestic emissions ... is not a useful basis for deciding whether or to what extent to consider climate change effects under NEPA'.[86]

---

[80] Eccleston (n 58) 111.

[81] *Center for Biological Diversity v Department of Fish and Wildlife* (n 42) 219 (references omitted). See also *Environment Council of Central Queensland Inc v Minister for the Environment and Water* [2024] FCAFC 56 [89].

[82] *Diné Citizens* (n 5) 1043–44.

[83] *350 Montana v Haaland* (9th Cir 2022) 50 F 4th 1254, 1266 (emphasis in the original). See also Madeleine Siegel and Alexander Loznak, 'Survey of Greenhouse Gas Considerations in Federal Environmental Impact Statements and Environmental Assessments for Fossil Fuel-Related Projects 2017–2018' (White Paper, Sabin Centre for Climate Change Law 2019) (showing the extensive reliance of the Trump administration on the radical exclusion thesis).

[84] On the underlying confusion between magnitude and significance, see text above at n 43.

[85] See *Southwestern Electric Power Company v US Environmental Protection Agency* (5th Cir 2019) 920 F 3d 999, 1032, cited in *350 Montana* (n 83) 1266–67.

[86] CEQ, 2023 Guidance (n 56) 1201.

At the other extreme, the radical inclusion thesis in based on the premise that virtually any amount of GHG emissions is significant because these emissions further exacerbate climate change. As '[a]ll GHG emissions from projects will contribute to climate change', the Institute of Environmental Management and Assessment suggested, they 'may be considered significant'.[87] This seems to suggest that any activity causing GHG emissions would need to undergo a comprehensive EA procedure, even when these emissions are 'essentially innocuous'.[88] Wald CJ, at the US Court of Appeal for the District of Columbia Circuit, appeared to endorse this radical inclusion thesis when arguing that 'we cannot afford to ignore even modest contributions to global warming', before asking: '[i]f global warming is the result of the cumulative contributions of myriad sources, any one modest in itself, is there not a danger of losing the forest by closing our eyes to the felling of the individual trees?'[89] Jacqueline Peel could be read as suggesting a similar, all-inclusive approach to significance determination in relation to climate impacts when she argued that the significance of a project's GHG emissions should be assessed based on 'how it contributes, together with other actions, to cumulative emissions and the overall climate change problem and its consequences'.[90] She referred to the decision in *Gloucester Resources Ltd v Minister for Planning*, where the Land and Environment Court of New South Wales had suggested that climate change was to be addressed by regulating 'myriad of individual sources' of GHG emissions.[91]

But as some of these quotes suggest, the argument is often ambiguous: saying that every climate impact '*may*' be considered significant, for instance, does not necessarily mean that it *should* be considered significant. The Council on Environmental Quality (CEQ) noted that, under NEPA, '[t]he rule of reason and the concept of proportionality caution against providing an in-depth analysis of emissions regardless of the insignificance of the quantity of GHG emissions that the proposed action would cause'.[92] Surely the most minute climate impact—say the GHG emissions embedded in the electricity consumption of the computer used to write this book—could not seriously be considered as 'significant', or, at

---

[87] James Blake and others, 'Assessing Greenhouse Gas Emissions and Evaluating their Significance' (2nd edn, IEMA 2022) 8. See also Clemens Kaupa, 'Is It Still Permissible under EU Law to Issue New Permits for Oil and Gas Extraction?' (2024) 33 Review of European, Comparative & International Environmental Law 236, 241.

[88] Eccleston (n 58) 110.

[89] *City of Los Angeles v National Highway Traffic Safety Administration* (DC Cir 1990) 912 F 2d 478, 501 (Wald, CJ, dissenting in part).

[90] Jacqueline Peel, 'The *Living Wonders* Case: A Backwards Step in Australian Climate Litigation on Coal Mines' (2024) 36 Journal of Environmental Law 125, 130.

[91] *Gloucester* (n 16) [516], cited in Jacqueline Peel, 'The Land and Environment Court of New South Wales and the Transnationalisation of Climate Law: The Case of *Gloucester Resources v Minister for Planning*', in Elizabeth Fisher and Brian Preston (eds), *An Environmental Court in Action* (Hart 2022) 73, 83.

[92] CEQ, 2023 Guidance (n 56) 1202. See also Mott McDonald and others, *Environmental Impact Assessment Guide to: Climate Change Resilience & Adaptation* (IEMA, 2020) 1 (noting that 'a focus on proportionate assessment is also important in avoiding undue burden to developers and regulators').

least, not significant enough to justify an extensive environmental assessment procedure.

Both the radical inclusion and exclusion theses ignore the essential, at times obvious distinction between some activities whose GHG emissions are clearly significant—for instance because they impede the achievement of a national mitigation goal, when the state has spent considerable resources trying to achieve this goal—and others that clearly are negligible. The lack of a clear line between these two categories does not justify denying the distinction,[93] but it calls for the development of a test that could be implemented in a predictable, consistent, and reasonable manner. At present, as Alexander Crockett observed, EA frameworks 'offer … virtually no guidance on how to gauge how much of an incremental contribution is too much'.[94] Consequently, the assessment of the significance of climate impacts has often fallen back on courts and agencies.

The following sections of this Chapter show that further analysis can assist with the determination of significance in relation to climate impacts. None of the methods presented here provides a mathematical formula that could be applied to make an objective determination of significance; nor should that much be expected from any methods aimed at guiding the determination of significance, which is irreducibly a value-based decision.[95] Rather than an autopilot, the methods described here can be compared with a driver-assistance system that ensures the safe operation of the vehicle: these methods allow decisionmakers and other stakeholders to make safer decisions—decisions that are better informed, more convincing, and likely more consistent with one another over time—but they should not replace political deliberation.

These methods seek to present climate impacts in a way that enables a meaningful debate on their significance. The main impediment to such meaningful comparison is the vast difference in scale and timeframe of the proposed activity's costs and the benefits.[96] No meaningful comparison is immediately possible, for instance, between the local benefits of a coal plant over a few decades and its global climate cost unfolding over millennia. The plant's climate impacts may appear vanishingly small on a planetary scale, but so would its potential benefits, if considered at the same scale, as the plant would only satisfy an infinitesimal proportion of the world's energy demand. Further, these benefits would mainly occur over a few decades, whereas the adverse climate effects would continue to unfold on a much longer timescale.

---

[93] But see eg *Audubon Naturalist Society of the Central Atlantic States Inc v US Department of Transportation* (D Maryland 2007) 524 F Supp 2d 642, 708–09.

[94] Alexander G Crockett, 'Addressing the Significance of Greenhouse Gas Emissions under CEQA: California's Search for Regulatory Certainty in an Uncertain World' (2011) 4 Golden Gate University Environmental Law Journal 203, 208.

[95] Glasson and Therivel (n 2) 115.

[96] See Peel, 'Issues in Climate Change Litigation' (n 3) 16–17.

One theoretical way to facilitate a meaningful determination of the significance of a proposed activity's climate impacts would be to imagine that an equivalent impact was to be concentrated within the geography and timeframe in which society expects to enjoy the project's benefits. In other words, one could ask whether the project's beneficiaries would still support the project if *they*—rather than foreign populations and future generations—were to suffer all the social, ecological, economic, and other losses and damages resulting from the project's GHG emissions. This intellectual exercise would allow one to consider whether the project's climate impact appears reasonable, proportionate, and justifiable when compared, at the same scale, with the project's benefits. This test could be adapted to reflect a society's values, including its likely preference for immediate rather than future value (ie a discount rate for future harm) and for its own utility over that of aliens (ie a discount rate for extraterritorial harm).[97]

The following sections identify and discuss three complementary methodological approaches to determine the significance of climate impacts, based respectively on magnitude, benchmarks, or valuation. While there is no single best method, some methods appear more appropriate at particular stages of the EA. At the scoping and screening stages, national agencies must promptly determine the potential significance of the climate impacts of many projects. This is best done by relying on simple metrics and assumptions. Thus, it is at this stage that agencies can make the most convincing use of magnitude-based tools, for instance when considering that the climate impact of a project is likely to be significant when the project's GHG emissions exceed a set amount. By contrast, a more thorough assessment of significance may be warranted at the substantive decision-making stages of the CA procedure. Benchmarks can sometimes be found in GHG emission inventories or emission-reduction objectives applicable at the scale of the proposed activity. Alternatively, identifying a social cost of GHG emissions (SC-GHG) enables the integration of climate impacts in a comprehensive cost-benefit analysis. Rather than a choice between these tools, decisionmakers may best be informed by a combination of these different approaches.

B. Magnitude

This section considers methods that focus on the magnitude of a proposed activity's GHG emissions to determine the significance of the activity's climate impact. The first subsection accounts for how magnitude has been used as a proxy for significance, while the second subsection highlights the multidimensionality of magnitude. It is argued that these approaches are relevant to a preliminary

---

[97] See further discussion in this chapter, subsection D.2.

determination of potential significance at the screening and scoping stages, but that, even then, they cannot be reduced to the mechanical application of a single quantitative threshold. In particular, consideration should be given not only for the amount of GHG emissions by sources causally linked with the proposed activity, but also to the potential for a decision to reduce emissions compared with a scenario without the proposed activity.

1. Magnitude as a proxy for significance

The magnitude of the impacts that a proposed activity would cause is certainly a relevant consideration in determining the significance of these impacts. 'Usually', as Hong Kong's Technical Memorandum on Environmental Impact Assessment Process puts it, 'the greater the magnitude of the environmental change ... the more significant ... the impact'.[98] Thus, EA frameworks frequently rely on magnitude-based criteria at the stages of screening and scoping, that is, as a basis to determine whether and how thoroughly an environmental impact needs to be assessed.[99] Quantitative thresholds of magnitude provide a convenient heuristic for national agencies to focus resources on projects of greater concern.[100] For instance, a power plant may be subjected to an EA process if its heat output exceeds 300 megawatts (under the Espoo Convention and EU law),[101] or when its production capacity exceeds 200 megawatts (Canada),[102] or 100 megawatt (Hong Kong).[103] Such thresholds are not necessarily applied rigidly. In California, for instance, 'thresholds ... only define the level at which an environmental effect "normally" is considered significant', and they 'do not relieve the lead agency of its duty to determine the significance of an impact independently'.[104]

Magnitude-based criteria have also been used in relation to GHG emissions, for the preliminary determination of the potential significance of climate impacts at the screening and scoping stages of the CA. At times, these criteria are purely qualitative and somewhat imprecise. Thus, the Asian Development Bank requires

[98] Environmental Protection Department (Hong Kong) (n 33) 33 (table 1).
[99] European Commission, 'Environmental Impact Assessment of Projects: Guidance on the Preparation of the Environmental Impact Assessment Report' (2017) 49.
[100] See Charles H Eccleston, *Environmental Impact Assessment: A Guide to Best Professional Practices* (CRC Press 2011) 14. See also John Glasson, 'Principles and Purposes of Standards and Thresholds in the EIA Process', in Michael Schmidt and others (eds), *Standards and Thresholds for Impact Assessment: Environmental Protection in the European Union* (Springer 2008) 3.
[101] Directive 2011/92/EU (n 35) consolidated as of 15 May 2014, Annex I para 2(a); Convention on Environmental Impact Assessment in a Transboundary Context (adopted on 1 March 1991, entered into force 10 September 1997) 1989 UNTS 309 (Espoo Convention) App I para 2.
[102] Physical Activities Regulations, SOR/2019-285, Sch para 30.
[103] Environmental Impact Assessment Ordinance (1997) Cap 499, Sch 2 pt I, s D.1.
[104] *Center for Biological Diversity v Department of Fish and Wildlife* (n 42) 214. See also California Code Regs (2024) title 14, § 15064.7(a).

an assessment of projects that are expected to cause 'significant quantities of green-house gases',[105] and the World Bank's policy does not require a CA '[f]or projects that have diverse and small sources of emissions'.[106] Under CEQA, an initial study assesses '[t]he extent to which the project may increase or reduce greenhouse gas emissions as compared to the existing environmental setting'[107] to inform the agency's decision on whether a fully fledged environmental impact report is required.

In the absence of clear statutory or regulatory criteria, national agencies and judges have had to determine whether the magnitude of climate impacts warranted a CA. In some cases, the finding appeared rather obvious. An EIA report for a gas-fired power plant in South Africa, for instance, noted that 4.6 Mt CO_2 equivalent (CO_2e) were 'very large quantities' of annual emissions.[108] In *Gloucester*, the New South Wales Land and Environment Court assessed that the Rocky Hill Coal Project's aggregate direct and indirect emissions, estimated at 38 Mt CO_2e, would constitute 'a sizeable individual source of GHG emissions', even though this 'may represent a small fraction of the global total of GHG emissions'.[109] A fortiori, the Land Court of Queensland observed that the mining project in *Waratah Coal Pty Ltd v Youth Verdict Ltd* would result in 1.58 Gt CO_2e of combustion emissions.[110] Noting that this was far more than the level of emissions in *Gloucester*,[111] the Land Court of Queensland concluded that preventing this amount of emissions would be 'in absolute terms … a meaningful contribution to achieving the long-term temperature goal' agreed upon by states.[112]

Likewise, the significance of the GHG emissions was beyond doubt in several leading US court cases, including *WildEarth Guardians v US Bureau of Land Management* (on the extension of the lease of coal mines accounting for 20 percent of the US's annual coal production)[113] and *WildEarth Guardians v Zinke* (on the issuance of 473 oil and gas leases).[114] By contrast, the Court of Appeal for the Ninth Circuit held in *Hapner v Tidwell* that a thorough assessment of climate impacts was not required for a project authorizing logging and prescribed burning on 810 acres (3.3 km²) of land, which the Court considered to be a 'relatively small amount of land'.[115] Subsequent decisions excluded a fully fledged CA requirement

[105] Asian Development Bank, 'Safeguard Policy Statement' (2009) 38.
[106] World Bank, 'Environmental and Social Framework' (2016) 41.
[107] California Code Regs (2024) title 14, § 15064.4(b)(1).
[108] Savannah Environmental on behalf of Eskom Holdings SoC Ltd, Revised Environmental Impact Assessment Report, 'Richards Bay Combined Cycle Power Plant (CCPP) Project' (July 2019) 216.
[109] *Gloucester* (n 16) [515].
[110] *Waratah Coal* (n 7) [649].
[111] Ibid [774].
[112] Ibid [775].
[113] *WildEarth Guardians v US Bureau of Land Management* (10th Cir 2017) 870 F 3d 1222, 1227.
[114] *WildEarth Guardians v Zinke* (D DC 2019) 368 F Supp 3d 41, 55.
[115] *Hapner v Tidwell* (9th Cir 2010) 621 F 3d 1239, 1242, 1245. See also Cour administrative d'appel de Bordeaux (Administrative Court of Appeal of Bordeaux), 28 March 2023, 22BX02010 [32] (finding GHG emissions from land clearing for the construction of a power plant to be negligible).

for the logging of 1,631 acres (6.6 km^2),[116] but required it when the logging project extended to 3,902 acres (15.8 km^2) of land.[117] It remains unclear where precisely courts would draw the line of potential significance between 6.6 and 15.8 km^2, or how courts might also consider other factors such as the forest density and the logging format.

The adoption of regulatory thresholds would considerably reduce this uncertainty.[118] Such thresholds can be expressed in activity data (eg the area covered by a logging project and the capacity of a power plant) or as an expected level of GHG emissions associated with the proposed activity. For instance, the European Investment Bank's policy requires a carbon footprint assessment for projects causing over 20 kt CO_2e per year.[119] The CEQ's 2010 and 2014 draft guidance suggested that a quantitative analysis of GHG emissions would be expected for projects likely to emit more than 25 kt CO_2e annually.[120] For the Environmental Protection Authority of Western Australia, a detailed assessment is 'generally' required for projects whose GHG emissions are 'reasonably likely to exceed' 100 kt CO_2e per year.[121] The Equator Principles adopts the same 100 kt CO_2e threshold.[122] And the state of Victoria only requires an EA study for projects whose emissions could exceed 200 kt CO_2e per year.[123]

Magnitude-based thresholds are an approximate attempt at finding a reasonable balance between the need to ensure that decisions are well informed and the need not to impose exceedingly burdensome procedures on minor activities. They vary significantly, and there is no objective ground for setting a threshold at, for instance, either 25 or 100 kt CO_2e of annual emissions.[124] During consultations on the CEQ's 2010 draft guidance, some commenters asserted that, 'when

---

[116] *Swomley v Schroyer* (D Colorado 2020) 484 F Supp 3d 970, 973, 977.

[117] *Center for Biological Diversity v US Forest Service* (D Montana 2023) 687 F Supp 3d 1053, 1065.

[118] While the Forest Service revised its NEPA compliance rules in 2020, these rules still do not define any criteria to determine the significance of climate impacts. See Forest Service, 'National Environmental Policy Act (NEPA) Compliance' (19 November 2020) 85 Federal Register 73620; 36 CFR (2024) § 220.

[119] European Investment Bank, 'Project Carbon Footprint Methodologies: Methodologies for the Assessment of Project Greenhouse Gas Emissions and Emission Variations' (version 11.3, 2023) 3.

[120] CEQ, 'Draft NEPA Guidance on Consideration of the Effects of Climate Change and Greenhouse Gas Emissions' (18 February 2010) <https://perma.cc/4P7U-RFYW>, 2; CEQ, 'Revised Draft Guidance for Federal Departments and Agencies on Consideration of Greenhouse Gas Emissions and the Effects of Climate Change in NEPA Reviews' (24 December 2014) 69 Federal Register 77802, 77827–28. The 2016 and 2023 guidance no longer included a threshold of magnitude.

[121] Environmental Protection Authority, 'Environmental Factor Guideline: Greenhouse Gas Emissions' (April 2023) 4.

[122] Equator Principles, version 4 (July 2020) principle 2. See also Equator Principles, 'Guidance Note on Climate Change Risk Assessment' (May 2023) 16.

[123] Department of Transport and Planning, 'Ministerial Guidelines for Assessment of Environmental Effects' (8th edn, 2023) <https://perma.cc/R392-26GN> 15.

[124] See Jennifer Stewart, 'American Petroleum Institute's Comments on the Council for Environmental Quality's Notice of Interim Guidance on Consideration of Greenhouse Gas Emissions and Climate Change under the National Environmental Policy Act (88 Fed Reg 1,196) (CEQ-2022-0005)' (American Petroleum Institute 2023) 11 (noting that 'no scientifically reliable method exists to attempt to establish any threshold for significance').

compared with nationwide or global GHG emissions, a 25,000 metric ton disclosure threshold is too low to be meaningful for the purposes of a NEPA analysis'; while others pleaded for a lower threshold, even perhaps a 'zero threshold standard', on the ground that 'any additional contribution of CO_2 ... would contribute to a significant cumulative effect'.[125] It is apparently this difficulty of determining where precisely to draw that line that eventually led the CEQ to refrain from 'establish[ing] any particular quantity of GHG emissions as "significantly" affecting the quality of the human environment'[126]—although this only displaced the problem by forcing agencies to define these thresholds for themselves, either in general and abstract terms or on a case-by-case basis.

In spite of the lack of a clear line, it is remarkable that the thresholds adopted in various jurisdictions are generally within the same order of magnitude, thus reflecting an agreement that emissions lower than 20 kt CO_2e pear year do not warrant detailed analysis, whereas emissions exceeding 200 kt CO_2e per year do require particular attention. These amounts correspond roughly to the GHG emissions from some of the smallest commercial fossil-fuel fired power plants.[127] Variations within this range could be justified by two main factors. First, a lower threshold of magnitude could apply in relation to less onerous EA frameworks, in particular those providing for a simplified EA procedure for proposed activity of lesser significance.[128] Second, developing countries, whose international law obligations on climate change mitigation are less demanding,[129] could rely on higher thresholds of magnitude.

2. The multidimensionality of magnitude

Magnitude-based thresholds can facilitate a preliminary assessment of significance in the early stages of a CA. However, the magnitude of a project's climate impact cannot be reduced to a single metric. In addition to a predicted level of annual emissions, a magnitude-based assessment of significance should consider

---

[125] CEQ, 2014 Guidance (n 120) 77811.

[126] CEQ, 2023 Guidance (n 56) 1200. See also CEQ, 'Final Guidance for Federal Departments and Agencies on Consideration of Greenhouse Gas Emissions and the Effects of Climate Change in National Environmental Policy Act Reviews' (1 August 2016) <https://perma.cc/2A9H-GPNH> (not including any threshold but, instead, recommending the use of economic valuation tools). See discussion in this chapter, Section D.

[127] Thus, 96 percent of carbon dioxide emissions from fossil-fuel fired power plants in the United States originate from power plants that emit more than 100 kt CO_2 per year, and nearly 100 percent from plants emitting more than 10 kt CO_2 per year. Calculation based on Environmental Protection Agency, 'eGRID' (database) (30 January 2024) <www.epa.gov/egrid/download-data>.

[128] See Glasson and Therivel (n 2) 86.

[129] United Nations Framework Convention on Climate Change (adopted 9 May 1992, entered into force 21 March 1994) 1771 UNTS 107 (UNFCCC) art 3(1) (principle of common but differentiated responsibilities and respective capabilities); Paris Agreement (adopted 12 December 2015, entered into force 4 November 2016) 3156 UNTS 79, art 4(4).

the duration of the proposed activity: the climate impact of a long-term project is more significant, all other things being equal, than the impact of a short-term project with the same level of annual GHG emissions.[130] Another relevant factor is the propensity of certain projects (eg transport infrastructure) to lock society into a GHG-intensive production and consumption development model that would hinder future efforts to mitigate climate change.[131]

One technical but important issue when assessing the magnitude of a project's climate impact regards the scope of the GHG emissions that are to be taken into account. A fossil-fuel extraction project, for instance, may have limited on-site emissions; yet, by providing fossil fuels, it facilitates other activities that may result in far larger GHG emissions. Some thresholds of magnitude expressed in terms of annual GHG emissions apply only to direct emissions,[132] while others apply to both direct and indirect emissions taken separately,[133] or to the sum of direct and indirect emissions.[134] None of these approaches is fully satisfying. On the one hand, the assessment of the impact of a project on climate change would not be complete without considering its indirect emissions.[135] On the other hand, some of these indirect emissions would likely occur even without the proposed activity: the rejection of a coal mine project, for instance, would likely prompt increased coal production from other projects to meet some of the demand, albeit at higher price.[136] As such, a preferable approach would be to define higher (ie less stringent) thresholds of magnitude in relation to indirect emissions.

Another issue with assessing the magnitude of an activity's climate impact is that different GHGs have different atmospheric lifetimes. Methane, for instance, has an estimated atmospheric lifetime of only about 12 years, compared with 109 years for nitrous oxide, and far longer for carbon dioxide.[137] EAs often compare emissions in different GHGs based on their global warming potential with a time horizon of a hundred years (GWP100),[138] the metric that states are required to

---

[130] See European Commission, 'Preparation of the Environmental Impact Assessment Report' (n 99) 49; Council on Environmental Quality, 'National Environmental Policy Act Implementing Regulations Revisions Phase 2' (1 May 2024) 89 Federal Register 35442, 35557 (to be codified as 42 USC §1501.3(d) ('In assessing context and intensity, agencies should consider the duration of the effect').

[131] Gregory Unruh, 'Understanding Carbon Lock-In' (2000) 28 Energy Policy 817.

[132] See eg Department of Transport and Planning (n 123) 15.

[133] See eg Environmental Protection Authority (n 121) 4.

[134] See eg Equator Principles, version 4 (n 122) principle 10 (applicable to 'combined Scope 1 and Scope 2 emissions').

[135] See Chapter IV, subsection A.3.

[136] See Chapter IV, subsection B.3.

[137] Piers Forster and others, 'The Earth's Energy Budget, Climate Feedbacks and Climate Sensitivity', in Masson-Delmotte and others (eds), *Climate Change 2021* (n 63) 923, 1017.

[138] See eg CEQ, 2023 Guidance (n 56) 1199 (fn 32) (encouraging agencies to use GWP100 '[t]o avoid potential ambiguity', on the ground that the US NDC uses this metric); *Gloucester* (n 16) [423]; referring to Pacific Environment Ltd and D McKenzie, Development Application No SSD 5156, 'Amended Rocky Hill Coal Project' (June 2016), pt 2, 2A-158; which, in turn, refers to Australian Government, Department of the Environment, 'National Greenhouse Accounts Factors' (August 2015) 57.

use when developing their national GHG inventories under climate treaties.[139] Yet, this metric does not account for the more immediate need to smoothen the pace of global warming in the coming few decades in order to give time to societies and ecosystems to adapt to climate change.[140] An alternative metric, global warming potential with a time horizon of 20 years (GWP20), gives more weight to GHGs with a more rapid warming effect like methane.[141] US courts have been divided on whether agencies could rely solely on GWP100,[142] and agencies have increasingly leaned towards including both GWP20 and GWP100 values on a conservative basis.[143] The simultaneous use of these two metrics is consistent with a general EA practice of considering the significance of environmental impacts on multiple time horizons.[144] It demands very little additional work but provides potentially relevant information, for instance to avoid the potential oversight of a project with a particularly significant medium-term climate impact.

Further, whether a proposed activity would cause emissions in excess of a threshold is not necessarily a straightforward question, especially if one considers indirect effects such as carbon leakage and market substitution, which can be difficult to predict.[145] The US Court of Appeal for the Eighth Circuit rightly noted in *Mid States Coalition for Progress v Surface Transportation Board* that an agency 'may not simply ignore' an effect of a proposed activity on the ground that 'the *extent* of the effect is speculative', as long as the '*nature* of the effect is reasonably foreseeable'.[146] It remains however that determining whether the proposed activity reaches a magnitude-based threshold of significance requires, precisely, a prediction of the extent of this effect. While a precautionary approach may justify a CA when there is a small risk of very high levels of GHG emissions,[147] it remains difficult to determine precisely at what point a preliminary finding of potential significance is warranted (ie what risk and what level of GHG emissions).

---

[139] Decision 18/CMA.1, 'Modalities, Procedures and Guidelines for the Transparency Framework for Action and Support Referred to in Article 13 of the Paris Agreement', FCCC/PA/CMA/2018/3/Add 2 (19 March 2019) 18 para 37. See also Paris Agreement (n 129) art 13(7)(a).

[140] See eg IPCC Core Writing Team and others, *Climate Change 2023: Synthesis Report* (WMO & UNEP 2023) 23.

[141] Thus, one tonne of methane is equivalent to 29.8 t CO_2 based on GWP100, but it is equivalent to 82.5 t CO_2 based on GWP20. See Forster and others (n 137) 1017.

[142] See *Western Organization of Resource Councils v US Bureau of Land Management* (D Montana 26 March 2018) CV-16-21-GF-BMM, 2018 WL 1475470, at *15 (an agency's 'unexplained decision to use the 100-year time horizon, when other more appropriate time horizons remained available, qualifies as arbitrary and capricious'); *Diné Citizens* (n 5) 1038 (suggesting that GWP100 could be considered a 'reliable' methodology given its adoption as a prevailing accounting convention).

[143] See eg *Diné Citizens* (n 5) 1039.

[144] Office of the Deputy Prime Minister (Scotland), 'A Practical Guide to the Strategic Environmental Assessment Directive: Practical Guidance on Applying European Directive 2001/42/EC on the Assessment of the Effects of Certain Plans and Programmes on the Environment' (2005) s 5.B.16.

[145] See discussion in Chapter IV, subsection D.1.

[146] *Mid States Coalition for Progress v Surface Transportation Board* (8th Cir 2003) 345 F 3d 520, 549 (emphasis in the original).

[147] Robert B Gibson and others, *Sustainability Assessment: Criteria and Processes* (Earthscan 2005) 33.

These observations show that even an apparently simple magnitude-based threshold does not always provide a straightforward method to assess significance. At times, magnitude-based approaches require more than the mechanical application of a simple quantitative threshold. In such non-obvious cases, quantitative thresholds of magnitude should be a starting point for a more qualitative assessment of other relevant factors and considerations, including consideration for the length of the proposed activity and its potential indirect effects, among other things. Yet, in many cases, thresholds of magnitude provide clear and convenient guidance to agencies. For instance, the US Court of Appeal for the District of Columbia Circuit had no difficulty finding that the GHG emissions resulting from changes in air traffic control procedures and flight paths, which was estimated at up to 42 t CO_2e per year, would be far below the 25 kt CO_2e threshold suggested by the CEQ.[148] This led the Court to uphold the agency's finding that 'the project would not have a significant effect on the climate'.[149]

On the other hand, magnitude-based approaches are far less relevant at the decision-making stage, when determining the actual significance of the activity's climate impacts and the acceptability of the proposed activity. This is because a large among of GHG emissions may simply reflect the size of the proposed activity. For instance, a residential project would attract populations, resulting in a large amount of GHG emissions (eg from transportation and electricity consumption), but plausibly a more limited overall climate impact when one considers that some of the same emissions would probably have occurred elsewhere if the project had not been implemented. In these circumstances, as the California Supreme Court noted in *Center for Biological Diversity v Department of Fish and Wildlife*, 'a significance criterion framed in terms of efficiency' can appear 'superior to a simple numerical threshold because [EA procedure] is not intended as a population control measure'.[150] More scrutiny is warranted with regard to small projects causing relatively large GHG emissions for limited benefits.

C. Benchmarks

Information on the magnitude of an activity's GHG emissions is important, but it is not sufficient to assess significance at the decision-making stage. Without further analysis, the US Court of Appeal for the District of Columbia Circuit noted, 'it is difficult to see how [an agency] could engage in "informed decision making" with respect to the greenhouse-gas effects of [a] project, or how "informed public

---

[148] *Vaughn v Federal Aviation Administration* (DC Cir 2018) 756 Fed Appx 8, 15.
[149] Ibid. See also *Village of Logan v US Department of the Interior* (D New Mexico 14 January 2013) 12-CV-401 WJ/LFG, 2013 WL 12084730, at *14.
[150] *Center for Biological Diversity v Department of Fish and Wildlife* (n 42) 220.

comment" could be possible.'[151] Benchmarks contextualize and give meaning to a level of GHG emissions, although they do not provide an objective test for the assessment of significance. It is often useful for a CA to use several or multiple concurrent benchmarks to better allow the public and decision-makers to assess the significance of a climate impact.

Two types of benchmark have been used to assess the significance of climate impacts. The first type is based on observation: the climate impact of an activity can be compared with overall levels of GHG emissions globally or at the relevant scale, or they can be put in perspective with relevant sectorial practice. These empirical benchmarks do not account for the changes implied by existing commitments, goals, and policies on climate change mitigation. Holding a new power plant to a level of carbon efficiency observed in power plants today, for instance, would not properly reflect the expectation of a reduction in GHG emissions in the energy sector during the plant's lifetime. By contrast, benchmarks of the second type are normative in nature: they include mitigation goals, adopted at the global, national, or local scale, in relation to the entire economy or focusing on particular sectors or categories of GHG emissions. A frequent issue in this regard concerns the relevance of non-binding goals. Imposing consistency with purely aspirational goals could lead to arbitrary treatments and to an unacceptable burden on proposed activities.

Among both types of benchmark, a distinction can be drawn between benchmarks that are more or less specific. The more useful benchmarks are those that situate the climate impact of the proposed activity at relevant scales, where this impact can be compared with the benefits of the project.[152] The climate impact of a large power plant may appear very small when compared with global historical or predicted GHG emissions or with estimated carbon budgets consistent with global mitigation goals. Yet, the contribution of this power plant to global electricity production would similarly appear very small if considered at this global scale within such time periods. The merits of the project are best assessed at a scale at which one can meaningfully compare the project's costs with its benefits. If the project's main objective is to improve energy access in a given region, its GHG emissions can usefully be assessed, for instance, by reference to other sources of GHG emissions in this region, to the region's total emissions, and to emission-reduction goals applicable to that region. Approaching a proposed activity's climate impact in its relevant context often transforms a drop in the ocean into a drop in a spoon, whose significance can more readily be assessed.[153]

---

[151] *Sierra Club v Federal Energy Regulatory Commission* (DC Cir 2017) 867 F 3d 1357, 1374 (references omitted).

[152] See also CEQ, 2023 Guidance (n 56) 1201 ('when considering GHG emissions and their significance, agencies should … place emissions in relevant context, including how they relate to climate action commitments and goals').

[153] See also Peel, 'Issues in Climate Change Litigation' (n 3) 17 (noting that the drop in the sea issue can be addressed by 'embrac[ing] climate change as a "multiscalar" environmental problem').

1. Empirical benchmarks

Current levels of GHG emissions may provide a benchmark to assess the significance of climate impacts. An obvious difficulty, however, is that any individual activity contributes only a tiny fraction of global GHG emissions. The Intergovernmental Panel on Climate Change (IPCC) estimates global net anthropogenic GHG emissions at 59 Gt CO_2e per year (in 2019), and global cumulative carbon dioxide emissions at 2.4 Tt between 1850 and 2019.[154] When compared with these global totals, the climate impact of any proposed activities will appear vanishingly small, thus seemingly insignificant. On this ground, Nelson CJ's dissenting opinion in *350 Montana* argued that an agency could lawfully find that 'the incremental effects of 0.04% of annual global … GHG … emissions were "minor"'.[155] Yet, the benefits of these activities will also appear vanishingly small if they are assessed from the same global, long-term perspective. Thus, Takafumi Ohsawa and Peter Duinker denounce a 'scale trick': that a proposed activity would only contribute a small percentage of global GHG emissions does 'not necessarily mean that the emissions are not environmentally influential'.[156]

A concrete example can illustrate this point. The James H Miller plant, a very large coal plant situated in Alabama, emitted 24 Mt CO_2e in 2022.[157] This is several orders of magnitude beyond typical quantitative thresholds of significance discussed in the previous section.[158] Yet, these emissions pale in comparison with global annual GHG emissions: the Miller plant emitted less than 0.04 percent of global GHG emissions in 2022.[159] Presented this way, the climate impact of this very large coal plant appears very small, even somewhat insignificant. On the other hand, the James H Miller coal-fired power plant generates about 20 GWh of electricity per year,[160] which—from the same global perspective—represents only 0.08 percent of electricity generation.[161] Thus, if the plant's carbon dioxide emissions are very small from a global perspective, its benefits are also very small from that same perspective. The fact that the power plant's electricity-generating

---

[154] IPCC Core Writing Team and others (n 140) 45.

[155] *350 Montana* (n 83) 1281 (R Nelson, CJ, dissenting). The majority observed that the agency had estimated the project's annual emissions at 0.44% of global annual GHG emissions, not at 0.04% as Nelson CJ suggested. See ibid 1270 (fn 24).

[156] Ohsawa and Duinker (n 4) 224.

[157] Environmental Protection Agency, 'eGRID' (n 127) sheet PLNT22, cell AW190.

[158] See Chapter III, subsection B.1.

[159] M Crippa and others, 'GHG Emissions of All World Countries' (JCR/IEA Report 2023) <https://edgar.jrc.ec.europa.eu/report_2023>, 4 (estimating global GHG emissions at 53.8 Gt CO_2e).

[160] Energy Information Administration, 'Form EIA-923 detailed data with previous form data (EIA-906/920)' (16 January 2024) <www.eia.gov/electricity/data/eia923/> (reporting a net generation from the James H Miller coal-fired power plant of around 20 GWh per year from 2018 to 2022).

[161] See Hanna Ritchie and Pablo Rosado, 'Electricity Mix', *Our World in Data* (10 June 2020) <https://ourworldindata.org/electricity-mix> (estimating global electricity production in 2022 at 29 PWh).

benefits are localized whereas its global-warming impacts are diffuse does not jus-tify a finding that the former are significant while the latter are not.

As a tool to inform decision-makers, EA aims precisely at dispelling illusions and at fostering a better understanding of the impacts of proposed activities. As the Institute of Environmental Management and Assessment noted, 'a key goal of EIA is to inform the decision-maker about the relative severity of environmental ef-fects', and this goal can only be achieved by 'provid[ing] context for the magnitude of GHG emissions ... in a way that aids evaluation of these effects'.[162] An EA is all the more needed for abstract and diffuse impacts, the significance of which cannot readily be assessed without further analysis.

In principle, one could compare the costs and benefits of a proposed activity on a global scale—that is, put the very small contribution of a coal plant to global GHG emissions in perspective with its very small contribution to global energy generation. Yet, the exercise would be hindered by the very small magnitude of a proposed activity's costs and benefits, reducing the determination of significance to an apparent hair-splitting exercise that may do little to inform the public and decision-makers about the significance of the climate impact. For instance, it is difficult to agree or disagree with the Land Court of Queensland's submission that an increase in global GHG emissions by 0.2 percent could not 'be dismissed as negligible' while an increase in by 0.002 percent would be 'infinitesimal'.[163] Such numbers are not immediately meaningful without an extensive and careful consid-eration of the context.

By contrast, a more meaningful assessment of significance can be achieved by comparing the activity's costs and benefits on a local scale, including national or local GHG emission inventories.[164] The James H Miller plant's GHG emissions in 2022 may represent only 0.04 percent of global GHG emissions, but it also con-stitutes 0.4 percent of the GHG emissions of the United States and 24 percent of Alabama's GHG emissions.[165] While the plant emitted half of Alabama's GHGs from electricity generation,[166] it contributed only a quarter of the state's electri-city generation and met only a quarter of the state's electricity demand.[167] This

---

[162] Blake and others (n 87) 24.

[163] *Hancock Coal Pty Ltd v Kelly (No 4)* [2014] QLC 12 [208]–[209].

[164] Peel, 'Issues in Climate Change Litigation' (n 14) 17.

[165] See Crippa and others (n 159) 5 (estimating the GHG emissions of the US at 6.0 Gt CO_2e in 2022); Environmental Protection Agency, 'State GHG Emissions and Removals', 'consolidated data for all states' (database) (3 October 2023) <www.epa.gov/ghgemissions/state-ghg-emissions-and-remov als> (reporting US and Alabama 2021 GHG emissions at, respectively, 5.98 Gt CO_2e and 101 Mt CO_2e). On the emissions from the Miller plant, see text at n 157.

[166] See Environmental Protection Agency, 'State GHG Emissions and Removals' (n 165) (reporting Alabama's 2021 GHG emissions for electricity generation at 48 Mt CO_2e).

[167] See Energy Information Administration, 'Form EIA-923' (n 160) (reporting that the Miller plant generated 21.1 TWh, and that all plants situated in Alabama produced 78.9 TWh, in 2022); Energy Information Administration, 'US Electricity Profile 2022' (November 2023) <https://www.eia.gov/elec tricity/state/> (reporting electricity retail sales at 87 TWh in Alabama in 2022).

contextualization of the GHG emissions of the Miller plant does a better work at informing value-based judgements on the significance and acceptability of the project's climate impact.

Thus, agencies and courts have sometimes relied on national or local GHG emission inventories to determine the significance of the climate impact of a proposed activity. For instance, the EA report for the Richards Bay power plant acknowledged that the plant's GHG emissions would be 'high for an individual source' as they would 'account for as much as 1% of the South African greenhouse gas inventory'.[168] In *350 Montana*, the US Court of Appeal for the Ninth Circuit questioned the relevance of global GHG emissions as a benchmark to assess the significance of the emissions from a proposed new coal mine.[169] The Court suggested more relevant benchmarks when it observed that the project's emissions would represent 3 percent of the US emissions and 519 percent of Montana's emissions.[170] The Court thus emphasized that 'a more complete comparison of the [project]'s GHG emissions against US- and Montana-based emissions would go a long way toward contextualizing the significance of the project's environmental consequences'.[171] Similarly, the Land and Environment Court of New South Wales noted in *Gray v Minister for Planning* that a coal mine's downstream emissions were 'clearly a potential major single contributor to GHG emissions deriving from [New South Wales] given the large size of the [project]'.[172]

A frequent issue is to identify the relevant benchmark for the proposed activity at issue. The Guidelines on the implementation of CEQA note that a project's GHG emissions may be significant even when they appear 'relatively small compared to statewide ... emissions'.[173] Thus, the California Court of Appeal for the Third District was rightly critical of the possibility of assessing significance merely by comparing a project's GHG emissions with those of 'the world's eighth largest economy'.[174] Similarly, the US Court of Appeal for the Tenth Circuit noted in *Diné Citizens against Ruining Our Environment v Haaland* that the GHG emissions resulting from the issuance of permits to drill for oil and gas in New Mexico would 'add only a small percentage to the annual GHG emissions in the nation and the state'[175]—the EA study suggested that the project would add 0.0008 percent to the US annual emissions and 7 percent to New Mexico's.[176] However, the Court added, 'this comparative analysis proves only that there are other, larger sources

---

[168] Savannah Environmental (n 108) 216.
[169] *350 Montana* (n 83) 1265–66.
[170] Ibid, 1269.
[171] Ibid.
[172] *Gray* (n 73) [98].
[173] California Code Regs (2024) title 14, § 15064.4(b).
[174] *Friends of Oroville v City of Oroville* (2013) 219 Cal App 4th 832, 842 (finding this comparison 'meaningless').
[175] *Diné Citizens* (n 5) 1042.
[176] Ibid, 1041.

of GHGs'.[177] The New South Wales Land and Environment Court pointed out in *Mullaley Gas and Pipeline Accord Inc v Santos NSW (Eastern) Pty Ltd* that 'whilst the relative extent of a project's GHG emissions can assist in understanding the scale of a project's contribution to climate change, it is not dispositive and can in fact, if blindly applied, lead to a misunderstanding of the project's contribution to climate change'.[178]

Another complicating factor is that some of the GHG emissions associated with proposed activities would unfold beyond the jurisdiction in which the activities would be implemented. Many of the cases mentioned above concern the downstream emissions resulting from the combustion of fossil fuels, which may occur in other jurisdictions. The coal mine project at issue in *350 Montana*, for instance, would not cause a five-fold increase in GHG emissions *in Montana*, as most of the coal would be exported to foreign markets.[179] The comparison might nonetheless help stakeholders to understand the scale of the climate impacts associated with a proposed activity, but it needs to be used cautiously.

Alternatively to current levels of emissions, relevant sectorial practice may provide useful empirical benchmarks to assess the significance of climate impacts.[180] For instance, the Environmental Protection Authority of Western Australia requests proponents to provide information on 'projected emission intensity ... for the proposal and international benchmarking against other comparable projects, best practice, industry standards and/or milestones and sector pathways, benchmarks and/or milestones'.[181] Similarly, the Environmental Assessment Office of British Columbia may request information on the measures that proponent may have identified to reduce GHG emissions, 'including through the use of Best Available Technologies'.[182]

Sectorial benchmarks often avoid the issue of scale by relying on criteria such as technology, processes, and emission intensity rather than absolute levels of emissions. Intensity-based standards can be particularly useful in relation to activities likely to have a strong market substitution effect. For instance, it could be found that the climate impact of a waste treatment facility is not significant, despite a high level of on-site GHG emissions, if this facility substitutes to other facilities with a higher emission intensity. On the other hand, difficulties are likely to occur in identifying relevant analogues. The best technologies available in a developed country, for instance, may not be accessible to a project proponent in a developing country.

---

[177] Ibid, 1042.

[178] *Mullaley Gas and Pipeline Accord Inc v Santos NSW (Eastern) Pty Ltd* [2021] NSWLEC 110 [66].

[179] *350 Montana* (n 83) 1287 (R Nelson, CJ, dissenting).

[180] See Ohsawa and Duinker (n 4) 231.

[181] Environmental Protection Authority (n 121) 6.

[182] Environmental Assessment Office, 'Effects Assessment Policy' (version 1.0, April 2020) 58.

A more fundamental limitation with any empirical benchmarks relates to the tension between the conservative nature of these benchmarks (being based on observation of past practice) and the transformative goal of mitigating climate change. Mitigating climate change requires a transition: future activities must break with past practice. In this regard, normative benchmarks provide an essential complementary standpoint to consider the significance of the climate impacts of proposed activities.

2. Normative benchmarks

Normative benchmarks are defined in documents such as treaties, declarations, statutes, and national policies, and they often define environmental protection standards more stringently than those observed in practice. EA law and practice often relies on normative benchmarks to assess the significance of environmental impacts, notwithstanding the legal force of the underlying norm. For what concerns CA, reference can be made to global as well as national (or subnational) goals on climate change mitigation. Reference to these goals may enrich public deliberations on the merits of a proposed activity, but, as for other benchmarks, it cannot substitute for, and should not conceal, value judgements.

a) Global goals

States have long agreed on the 'ultimate objective' of the international climate regime to prevent 'dangerous anthropogenic interference with the climate system'.[183] In the Paris Agreement, they proclaimed the goal of 'holding the increase in the global average temperature to well below 2°C above pre-industrial levels' and of 'pursuing efforts to limit [it] to 1.5°C'.[184] This, they specified, required them collectively 'to reach global peaking of greenhouse gas emissions as soon as possible ... and to undertake rapid reductions thereafter ... so as to achieve a balance between anthropogenic emissions by sources and removals by sinks of greenhouse gases in the second half of this century'.[185] The Meeting of the Parties to the Paris Agreement has subsequently 'resolve[d] to pursue efforts to limit the temperature increase to 1.5°C'[186] while also noting with 'utmost concern' that 'carbon budgets consistent with achieving the Paris Agreement temperature goal are now small and being rapidly depleted'.[187]

---

[183] UNFCCC (n 129) art 2.
[184] Paris Agreement (n 129) art 2(1)(a).
[185] Ibid, art 4(1).
[186] Decision 1/CMA.5, 'Outcome of the First Global Stocktake', FCCC/PA/CMA/2023/16/Add 1 (15 March 2024) 2, para 4 (emphasis removed).
[187] Decision 1/CMA.3, 'Glasgow Climate Pact', FCCC/PA/CMA/2021/10/Add.1 (8 March 2022) 2 para 3. See also Decision 1/CMA.5 (n 186) para 25.

These goals and carbon budgets can inform the determination of the significance of an activity's climate impact. Thus, the Institute of Environmental Management and Assessment submits that, '[w]hen setting [a project's climate] impact into context to determine significance, it is important to consider the net zero trajectory in line with the Paris Agreement's 1.5°C pathway'.[188] Similarly, the plaintiffs in *Diné Citizens against Ruining Our Environment v Haaland* argued that the US Bureau of Land Management should 'have compared the expected emissions from the [applications for permits to drill] to the remaining carbon budget to determine the cumulative impact of the expected emissions'.[189] The Court of Appeal for the Tenth Circuit largely agreed: where the agency 'neither applied the carbon budget method nor explained why it did not', it 'acted arbitrarily and capriciously by failing to consider the impacts of the projected GHGs'.[190] And the decision of the Queensland Land Court against a coal mining project, in *Waratah Coal*, relied in part on the finding that, '[o]n the evidence in this hearing, there is no modelled scenario that demonstrates all the Project coal could be combusted, unabated, while still meeting the temperature goal'.[191]

A limitation of this approach, however, relates to the lack of legal force of the global mitigation goals. While states have agreed on the adoption of temperature goals, they have not explicitly agreed upon an obligation to act consistently with this goal, and there is little evidence of a subsequent practice of states from which such an obligation could have arisen.[192] Nor do the temperature goals reflect a scientific finding on 'safe limits of warming'[193]—as Reto Knutti and colleagues note, 'no scientific assessment has clearly justified or defended the 2°C target as a safe level of warming, and indeed, this is not a problem that science alone can address'.[194]

The non-binding nature of global mitigation goals does not necessarily preclude their relevance as benchmarks to inform political decisions on the significance of the climate impact of proposed activities. EA law and practice often use non-binding goals, and decision-makers are not necessarily required to reject a

[188]   Blake and others (n 87) 24.

[189]   *Diné Citizens* (n 5) 1043.

[190]   Ibid, 1044. See also *WildEarth Guardians v Bernhardt* (D DC 2020) 502 F Supp 3d 237, 255 (holding that the Bureau of Land Management should either have conducted a carbon budget analysis or explained why it considered that such an analysis 'would not contribute to informed decision making').

[191]   *Waratah Coal* (n 7) [1025].

[192]   See Benoit Mayer, *International Law Obligations on Climate Change Mitigation* (OUP 2022) 231–42; *R (Friends of the Earth) v Heathrow Airport Ltd* [2020] UKSC 52, [2021] 2 All ER 967 [71].

[193]   Centre for Environmental Rights, 'Comments on the Intended Draft Guideline for Consideration of Climate Change Implications in Applications for Environmental Authorisations, Atmospheric Emissions Licenses and Waste Management Licences' (23 July 2021) <https://cer.org.za/wp-content/uploads/2021/07/LAC-Submission-National-Guideline-for-Climate-Change-Considerations_26-July-2021.pdf> para 16.

[194]   Reto Knutti and others, 'A Scientific Critique of the Two-Degree Climate Change Target' (2016) 9 Nature Geoscience 13, 13.

project inconsistent with these goals.[195] Yet, the emphasis is generally on national rather than international goals, with exceptions mainly in European countries.[196] In dualist jurisdictions, where even binding treaty provisions are not automatically applicable as the law of the land, one could raise further questions about the possibility of relying on non-binding treaty objectives and other non-binding goals adopted on the international plane as part of national EA frameworks.

Another limitation of global mitigation goals as benchmark of significance is that these goals currently provide little information about the significance of a local activity. For one thing, these goals are vaguely defined.[197] When advocates suggest that '[t]he carbon budget derives from science',[198] they omit to mention that any estimate of a carbon budget depends on far-reaching assumptions. To estimate a carbon budget consistent with the temperature goals of the Paris Agreement, one needs not only to decide whether the goal is actually to limit global warming to 1.5 *or* to 2°C, what precisely the 'pre-industrial' reference point is, what the time horizon is (2100 or peak temperature?), and what level of certainty one must have to achieve this the carbon budget must give to achieving the goal.[199]

Thus, carbon budget analyses can point to different conclusions. In *Waratah Coal*, the Queensland Land Court found that the project would cause 1.58 Gt CO_2, which the Court considered to be a 'material' contribution to 'a remaining carbon budget of between 320 Gt and 620 Gt'.[200] The Court thus relied on the carbon budgets estimated by the IPCC for a 67 percent chance of holding the temperature limit (ie the peaking in global average temperature) to, respectively, 1.5 or 1.7°C relative to 1850–1900.[201] Relying on the same IPCC report, another plausible

[195] See eg Environmental Impact Assessment Ordinance (1997) Cap 499 (Hong Kong), s 10(2)(b) (requiring the Director of Environmental Protection to take into account 'the attainment and maintenance of an acceptable environmental quality' when deciding whether to issue an environmental permit); Environmental Protection Department, 'Technical Memorandum on Environmental Impact Assessment Process' (30 June 2023), Annex 4 para 1.1(a), Annex 6 para 1.2.1 (referring respectively to air and water quality objectives).

[196] See eg Directive 2011/92 (n 35) consolidated as of 15 May 2014, Annex IV para 5 ('the environmental protection objectives established at Union or Member State level which are relevant to the project'); Directive 2001/42/EC of the European Parliament and of the Council of 27 June 2001 on the assessment of the effects of certain plans and programmes on the environment, [2001] OJ L 197/30, Annex I para (e) ('the environmental protection objectives, established at international, Community or Member State level, which are relevant to the plan or programme').

[197] See Benoit Mayer, 'Temperature Targets and State Obligations on the Mitigation of Climate Change' (2021) 33 Journal of Environmental Law 585, 590–93. See also Joeri Rogelj and others, 'Mitigation Pathways Compatible with 1.5°C in the Context of Sustainable Development', in Valérie Masson-Delmotte and others (eds), *Global Warming of 1.5°C: An IPCC Special Report* (IPCC 2019) 93, 104–108.

[198] *Diné Citizens* (n 5) 1043.

[199] MR Allen and others, 'Framing and Context' in Masson-Delmotte and others (eds), *Global Warming of 1.5°C* (n 197) 49, 56–68; Mayer, 'Temperature Targets' (n 197) 590–93.

[200] *Waratah Coal* (n 7) [1937].

[201] Hans-Otto Pörtner and others, 'Summary for Policymakers', in Hans-Otto Pörtner and others (eds), *Climate Change 2022: Impacts, Adaptation and Vulnerability. Working Group II Contribution to the Sixth Assessment Report of the Intergovernmental Panel on Climate Change* (CUP 2022) 3, 29 reproduced in *Waratah Coal* (n 7) [761]. The carbon budgets are adjusted for the emission of 80 Gt CO_2 over the following two years.

interpretation of the global temperature goals—a 50 percent chance of limiting global warming to 1.9°C—would allow 1,120 Gt CO_2 emissions.[202] Moreover, considerable uncertainty surrounds each of these estimates. A slightly more recent study suggests that the remaining carbon budget for a 50 percent chance of achieving the 1.5°C goal was likely between –200 and 830 Gt CO_2 (with the lower range implying that the budget was already exceeded); for a 50 percent chance of achieving the 2°C goal, it was likely between 600 and 2,240 Gt CO_2.[203]

Overall, it is far from clear how a global, long-term carbon budget should be divided among states and through time, or apply in relation to any particular sector or activity. States have agreed in principle on the relevance of 'equity', 'common but differentiated responsibilities and respective capabilities', and 'national circumstances' as grounds for differentiation,[204] but they have not agreed upon a particular formula that could be used to divide a global carbon budget into national budgets. While states acknowledged the relevance of 'intergenerational equity',[205] they have not determined precisely how states should use carbon budgets through time, or how far they can rely on the expectation that they will be able to deploy negative emission technologies at scale to make up for excess emissions. The French State Council thus observed that the temperature goals of the Paris Agreement 'do not prevent as a matter of principle any new highway project'.[206]

Carbon budgets might be useful as a general reference to assess the significance of a climate impact, but they certainly do not provide the ground for an objective determination of the merits of a proposed activity. In abstraction from a discussion of the benefits of the project, ascertaining that the Galilee coal mine, in Queensland, could use anywhere between 0.1 and 0.5 percent of a global carbon budget says little about the opportunity of approving this project. In particular, the formal compatibility or incompatibility of a proposed activity with carbon budgets should not be taken as a ground to approve or reject this activity.

b) National goals

National goals on climate change mitigation may provide more directly relevant benchmarks to assess the significance of a climate impact.[207] For instance, the CEQ

---

[202] Pörtner and others, 'Technical Summary' in Pörtner and others (eds) (n 201) 35, 93 (table TS.3). As for the previous values, 80 Gt CO_2 were deducted from the IPCC's carbon budget to reflect emissions in the years since the IPCC's assessment.

[203] Robin D Lamboll and others, 'Assessing the Size and Uncertainty of Remaining Carbon Budgets' (2023) 13 Nature Climate Change 1360, 1365.

[204] See eg UNFCCC (n 129) art 3(1); Paris Agreement (n 129) art 4(4).

[205] Paris Agreement (n 129) preamble para 12.

[206] Conseil d'État (State Council) (10ème–9ème chambres réunies), 30 December 2021, 438686, ECLI:FR:CECHR:2021:438686.20211230 [23]. See also Conseil d'État (State Council) (6ème–5ème chambres réunies), 19 November 2020, 417362, ECLI:FR:CECHR:2020:417362.20201119 [57].

[207] See Ohsawa and Duinker (n 4) 230 ('each provincial government might need to check whether approaches to GHG emissions in EIAs are consistent with their targets ... and their opinions should be explicitly written in comprehensive reports').

notes in its latest Guidance that one of the goals of a CA is to '[e]nable agencies to make informed decisions to help meet applicable Federal, State, Tribal, regional, and local climate action goals'.[208] Accordingly, the CEQ guides national agencies to 'place emissions in relevant context, including how they relate to climate action commitments and goals', so as to 'present the environmental and public health effects of a proposed action in clear terms and with sufficient information to make a reasoned choice'.[209] Similarly, the European Commission suggests that CA 'should take relevant greenhouse gas reduction targets at the national, regional, and local levels into account, where available', including by 'identify[ing] opportunities to reduce emissions through alternative measures'.[210] The French Energy Code requires consideration for national climate change mitigation objectives in the EA of relevant projects, for instance power plants.[211]

Compared with global goals, the main added value of national goals is that they provide a benchmark that is more directly relevant to the proposed activity. As the Technical Manual for New York City's Environmental Quality Review notes, '[a]lthough the contribution of a proposed project's GHG emissions to global GHG emissions is likely to be considered insignificant when measured against the scale and magnitude of global climate change, certain projects' contribution of GHG emissions still should be analyzed to determine their consistency with the City's Citywide GHG reduction goal'.[212] Similarly, the Department of Environmental Conservation of New York State may determine that an action would have a significant impact if this action 'prevents or makes it more difficult or more expensive for the State to reduce GHG emissions'.[213] When society invests massive resources in the pursuance of a national mitigation goal, it certainly should be interested to know how a proposed activity may contribute or hinder the achievement of this goal.

Nonetheless, it remains at times difficult to assess the significance of the climate impact of a specific activity by reference to a broad national goal. In *R (Boswell) v Secretary of State for Transport*, the High Court of England and Wales dismissed the claim that the defendant should have used national carbon budgets as a benchmark to determine the significance of local road schemes: the Court observed that the carbon dioxide emissions from these schemes would represent only 0.5 percent of the national carbon budget and, thus, could not meaningfully be compared

[208] CEQ, 2023 Guidance (n 56) 1197.
[209] Ibid 1201.
[210] European Commission, 'Preparation of the Environmental Impact Assessment Report' (n 99) 39.
[211] Code de l'environnement, art L181-3(II)(8).
[212] New York City Executive Order 91 of 1977, ch 6 §6-12. See also Mayor's Office of Environmental Coordination, CEQR Technical Manual (December 2021) <www.nyc.gov/site/oec/environmental-quality-review/technical-manual.page>, ch 18, 2.
[213] New York State Department of Environmental Conservation, Commissioner Policy 49, 'Climate Change and DEC Action' (issued 22 October 2010, revised 14 December 2022) <https://extapps.dec.ny.gov/docs/administration_pdf/cp492022.pdf>, 6–7.

with national emissions.[214] In a 2018 Airports National Policy Statement, the UK Department for Transport suggested that a development consent should only be refused when a project's carbon emissions 'would have a *material impact* on the ability of Government to meet its carbon reduction targets'.[215] Similarly, French courts have regularly held that applicants had not demonstrated a project's 'incompatibility'[216] or 'direct contradiction'[217] with France's climate goals.

These authorities fail to distinguish between a test of compatibility and a test of consistency. A test of compatibility would check that the proposed activity does not make it impossible to achieve an existing goal on climate change mitigation. This test would generally be ineffective because, when taken in isolation, a single activity would generally not preclude the achievement of the goal. Rather than compatibility, a more promising approach assesses the significance of a climate impact by assessing the consistency of an activity with a national goal. A proposed activity is consistent with a national goal if it is based on principles that, if applied to all other activities, would ensure the realization of the mitigation goal. A significant climate impact, in this approach, is one that is in tension with the realization of the goal, even if not formally incompatible with this goal.

Various national goals can be used as benchmarks for significance. Some EA instruments emphasize the need for a proposed activity to 'support meeting' nationally determined contributions (NDCs) under the Paris Agreement[218] or, more broadly, to assess the extent to which the proposed activity may 'hinder or contribute to [the state's] ability to meet its environmental obligations and its commitments in respect of climate change'.[219] EA instruments adopted by federated states and local governments can refer to local goals and policies on climate change mitigation.[220]

Long-term strategic goals can also be particularly relevant, especially with regard to activities with long-term implications. For instance, the Institute of Environmental Management and Assessment suggests that '[t]he crux of significance' in the context of a CA is whether the project 'contributes to reducing GHG emissions' in a way that is 'consistent with a trajectory towards net zero by 2050'.[221]

[214] R *(Boswell) v Secretary of State for Transport* [2023] EWHC 1710 (Admin) [84], affirmed in *R (Boswell) v Secretary of State for Transport* [2024] EWCA Civ 145, [2024] All ER (D) 161 (Feb).

[215] Department for Transport, 'Airport National Policy Statement: New Runway Capacity and Infrastructure at Airports in the South East of England' (June 2018) para 5.82 (emphasis added).

[216] See eg Conseil d'État, 19 November 2020 (n 206) [57]; Cour administrative d'appel de Lyon (Administrative Court of Appeal of Lyon), 12 October 2022, 20LY00136 [18].

[217] Cour administrative d'appel de Nantes (Administrative Court of Appeal of Nantes), 21 May 2019, 17NT03927 [25].

[218] Asian Infrastructure Investment Bank, 'Environmental and Social Framework' (2019) 33.

[219] Impact Assessment Act (Canada) (n 31) s 22(1)(i). See also Equator Principles, version 4 (n 122) Annex A (suggesting that EA 'should ... consider the Project's compatibility with the host country's national climate commitments, as appropriate').

[220] Eg Environmental Assessment Act, SBC 2018, c 51, s 25(2)(h) ('The following matters must be considered in every assessment: ... (h) greenhouse gas emissions, including the potential effects on the province being able to meet its targets under the Greenhouse Gas Reduction Targets Act').

[221] Blake and others (n 87) 24.

Consistently with this recommendation, the Environmental Protection Authority of the Northern Territory considers that EIAs should seek to '[m]inimize greenhouse gas emissions so as to contribute to the NT Government's goal of achieving net zero greenhouse gas emissions by 2050',[222] whereas the Guidelines for implementation of CEQA authorize an agency conducting an EA to 'consider a project's consistency with the State's long-term climate goals or strategies'.[223] Absent any clear guidance, the High Court of South Africa held that a CA 'is necessary and relevant to ensuring that [a] proposed coal-fired power station fits South Africa's peak, plateau and decline trajectory as outlined in the NDC and its commitment to build cleaner and more efficient than existing power stations'.[224]

National goals are more helpful when they apply at a scale and to a sectorial scope coinciding with the benefits expected from the proposed activity.[225] In *North Coast Rivers Alliance v Marin Municipal Water Distribution Board of Directors*, the California Court of Appeal for the First District considered a desalinization plant project aimed at supplying fresh water to the Marin county. The Court found that the agency had lawfully concluded that the project's climate impact would not be significant on the ground, among other things, that this project would be consistent with the county's mitigation goal.[226] In *Tsakopoulos Investments v County of Sacramento*, likewise, the California Court of Appeal for the Third District upheld an agency's reliance on 'county-specific thresholds of significance for different sectors' while suggesting that statewide benchmarks would not have been appropriate.[227] At times, significance can be determined by reference not just to policy goals, but also to more specific regulations or requirements.[228] Other times, the preliminary strategic environmental assessment (SEA) of a policy, plan, or programme can define objectives that can inform subsequent assessments of significance in relation to more specific projects. The CEQA guidelines, for instance, provide that a project-specific CA may rely on a previous CA report 'containing a programmatic analysis of greenhouse gas emissions'.[229]

---

[222] Environment Protection Authority, 'Referring a Proposal to the NT EPA: Environmental Impact Assessment Guidance for Proponents' (2021) 27–28.

[223] California Code Regs (2024) title 14, § 15064.4(b)(3). See also *Friends of Oroville* (n 174) 842 ('The relevant question to be addressed in the [environmental impact report] is ... whether the Project's GHG emissions should be considered significant in light of the threshold-of-significance standard of Assembly Bill 32, which seeks to cut about 30 percent from business-as-usual emission levels projected for 2020, or about 10 percent from 2010 levels').

[224] *Earthlife Africa Johannesburg v Minister of Environmental Affairs* [2017] 2 All SA 519 (GP) [90].

[225] See Ohsawa and Duinker (n 4) 230. (noting that 'targets at small scales can ... avoid the scale trick').

[226] *North Coast Rivers Alliance v Marin Municipal Water District Board of Directors* (2013) 216 Cal App 4th 614, 652.

[227] *Tsakopoulos Investments v County of Sacramento* (2023) 95 Cal App 5th 280, 307.

[228] See also California Code Regs (2024) title 14, § 15064.4(b)(3). See *Center for Biological Diversity v Department of Fish and Wildlife* (n 42) 229.

[229] California Code Regs (2024) title 14, § 15183.5. See also Conseil d'État (State Council) (6ème–5ème chambres réunies), 15 November 2021, 432819, ECLI:FR:CECHR:2021:432819.20211115, para 3 (finding that the EIA for a fossil-fuel fired power plant did not need to consider alternative sources of energy since these alternatives had been considered in a previous policy decision); and,

National targets do not generally provide a clear determination of whether the proposed activity can be approved. This does not mean that these benchmarks are ineffective. National targets can inform decision-makers even while allowing them some discretion in determining whether the proposed activity's climate impact is something they can approve in light of the existing goals and policies. As the High Court of England and Wales noted in *R (Goesa Ltd) v Eastleigh Borough Council*, 'it is permissible for a planning authority to look at the scale of the GHG emissions relative to a national target and to reach a judgment, which may inevitably be of a generalized nature, about the likelihood of the proposal harming the achievement of that target'.[230] The substantive decision to be taken based on this information, the Court emphasized, 'is a matter of judgment ... not a hard-edged point of law'.[231] Whether courts can review these decisions thus depends on the level of judicial deference.[232]

The California Supreme Court's decision in *Center for Biological Diversity v Department of Fish and Wildlife* illustrates the persistence of a subjective judgement even when agencies can rely on relatively clear and specific goals. In that case, the Department of Fish and Wildlife had determined that the climate impact of a large real-estate development project was not significant on the ground that the project was consistent with the state's (implied) policy goal of reducing GHG emission intensity in specified sectors by 29 percent compared with a business-as-usual scenario.[233] However, the California Supreme Court held that the agency 'erred in failing to substantiate its assumption that [this] statewide measure of emissions reduction can also serve as a criterion for an individual land use project'.[234] The Court thus appeared receptive to the plaintiffs' argument that, in order to achieve the statewide goal, more substantial emission reductions would be expected from

by analogy, RenewableUK, 'Cumulative Impact Assessment Guidelines: Guiding Principles for Cumulative Impacts Assessment in Offshore Wind Farms' (1 June 2013) <https://tethys.pnnl.gov/publications/cumulative-impacts-assessment-guidelines-guiding-principles-cumulative-impacts>, 10 ('Regulators need to undertake an SEA to identify thresholds at which impacts are likely to become significant, and in turn should help determine the focus of the individual project-level EIAs. SEAs can help to scope project-level CIAs [climate impact assessments], identify strategic research needs, and provide a clear steer on mitigation and monitoring requirements').

[230] *R (Goesa Ltd) v Eastleigh Borough Council* [2022] EWHC 1221 (Admin), [2022] PTSR 1473 [123].
[231] Ibid. See also Cour administrative d'appel de Nantes (n 217) [25] (finding that a new fossil-fuel fired power plant could be considered to be consistent with France's mitigation goals, considering that it would operate as a load-following power plant aimed at enabling the development of renewable energy capacity and that it would substitute to less efficient fossil-fuel fired power plants).
[232] See discussion in Chapter V, subsection A.3.
[233] *Center for Biological Diversity v Department of Fish and Wildlife* (n 42) 225. See also California Air Resources Board, 'Climate Change Scoping Plan: A Framework for Change' (2008) <https://ww2.arb.ca.gov/sites/default/files/classic/cc/scopingplan/document/adopted_scoping_plan.pdf>, executive summary, 1.
[234] *Center for Biological Diversity v Department of Fish and Wildlife* (n 42) 228. See also *Cleveland National Forest Foundation v San Diego Association of Governments* (2017) 3 Cal 5th 497, 503.

new buildings and infrastructures, given the difficulty of retrofitting existing ones.[235]

Two justices disagreed with the suggestion that an individual project should be held to a higher standard than applicable statewide standards. Corrigan J highlighted that 'no CEQA provision places the responsibility on developers to mitigate environmental impacts caused entirely by *other* projects'.[236] Chin J likewise noted that '[n]o legal basis exists to determine that [the project's departure from a business-as-usual emission scenario] is insufficient'.[237] While the majority points to valid grounds to impose more stringent standards on new real-estate developments than on existing ones, one could question the role of the judge, rather than the agency, in making this determination. English courts, by contrast, have affirmed that, '[a]s a matter of principle', decision-makers have discretion in 'using benchmarks [they] consider . . . to be appropriate'.[238]

Another occasional critique of the use of national goals as benchmarks for significance is that these goals may lack ambition. For instance, the Centre for Environmental Rights argued that South Africa's NDC was 'not an adequate benchmark against which an activity's impacts should be assessed in terms of climate change mitigation'.[239] The Centre for Environmental Rights emphasized that NDCs 'are ultimately the outcome of the discretion of the executive arm of the state, and not necessarily a reliable indicator of a safe emission reduction pathway'.[240] A limitation of this line of argument concerns the ability and legitimacy of other actors—national agencies in the course of implementing EAs or courts when reviewing EA decisions—to substitute their view to that of democratic institutions in charge of devising national mitigation goals.[241] A distinct and perhaps more persuasive argument might invoke inconsistencies among national goals, for instance to exclude reliance on a more specific goal incompatible with the state's NDC or on an NDC inconsistent with the state's long-term emission-reduction strategy.

D. Economic Valuation

A third type of approach to determining the significance of climate impacts relies on an economic valuation of GHG emissions. The premise of valuation-based approaches is that one can put a price tag on each unit of GHG emissions to facilitate a comparison between the proposed activity's climate impacts and its other costs

---

[235] *Center for Biological Diversity v Department of Fish and Wildlife* (n 42) 226.
[236] Ibid 242 (Corrigan J, concurring and dissenting).
[237] Ibid 249 (Chin J, dissenting).
[238] *Goesa Ltd* (n 230) [122].
[239] Centre for Environmental Rights (n 193) para 11.
[240] Ibid para 12.
[241] See generally Benoit Mayer, 'Prompting Climate Change Mitigation through Litigation' (2023) 72 International & Comparative Law Quarterly 233, 242–43.

and benefits.[242] The first subsection presents an overview of CA law and practice endorsing valuation-based approaches. The second subsection examines existing valuation methodologies. The third subsection argues that valuation-based approaches allow a practical and consistent comparison between an activity's costs and benefits, but that this comes at a risk of hindering public engagement with CA decision-making.

1. Law and practice

Valuation-based approaches to significance have been implemented in several jurisdictions, in particular in Northern America. From the 1970s onwards, US courts have held that, in order to allow agencies to make well-informed decisions, 'NEPA, in effect, requires a broadly defined cost-benefit analysis of major federal activities'.[243] More precisely, courts have suggested that, when environmental or other factors 'are reasonably susceptible of being quantified in economic terms (dollars), such must be done'.[244]

In this context, issues arose when agencies compared costs that they did not consider susceptible of being quantified in economic terms, including climate impacts, with other costs and benefits. In 2006, the National Highway Traffic Safety Administration conducted a NEPA review to impose a fuel efficiency standard on new light trucks in which it sought to assess the 'maximum feasible' standard by comparing all costs and benefits.[245] Petitioners observed that the agency had not accounted for GHG emissions because it 'view[ed] the value of reducing emissions of CO_2 and other greenhouse gas as too uncertain to support their explicit valuation'.[246] In *Center for Biological Diversity v National Highway Traffic Safety Administration*, the Court of Appeal for the Ninth Circuit held that the agency's 'decision not to monetize the benefit of carbon emissions reduction was arbitrary and capricious'.[247] While acknowledging uncertainties as to the social cost of carbon dioxide emissions, the Court insisted that this value 'is certainly not zero'.[248] The Court thus suggested that it would be preferable for an agency to rely

---

[242] On the other hand, a cost-benefit analysis does not necessarily rely on an explicit economic valuation of costs and benefits. See eg *Bulga Milbrodale Progress Association Inc v Minister for Planning and Infrastructure* (2013) 194 LGERA 347 (NSWLEC) [20].

[243] *Chelsea Neighborhood Associations v US Postal Service* (2d Cir 1975) 516 F 2d 378, 386; *County of Suffolk v Secretary of Interior* (2d Cir 1977) 562 F 2d 1368, 1384. See also 40 CFR (2024) § 1502.22.

[244] *Alabama v US Army Corps of Engineers* (ND Alabama 1976) 411 F Supp 1261, 1268. But see *Sierra Club v Sigler* (5th Cir 1983) 695 F 2d 957, 978.

[245] National Highway Traffic Safety Administration, 'Average Fuel Economy Standards for Light Trucks Model Years 2008–2011' (6 April 2006) 71 Federal Register 17566, 17566.

[246] Ibid 17638, cited in *Center for Biological Diversity v National Highway Traffic Safety Administration* (n 53) 1200.

[247] *Center for Biological Diversity v National Highway Traffic Safety Administration* (n 53) 1203.

[248] Ibid 1200.

on a rough approximation of the value of GHG emissions rather than failing to account for these emissions at all.

In 2008, the Court of Appeal for the Ninth Circuit observed the existence of multiple valuations of GHG emissions ranging from 3 to 50 USD/t CO_2e.[249] The next year, the White House convened the Interagency Working Group on the Social Cost of Carbon (IWG-SCC) to harmonise agencies' valuation of climate impacts.[250] Since then, the IWG-SCC has published successive 'technical support documents' aimed at developing 'a range of SCC [social cost of carbon] values using a defensible set of input assumptions grounded in the existing scientific and economic literature'.[251] The IWG-SCC was disbanded in 2017 but re-established four years later as the Interagency Working Group on the Social Cost of Greenhouse Gases (IWG-SCGHG),[252] and it adopted a new 'interim estimate' of the social cost of GHG emissions (SC-GHG) in 2021.[253] In 2023, the IWG-SCGHG issued a memorandum inviting agencies to 'use their professional judgment to determine which estimates of the SC-GHG reflect the best available evidence, are most appropriate for particular analytical contexts, and best facilitate sound decision-making'.[254]

Along with the decision in *Center for Biological Diversity v National Highway Traffic Safety Administration*, the successive estimates of the social cost of carbon dioxide and other GHG emissions made it more difficult for national agencies to contend that they could not consider GHG emissions as part of a quantitative cost-benefit analysis of a proposed activity. Thus, in *High Country Conservation v US Forest Service*, the District Court for the District of Colorado rejected an agency's 'categorical explanation' that an analysis of the impacts of GHG emissions was impossible: referring to the IWG-SCC's technical support, the Court observed that

[249] Ibid 1199.

[250] Kevin Rennert and Cora Kingdon, 'Social Cost of Carbon 101: A Review of the Social Cost of Carbon, from a Basic Definition to the History of its use in Policy Analysis' (Resources for the Future, 1 August 2019) <https://media.rff.org/documents/SCC_Explainer.pdf> 4.

[251] IWG-SCC, 'Technical Support Document: Social Cost of Carbon for Regulatory Impact Analysis under Executive Order 12866' (February 2010) <www.epa.gov/sites/default/files/2016-12/documents/scc_tsd_2010.pdf>, 1. See also IWG-SCC, 'Technical Support Document: Technical Update of the Social Cost of Carbon for Regulatory Impact Analysis under Executive Order 12866' (May 2013, revised November 2013) <https://obamawhitehouse.archives.gov/sites/default/files/omb/assets/inforeg/technical-update-social-cost-of-carbon-for-regulator-impact-analysis.pdf>;   IWG-SCC, 'Technical Support Document: Technical Update of the Social Cost of Carbon for Regulatory Impact Analysis under Executive Order 12866' (August 2016) <www.epa.gov/sites/default/files/2016-12/documents/sc_co2_tsd_august_2016.pdf>.

[252] See Executive Order 13783 (28 March 2017) 82 Federal Register 16093; Executive Order 13990 (20 January 2021) 86 Federal Register 7037.

[253] IWG-SCGHG, 'Technical Support Document: Social Cost of Carbon, Methane, and Nitrous Oxide Interim Estimates under Executive Order 13990' (February 2021) <www.whitehouse.gov/wp-content/uploads/2021/02/TechnicalSupportDocument_SocialCostofCarbonMethaneNitrousOxide.pdf>.

[254] IWG-SCGHG, 'Memorandum' (22 December 2023) <www.whitehouse.gov/wp-content/uploads/2023/12/IWG-Memo-12.22.23.pdf>.

'a tool ... was available'.[255] While courts emphatically refrained from 'prescribing a specific metric for the agency to use',[256] they imposed a stringent requirement on those agencies to explain their decision if they had decided not to apply an economic analysis.[257]

Courts have seldom been satisfied by the justifications that national agencies had given for their decision not to use an SCC analysis. The District Court in *High Country Conservation Advocates* expressed doubts about the possibility of 'quantifying the effect of greenhouse gases in dollar terms' and suggested that 'the agencies might have justifiable reasons for not using' an SCC analysis.[258] Yet, it also noted that the agencies 'did not provide those reasons' and, thus, that their failure to carry out this analysis was 'arbitrary and capricious'.[259] By contrast, in *EarthReports Inc v Federal Energy Regulatory Commission*, the Court of Appeal for the District of Columbia Circuit accepted the defendant's explanation that an SCC analysis would not be appropriate due to a lack of consensus on its value.[260] Yet, the Court declined to apply this precedent in *Vecinos para el Bienestar de la Comunidad Costera v Federal Energy Regulatory Commission*, on the ground that, in the latter case, the petitioners had specifically cited a regulatory provision requiring the agency to use 'theoretical approaches and research methods generally accepted in the scientific community'.[261] And, in *Utah Physicians for a Healthy Environment v US Bureau of Land Management*, the District Court for the District of Utah found that the agency had properly justified its decision not to use an SCC analysis, but had failed to use alternative means to ensure balanced weighing of climate impacts with the project's socio-economic benefits.[262]

The CEQ has also encouraged the use of SCC and SC-GHG in NEPA reviews. The Council's 2014 Draft Guidance observed that 'the Federal social cost of carbon ... offers a harmonized, interagency metric that can provide decisionmakers and the public with some context for meaningful NEPA review'.[263] The 2016 Final Guidance contained a similar language.[264] Admittedly, the Tump administration's

---

[255] *High Country Conservation Advocates v US Forest Service* (D Colorado 2014) 52 F Supp 3d 1174, 1190. See also *California v Bernhardt* (ND California 2020) 472 F Supp 3d 573, 623 ('It is arbitrary for an agency to quantify an action's benefits while ignoring its costs where tools exist to calculate those costs.').
[256] *350 Montana* (n 83) 1271.
[257] See eg *Sierra Club v Federal Energy Regulatory Commission* (n 151) 1375; *Vecinos para el Bienestar de la Comunidad Costera v Federal Energy Regulatory Commission* (DC Cir 2021) 6 F 4th 1321, 1329–30; *Diné Citizens against Ruining Our Environment* (n 5) 1042 ('if an accurate method exists to determine the effect of the proposed action, BLM must perform that analysis or explain why it has not').
[258] *High Country Conservation Advocates* (n 255) 1193
[259] Ibid.
[260] *EarthReports Inc v Federal Energy Regulatory Commission* (DC Cir 2016) 828 F 3d 949, 956. See also *Delaware Riverkeeper Network v Federal Energy Regulatory Commission* (DC Cir 2022) 45 F 4th 104, 111–12; *Center for Biological Diversity v Federal Energy Regulatory Commission* (DC Cir 2023) 67 F 4th 1176, 1184–85.
[261] 40 CFR (2024) § 1502.21(c)(4). See *Vecinos para el Bienestar de la Comunidad Costera* (n 257) 1329–30.
[262] *Utah Physicians for a Healthy Environment v US Bureau of Land Management* (D Utah 2021) 528 F Supp 3d 1222, 1231–32.
[263] CEQ, 2014 Guidance (n 120) 77827.
[264] CEQ, 2016 Guidance (n 126) 33.

2019 guidance suggested that 'an agency need not weigh the effects of the various alternatives in NEPA in a monetary cost-benefit analysis using any monetized ... SCC ... estimates'.[265] Yet, the 2023 Interim Guidance observed again that '[t]he SC-GHG translates metric tons of emissions into the familiar unit of dollars, allows for comparisons to other monetized values, and estimates the damages associated with GHG emissions over time'.[266] The Interim Guidance further noted that this tool 'can assist agencies and the public in assessing the significance of climate impacts', emphasizing that it only requires 'a simple and straightforward calculation'.[267] Therefore, the CEQ recommended that agencies should conduct a SC-GHG analysis 'in most circumstances' when they have quantified GHG emissions.[268]

Several US states have followed the lead of the federal government in encouraging SCC analyses of climate impacts. In 2019, for instance, the State of New York adopted the Climate Leadership and Community Protection Act, which directed the Department of Environmental Conservation to 'establish a social cost of carbon for use by state agencies'[269] as 'a monetary estimate of the value of not emitting a ton of greenhouse gas emissions'.[270] Similarly, the California Air Resources Board is required to consider SCC analyses when adopting rules and regulations and developing a scoping plan,[271] and other agencies have also conducted similar analyses in CEQA reviews.[272] In Colorado, public utilities are required to consider SC-GHG when proposing specified projects, including 'the acquisition of new electric generating resources'.[273]

The development of the SCC in the United States has also influenced foreign other jurisdictions. Since 2011, Environment and Climate Change Canada has formally recommended the use of SCC as a tool to determine the significance of GHG emissions, building extensively on the IWG-SCC and IWG-SCGHG's

---

[265] CEQ, 'Draft National Environmental Policy Act Guidance on Consideration of Greenhouse Gas Emissions' (26 June 2019) 84 Federal Register 30097, 30098.

[266] CEQ, 2023 Guidance (n 56) 1202.

[267] Ibid 1202–03.

[268] Ibid 1202. See also Secretary of the Interior, Order 3399, 'Department-Wide Approach to the Climate Crisis and Restoring Transparency and Integrity to the Decision-Making Process' (16 April 2021) 2021 WL 1584759, *4.

[269] New York Env't Conserv Law (McKinney 2024) § 75-0113(1).

[270] Ibid, § 75-0113(2). See also New York State Department of Environmental Conservation, 'Establishing a Value of Carbon: Guidelines for Use by State Agencies' (2023) <https://extapps.dec. ny.gov/docs/administration_pdf/vocguide23final.pdf> (reflecting the IWG estimates).

[271] California Health & Safety Code (2024) § 38562.5.

[272] See California Natural Resources Agency, 'Guidelines for the Implementation of the California Environmental Quality Act', California Code Regs (2024) title 14 § 15064.4(c) ('A lead agency may use a model or methodology to estimate greenhouse gas emissions resulting from a project. The lead agency has discretion to select the model or methodology it considers most appropriate to enable decision makers to intelligently take into account the project's incremental contribution to climate change'). See also Policy Matters: California Senate Office of Research, 'Social Cost of Carbon: Federal and California Activity' (February 2018) <https://sor.senate.ca.gov/sites/sor.senate.ca.gov/files/Policy%20 Matters%20SCC%20Final.pdf>, 6.

[273] Colorado Rev Stat (2024) § 40-3.2-106(1). See also Washington Rev Code (2023) § 19.280.030(3) (a) (requiring public utilities to consider the SC-GHG when developing mandatory 'resource plans'); Minnesota Stat (2024) § 216B.2422(3)(b) (establishing a similar regime).

successive technical support documents.[274] Several Australian courts have also showed some interest in SCC as evidence of the possibility of ascertaining the significance of an activity's climate impact. For instance, the Land and Environment Court of New South Wales observed that 'it is methodologically possible to apply data from single ... projects in a climate model to quantify to some extent at least the social cost of carbon'.[275] Further, the Australian Capital Territory decided in 2019 that it would consider SCC 'in all ACT Government policies, budget decisions, capital works projects and procurements'.[276] Plaintiffs have also invoked the SC-GHG as a method for assessing significance in South Africa, although courts have so far refrained from imposing this methodology on national agencies.[277] And several multilateral development banks use their own economic valuation tools, such as the shadow price of carbon, to integrate climate impacts in comprehensive cost-benefit analyses of proposed actions.[278]

2. Methodologies

Three alternative methods can be used to assess the economic value of GHG emissions. The first and most commonly used method relies on a valuation of damages caused by climate change. A second method considers the willingness of a society to pay for emission abatements. A third method estimates the price that should be imposed on emissions in order to achieve a given global mitigation objective. Similar to the national and global benchmarks described in the previous section, a

---

[274] See Environment and Climate Change Canada, 'Technical Update to Environment and Climate Change Canada's Social Cost of Greenhouse Gas Estimates' (March 2016) <https://publications. gc.ca/collections/collection_2016/eccc/En14-202-2016-eng.pdf>; Environment and Climate Change Canada, 'Social Cost of Greenhouse Gas Estimates: Interim Updated Guidance for the Government of Canada' (April 2023) <www.canada.ca/en/environment-climate-change/services/climate-change/scie nce-research-data/social-cost-ghg.html>.

[275] *Hunter Environment Lobby Inc v Minister for Planning* [2011] NSWLEC 221 [96]. See also *Hunter Environment Lobby Inc v Minister for Planning (No 3)* [2012] NSWLEC 102 [7], [23]; *Hancock Coal* (n 163) [239]–[249]; *Environment Victoria Inc v AGL Loy Yang Pty Ltd* (2022) 71 VR 1 [103], [105].

[276] ACT Climate Change Strategy 2019–25 (2019) <www.environment.act.gov.au/__data/assets/ pdf_file/0003/1414641/ACT-Climate-Change-Strategy-2019-2025.pdf> 70 para 5.5 (Goal 5B). See also Nicki Hutley, *A Social Cost of Carbon for the ACT* (ACT Government, 2021) 34 (recommending the adoption of the SCC estimates adopted by the IWG).

[277] *South Durban Community Environmental Alliance v Minister of Forestry, Fisheries and the Environment* [2022] ZAGPPHC 741 [48]–[50].

[278] European Bank for Reconstruction and Development, 'Methodology for the Economic Assessment of EBRD Projects with High Greenhouse Gas Emissions' (22 January 2019) <www.ebrd. com/news/publications/institutional-documents/methodology-for-the-economic-assessment-of- ebrd-projects-with-high-greenhouse-gasemissions.html> 5; World Bank, 'Shadow Price of Carbon in Economic Analysis: Guidance Note' (12 November 2017) <https://pubdocs.worldbank.org/en/ 911381516303509498/2017-Shadow-Price-of-Carbon-Guidance-Note-FINAL-CLEARED.pdf>, 3. See also European Investment Bank, 'EIB Group Climate Bank Roadmap 2021–2025' (November 2020) <www.eib.org/attachments/thematic/eib_group_climate_bank_roadmap_en.pdf>, 47–48, 121; Asian Development Bank, 'Guidelines for the Economic Analysis of Projects' (March 2017) <www.adb. org/documents/guidelines-economic-analysis-projects> para 162 (reflecting values from an unspecified section of the Fifth Assessment Report of the IPCC, based on an unspecified methodology).

distinction can be drawn between empirical and normative methods. The first two methods are primarily empirical, as they rely on an observation of, respectively, the value society attaches to climate harm and the willingness of society to pay for emission abatement. By contrast, the third method is mainly normative, as it relies on the agreed-upon objectives rather than on actual climate action.

The first, damage-based method was mainly developed by the IWG-SCC and IWG-SCGHG. The latter defines SC-GHG as 'the monetary value of the net harm to society associated with adding a small amount of that GHG to the atmosphere in a given year'.[279] In this approach, estimating SC-GHG requires, '[i]n principle', an economic valuation of 'all climate change impacts, including (but not limited to) changes in net agricultural productivity, human health effects, property damage …, disruption of energy systems, risk of conflict, environmental migration, and the value of ecosystem services'.[280] The value of SC-GHG depends on the year in which GHG emissions take place 'because future emissions produce larger incremental damages as physical and economic systems become more stressed in response to greater climatic change, and because GDP is growing over time and many damage categories are modeled as proportional to GDP'.[281] The social cost of each GHG (eg carbon dioxide, methane, and nitrous oxide) further reflects its global warming potential and atmospheric lifetime.[282]

One preliminary issue, when estimating SC-GHG, relates to the nature of the damage to be considered. As climate impacts are global in nature, the SC-GHG is usually defined as 'the monetary value of world-wide damage done by' these emissions.[283] Yet EA procedures do not always require agencies to give the same weight to an activity's territorial and extraterritorial impacts. For one, NEPA clearly directs agencies to 'recognize the worldwide and long-range character of environmental problems',[284] and the Court of Appeal for the Seventh Circuit noted that 'global effects are an appropriate consideration' in the assessment of an agency's proposed action with implications for GHG emissions.[285] While the Trump administration briefly directed agencies to 'focus on the impacts that accrue to citizens and residents of the United States',[286] Executive Order 13990 of 20 January 2021 reaffirmed that agencies should 'capture the full costs of [GHG] emissions … including by taking global damages into account'.[287] Accordingly, the IWG-SCC and

---

[279] IWG-SCGHG, 'Technical Support Document' (2021) (n 253) 2.

[280] Ibid.

[281] Ibid, 24. This assumption of increasing marginal impact with regard to large increments (ie when adding hundreds of billion tonnes of carbon dioxide) is not incompatible with the assumption of a 'near linear' relation between GHG emissions and climate impacts with regard to smaller increments. See references above, n 63.

[282] IWG-SCGHG, 'Technical Support Document' (2021) (n 253) 2.

[283] David Pearce, 'The Social Cost of Carbon and its Policy Implications' (2003) 19 Oxford Review of Economic Policy 362, 363.

[284] 42 USC (2024) § 4332(I).

[285] Zero Zone Inc v Department of Energy (7th Cir 2016) 832 F 3d 654, 678–679.

[286] CEQ, 2019 Guidance (n 265) 30099. See also Executive Order 13783 (28 March 2017) 82 Federal Register 16093, s 5(c).

[287] Executive Order 13990 (n 252) s 5. See also CEQ, 2023 Guidance (n 56) 1203.

IWG-SCGHG's estimates reflect the global value of climate impacts.[288] Similarly, in the State of New York, a specific statutory provision clarifies that the SCC may reflect 'the global ... impacts' of marginal GHG emissions.[289] By contrast, valuation instruments have not generally been implemented under EA frameworks that are limited to consideration of local impacts.[290] As noted above, this focus on local impacts should logically impede any finding of the significance of an activity's climate impact in all but the most egregious cases.[291]

Other difficulties concern the actual methodology used for the valuation of climate change. There is no question that, as the District Court for the District of Colorado put it, 'quantifying the effect of [GHGs] in dollar terms is difficult at best'.[292] For one thing, this involves a prediction of future harm, the occurrence of which depends in part on conjectures about the ability of society to adapt.[293] Further, the exercise involves an economic valuation of various harms that do not have a market value.[294] Yet perhaps the greatest impediment to an objective valuation of the SC-GHG is that many of the adverse consequences of today's GHG emissions occur in a far-off future.

The present value of these future impacts is typically calculated using a discount rate reflecting our preference for immediate benefits over future ones (and for future costs over immediate ones).[295] Economists are divided on the value of the discount rate[296] and the method (eg empirical or normative) for identifying it.[297] As many effects of climate change will occur in a distant future, any small variation in the discount rate has major implications for the valuation of climate change: using a slightly higher discount rate would dramatically reduce the present valuation of future effects of climate change.[298] The IWG-SCGHG's latest interim estimate presents SC-GHG for discount rates ranging from 2.5 to 5 percent, while also encouraging the use of 'discount rates appropriate for intergenerational analysis in the context of climate change that are lower than 3 percent'.[299] A subsequent report by the US Environmental Protection Agency considers discount rates from 1.5 to

---

[288] IWG-SCGHG, 'Technical Support Document' (2021) (n 253) 14–16.
[289] New York Env't Conserv Law (McKinney 2024) § 75-0113(2).
[290] See eg *Montana Environmental Information Center v Montana Department of Environmental Quality* (Montana Dist Ct 6 April 2023) DV21-01307 <https://climatecasechart.com/wp-content/uploads/case-documents/2023/20230406_docket-DV21-01307_order.pdf> *29; *Wildlife Preservation Society of Queensland Proserpine v Minister for the Environment and* Heritage (2006) 232 ALR 510 [72].
[291] See Chapter II, subsection C.3(b).
[292] *High Country Conservation Advocates* (n 255) 1193.
[293] Steven Rose and others, 'Cross-Working Group Box ECONOMIC: Estimating Global Economic Impacts from Climate Change', in Pörtner and others (eds) (n 201) 2495, 2495–96.
[294] IWG-SCGHG, 'Technical Support Document' (2021) (n 253) 26.
[295] Ibid 16–17.
[296] William Nordhaus, 'Climate Change: The Ultimate Challenge for Economics' (2019) 109 American Economic Review 1991, 2004.
[297] See eg Stefano Giglio, Matteo Maggiori, and Johannes Stroebel, 'Very Long-Run Discount Rates' (2015) 130 Quarterly Journal of Economics 1; David Weisbach and Cass R Sunstein, 'Climate Change and Discounting the Future: A Guide for the Perplexed' (2008) 27 Yale Law & Policy Review 433.
[298] Sydney Hofferth, 'Significant Impacts under NEPA: The Social Cost of Greenhouse Gases as a Tool to Mitigate Climate Change' (2022) 11 Michigan Journal of Environmental & Administrative Law 333, 360.
[299] IWG-SCGHG, 'Technical Support Document' (2021) (n 253) 4.

2.5 percent.[300] Similarly, Environment and Climate Change Canada recommends that Canadian agencies rely on a 2 percent discount rate.[301] By contrast, multilateral development banks occasionally use a discount rate as high as 9 percent.[302]

A second method to value climate impacts, alternative to the estimation of SC-GHG, relies on the observation of the price that society appears willing to pay to avoid an additional unit of GHG emission. The United Kingdom relies on this approach, in 2007, when determining a 'Shadow Price of Carbon' as a policy appraisal tool relying on 'a measure of our willingness to pay for carbon abatement'.[303] In an ideal world, this willingness of society to pay for the abatement of marginal GHG emission should be equal to the SC-GHG, since a rational society should avoid GHG emissions so far as the benefit of doing so exceeds the cost.[304] In reality, however, society may be implementing more or less climate ambitious than in its rational interest. For a number of reasons (eg misinformation), a state's climate ambition may not align fully with its global and long-term interests.[305]

With the exception of some multilateral development banks,[306] authorities generally have shown little interest in this approach due to the practical difficulty of estimating society's willingness to pay for avoiding GHG emissions in various contexts. A particular difficulty is that society has likely a different willingness to pay to reduce GHG emissions in different contexts, based for instance on the social acceptability of different sources of GHG emissions. The UK stopped relying on a shadow price of carbon just two years after the introduction of this policy.[307] And when the lawmaker of New York State requested the Department of Environmental Conservation to decide whether to value climate impacts 'based on marginal greenhouse gas abatement costs or on the global … impact' of marginal GHG emissions,[308] the Department chose the latter approach, noting that a damage-based approach was more consistent with the aim of providing 'a set of values that can be used by any State entity in a wide variety of contexts'.[309]

[300] Environmental Protection Agency, 'EPA Report on the Social Cost of Greenhouse Gases: Estimates Incorporating Recent Scientific Advances' (31 October 2023) <www.regulations.gov/document/EPA-HQ-OAR-2023-0434-0044> 2.

[301] Environment and Climate Change Canada, 'Social Cost of Greenhouse Gas Estimates' (2023) (n 274).

[302] Eg Asian Development Bank, 'Guidelines' (n 278) para 194.

[303] Richard Price, Simeon Thornton, and Stephen Nelson, 'The Social Cost of Carbon and the Shadow Price of Carbon: What They Are, and How to Use Them in Economic Appraisal in the UK' (Department for Environment, Food & Rural Affairs 2007) 14.

[304] Ibid 2–3; Nordhaus (n 296) 2004. On the definition of SC-GHG in marginal rather than average terms (ie as the cost resulting from an additional tonne of GHG emissions rather than the average cost of all emissions), see eg IWG-SCGHG, 'Technical Support Document' (2021) (n 253) 9.

[305] Even in jurisdictions where EA is clearly to consider global impacts, one can question whether the state is expected to attribute as much weight to impacts occurring overseas as it does to domestic impacts. By analogy, states certainly spend more to promote human rights within their territory (eg by running schools and public hospitals) than abroad.

[306] See references above at n 278.

[307] Department of Energy and Climate Change, 'Carbon Valuation in UK Policy Appraisal: A Revised Approach' (July 2009) <https://perma.cc/D2EM-QB27> 17.

[308] New York Env't Conserv Law (McKinney 2024) § 75-0113(2).

[309] New York State Department of Environmental Conservation, 'Establishing a Value of Carbon' (n 270) 11.

A third method to the valuation of climate impacts seeks to define a value that would ensure the achievement of a global mitigation goal, for instance the temperature goals of the Paris Agreement. Thus, Nicholas Stern and Joseph Stiglitz promoted an alternative definition of SCC as a value that would impose 'a path where temperature is constrained below 2 degrees C'[310] with a 66 percent probability.[311] Alternatively, Noah Kaufman and colleagues pleaded for a redefinition of SCC consistent with a mid-century net-zero carbon dioxide emission target.[312] They submitted that the goal-based valuation approach to the valuation of climate impact would avoid some of the uncertainty associated with a damage-based assessment and ensure '[c]onsistency with policy objectives'.[313] However, significant sources of uncertainties are associated with the objective that is to be pursued[314] and with the cost of achieving this objective, which depends, among other things, on economic, demographic, and technological factors.

3.   Usefulness

Economic valuation can be a convenient tool for a meaningful appraisal of the significance of a project's climate impact. Once an activity's GHG emissions have been estimated, the application of economic valuation tools 'should not require additional time or resources'.[315] While the general public and even the decision-maker might struggle to understand how much weight to give to a number of tonnes of carbon dioxide emissions, a dollar value might appear more immediately meaningful. By design, these tools enable consideration of the climate impacts of 'adding a small amount of … GHG to the atmosphere in a given year'.[316] Thus, economic valuation tools facilitate the comparison of an activity's concrete economic benefits with its diffuse climate impact. As Pain J noted in *Hunter Environment Lobby Inc v Minister for Planning*, these methodologies can bring some nuance on 'the submission that a particular project is but one of many contributors to a local, regional and global problem'.[317]

The main limitation of economic valuation tools, however, is the difficulty of determining the value of climate impacts at the first place, as discussed in the previous subsection. Estimates of the value of avoiding a tonne of carbon dioxide

---

[310] Nicholas Stern and Joseph E Stiglitz, 'The Social Cost of Carbon, Risk, Distribution, Market Failures: An Alternative Approach' (National Bureau of Economic Research working paper 28472, February 2021) 59.

[311] Joseph Stiglitz and others, 'Report of the High-Level Commission on Carbon Prices' (Carbon Pricing Leadership Coalition 2017) 33.

[312] Noah Kaufman and others, 'A Near-Term to Net Zero Alternative to the Social Cost of Carbon for Setting Carbon Prices' (2020) 10 Nature Climate Change 1010.

[313] Ibid 1010.

[314] See this chapter, subsection C.2(a).

[315] CEQ, 2023 Guidance (n 56) 1203.

[316] IWG-SCGHG, 'Technical Support Document' (2021) (n 253) 3.

[317] *Hunter Environment Lobby Inc* [2011] (n 275) [96].

range from slightly below zero to several thousands of USD.[318] Yet, some of the extremes in this range are studies that have not been peer-reviewed,[319] and the literature seems to be converging towards narrower ranges.[320] Thus, the most influential institutional reports, using widely different methods and assumptions, suggest that the 2020 value of a tonne of carbon dioxide emissions emitted in 2030 should be between USD 19 and 450 (see Table 3.1).

Table 3.1 Economic valuations of avoiding one tonne of carbon dioxide emissions in 2030

Methodology	Author (date)	Valuation (USD$_{2020}$/ t CO$_2$ 2030)[321]
Damage-based	IWG-SCGHG (2021)[322]	19–89
	EPA (2023)[323]	140–380
	ECCC (2023)[324]	202
	New York State Department of Environmental Conservation (2023)[325]	64–450
	Hutley report (ACT) (2021)[326]	58–85
Willingness to pay	UK (2007)[327]	98
Goal-based	World Bank (2017)[328]	65–129
	European Investment Bank (2020)[329]	54
	Kaufman and others (2020)[330]	65–181

[318] Pei Wang and others, 'Estimates of the Social Cost of Carbon: A Review Based on Meta-Analysis' (2019) 209 Journal of Cleaner Production 1494, 1494; Tol (n 63) 14–15.

[319] Wang and others (n 318) 1505.

[320] Tol (n 63) 17.

[321] Converted using currency conversion rates from 'Yearly average rates' *OFX.com* (nd) <www.ofx.com/en-au/forex-news/historical-exchange-rates/yearly-average-rates/>, and USD GDP deflator calculation from 'Convert Current to Real US Dollars', *Areppim.com* (nd) <https://stats.areppim.com/calc/calc_usdlrxdeflator.php>.

[322] IWG-SCGHG, 'Technical Support Document' (2021) (n 253) 5. The range reflects discount rates from 2.5 to 5%.

[323] Environmental Protection Agency, 'EPA Report on the Social Cost of Greenhouse Gases' (n 300) 4. The range reflects discount rates from 2.5–1.5%.

[324] Environment and Climate Change Canada, 'Social Cost of Greenhouse Gas Estimates' (2023) (n 274) Table 1 (294 CAD$_{2021}$ for a 2% near-term Ramsey discount rate).

[325] New York State Department of Environmental Conservation, 'Establishing a Value of Carbon' (n 270) 16, 21 (64–450 USD$_{2020}$ for a range of discount rates from 3 to 1%).

[326] Hutley (n 276) 6–7 (84–123 AUD$_{2020}$ for a range of discount rates from 3 to 2.5%).

[327] Price, Thornton, and Nelson (n 303) 15, 21 (40.1 GBP$_{2007}$).

[328] European Bank for Reconstruction and Development (n 278) 5; World Bank, 'Shadow Price' (n 278) 3 (50–100 USD$_{2005}$ for a 66% chance of achieving a 2°C goal). See also Stiglitz and others (n 311) 33, 50.

[329] European Investment Bank, 'Roadmap 2021–2025' (n 278) 47–48, 121 (250 EUR$_{2016}$ for an unspecified '1.5°C target').

[330] Kaufman and others (n 312) (source data for fig. 1) ('NT2NZ Pathway 2060', 65 or 181 USD$_{2020}$ for a global net-zero carbon dioxide emission goal respectively by 2060 and 2040).

While there remains a wide range of uncertainty, the IWG-SCGHG rightly notes that 'scientific and economic analysis can provide valuable information to the public and decision makers'.[331] In particular, these valuation tools may help to exclude some extreme arguments from the scope of public deliberations on the merits of proposed activities. Accordingly, US courts have rightly noted that the use of approximate values is preferable to not providing any value at all for climate impacts.[332] On the other hand, the IWG-SCGHG recommends that uncertainty 'be acknowledged, communicated as clearly as possible, and taken into account in the analysis whenever possible'.[333] Providing a range of values rather than a single value is essential to informing the public and decision-makers about the wide uncertainties associated with SC-GHG.

The values of climate impacts listed in Table 3.1, in turn, justify the thresholds of magnitude that are commonly used to determine the need for a CA. These thresholds, typically around 50 kt CO_2e GHG emissions per year,[334] correspond to annual climate impacts that would be valued between USD1 to 23 million. Requiring agencies to assess an activity's climate impact when it exceeds such monetary values seems about right. On the other hand, the climate impact of some large industrial activities, whose GHG emissions reach 1 Gt CO_2e during their lifetime,[335] would be valued between USD19 and 450 billion.

Conclusion

This chapter has shed light on three broad approaches to assess the significance of the climate impact of a proposed activity. These methods can be relevant at different stages of the EA process. Magnitude-based approaches are most relevant at the EA's preliminary stages: it is reasonable for an EA procedure to focus limited resources on those proposed activities that would likely be causing a higher level of GHG emissions. Benchmarks can be used at this preliminary stage, but they are also relevant at the substantive stage: the public and the decision-maker may find it useful to understand how a proposed activity would contribute to total GHG emissions at the relevant scale and whether it would interfere with the realization of local goals on climate change mitigation. Economic valuation, likewise, can help

---

[331] IWG-SCGHG, 'Technical Support Document' (2021) (n 253) 26.
[332] See eg *Border Power Plant Working Group v Department of Energy* (SD California 2003) 260 F Supp 2d 997, 1029; *Center for Biological Diversity v National Highway Traffic Safety Administration* (n 53) 1200; *High Country Conservation Advocates* (n 255) 1192; *WildEarth Guardians v Zinke* (D Montana 11 February 2019) CV-17-80-BLG-SPW-TJC, 2019 WL 2404860, at *12. But see *Utah Physicians for a Healthy Environment* (n 262) 1231–32 (noting that '[d]ecisions that implicate an agency's technical or scientific expertise are entitled to "especially strong" deference').
[333] IWG-SCGHG, 'Technical Support Document' (2021) (n 253) 26.
[334] See text above at nn 119–23.
[335] See Chapter II, subsection A.3(b).

the public and decision-makers to understand the weight to give to climate impacts when appraising the costs and benefits of a proposed activity and, ultimately, to assess the merits of this activity.

In conclusion, it needs to be reiterated that none of the approaches mentioned in this chapter can be used, singly or in combination with others, as a mechanical test to determine whether a proposed activity can be approved. Whether an activity is to be approved is inherently a political decision that involves far-reaching value-based judgements.[336] The aim of any approach discussed in this chapter is to provide the best possible information to the public and the decision-makers, not to substitute to them. The methods described in this chapter, especially when they are used in combination with one another, allow project proponents, national agencies, the public, decision-makers, and courts to make well informed decisions on whether to approve projects that have potentially significant climate impacts.

---

[336] See also *Goesa Ltd* (n 230) [122] (noting that the 'acceptability' of a particular proposal is ultimately 'for the judgment of the decision-maker').

IV

Assessing Indirect Climate Impacts

Introduction

As Chapter II has shown, most environmental assessment (EA) frameworks across the world have been applied as a tool to assess climate impacts and mitigate climate change. However, a more granular analysis shows that not all climate assessment (CA) frameworks consider the same climate impacts. One particularly difficult question is whether CA should consider the greenhouse gas (GHG) emissions that are only the remote or indirect consequences of the proposed activity—for instance, in fossil-fuel production projects, the downstream emissions from the combustion of fossil fuels. This question can be raised at the stage of deciding what potential impacts are to be assessed in an EA procedure (ie at the scoping stage), but also, earlier on, when deciding whether an assessment is required at all (ie at the screening stage).[1]

The GHG Protocol is at the origin of a widely used distinction between three 'scopes' of GHG emissions that can be associated with an activity. Scope 1 emissions are direct emissions 'from sources that are owned or controlled by the company,' such as emissions 'from combustion in owned or controlled boilers, furnaces [and] vehicles'.[2] Scope 2 emissions are indirect emissions 'from the generation of purchased electricity consumed by the company'.[3] And scope 3 includes 'all other indirect emissions' that 'are a consequence of the activities ... but occur from sources not owned or controlled by the company', such as emissions from the 'extraction and production of purchased materials', from the 'transportation of purchased fuels', and from the 'use of sold products and services'.[4] A project's direct climate impact (ie scope 1 emissions) is often the indirect climate impact (ie its scope 2 and scope 3 emissions) of another project. For instance, the GHG emitted from a coal plant are the scope 1 emissions of this coal plant, the scope 2 emissions

---

[1] On the distinction between the two stages, see John Glasson and Riki Therivel (eds), *Introduction to Environmental Impact Assessment* (5th edn, Routledge 2019) 83.

[2] See eg The Greenhouse Gas Protocol, 'A Corporate Accounting and Reporting Standard' (revised edn, World Business Council for Sustainable Development and World Resources Institute 2015) https://ghgprotocol.org/corporate-standard 25.

[3] Ibid (reference omitted). See also The Greenhouse Gas Protocol, 'GHG Protocol Scope 2 Guidance: An Amendment to the GHG Protocol Corporate Standard' (World Resources Institute 2019) <https://ghgprotocol.org/scope-2-guidance> 35.

[4] GHG Protocol, 'A Corporate Accounting and Reporting Standard' (n 2) 25. See also GHG Protocol, 'GHG Protocol Scope 2 Guidance' (n 3) 5–6.

Environmental Assessment as a Tool for Climate Change Mitigation. Benoit Mayer, Oxford University Press.
© Benoit Mayer 2024. DOI: 10.1093/oso/9780198939184.003.0004

of the companies purchasing electricity from this coal plant, and the scope 3 emissions of (among other things) any mine supplying coal to the plant or railways transporting the coal from the mine to the plant.

Even in jurisdictions where a CA requirement is firmly established, questions have repeatedly been asked about the need to extend this assessment to indirect climate impacts. The argument in favour of considering indirect climate impacts is rather straightforward: to allow an informed decision, CA should, in principle, consider all important consequences of a proposed activity, and indirect climate impacts can be an important consequence of a proposed activity. On the other hand, one reason frequently invoked against assessing indirect climate impacts is that scope 2 and scope 3 emissions are difficult to predict, as they are contingent on decisions extraneous to the proposed activity. Another oft-heard argument is that scope 2 and scope 3 emissions may take place in other jurisdictions, whereas it is often assumed that a state's obligation to mitigate climate change is territorial in nature.

To illustrate these arguments, consider a hypothetical natural gas extraction project. This project would likely have a direct climate impact due to the leakage, venting, and flaring of natural gas during production processes,[5] but its main consequence for climate change would result from the combustion of natural gas outside the project boundaries, possibly in other countries. This indirect climate impact would depend on future actions extraneous to the proposed activity, which would affect, for instance, the amount of gas leaked during the transportation of the natural gas[6] and, conceivably, the amount of carbon dioxide captured and sequestered after combustion. From a more holistic perspective, the overall impact of the project on global GHG emissions depends on the extent to which it would replace production by other existing or future projects. It is even conceivable that the project could achieve a net reduction in global GHG emissions if its gas were to substitute to less efficient fossil fuels (eg coal) or to natural gas produced in more polluting ways (eg with more fugitive emissions). On the other hand, one might be concerned that, in the long term, increasing natural gas production capacity could make it more difficult for society to agree on a more complete phase-out of fossil-fuel consumption.[7]

Scientific research can foster a better understanding of these issues. For instance, experts in GHG emission inventories can help assess the effect of project

---

[5] See Leon Clarke and others, 'Energy Systems', in Priyadarshi R Shukla and others (eds), *Climate Change 2022: Mitigation of Climate Change. Working Group III Contribution to the Sixth Assessment Report of the Intergovernmental Panel on Climate Change* (CUP 2022) 613, 620 (noting that fugitive emissions from oil and gas production amounted to 2.6 Gt CO_2e in 2019).

[6] This is because the methane contained in the natural gas has a far higher global warming potential than the carbon dioxide produced during the combustion of the gas. See Piers Forster and others, 'The Earth's Energy Budget, Climate Feedbacks and Climate Sensitivity', in Valerie Masson-Delmotte and others (eds), *Climate Change 2021: The Physical Science Basis* (CUP 2021) 923, 1017.

[7] See Clarke and others (n 5) 696–98.

substitution by comparing fugitive emissions from the proposed project with those of existing projects.[8] Many other economic, social, and political effects are difficult to ascertain because they depend on social and political variables, from market substitution effects to the moral hazard of extending fossil-fuel production capacity. In this regard, economic and other social scientific models can make predictions that are informative and useful even though they come with a significant level of uncertainty.

In this context, this chapter argues for the assessment of any potentially significant indirect climate effect from a proposed activity. This broad scope of assessment is necessary to ensure that decision-makers are properly informed before they make a decision liable to cause environmental harm. The uncertainty associated with indirect climate impacts does not justify their complete exclusion from the scope of CA, although such uncertainty should be clearly acknowledged as part of the assessment.

Admittedly, there needs to be a limit to the scope of indirect effects that are considered in any EA procedure. A frequent concern with the assessment of indirect climate impacts is that, as the Irish Supreme Court put it in *An Taisce v An Bord Pleanála*, there could be 'hardly any limits but the sky' for causal claims about the remote, unintended, and often unpredictable consequences of human actions.[9] Any proposed activity could have ripple effects extending *ad infinitum*.[10] This chapter argues that a CA should consider every impact to the extent that this impact can reasonably be anticipated to be potentially significant. This definition excludes the assessment of indirect climate impacts that are purely speculative or extremely unlikely as well as those that would not reach a threshold of significance. As indirect climate impacts are not fully under the control of the proposed activity, higher thresholds of significance could be used for indirect climate impacts than for direct climate impacts. Thus, in a jurisdiction requiring a CA for projects causing more than 50 kt CO_2 per year in scope 1 emissions, a coal mine project with 75 kt CO_2 per year in downstream scope 3 emissions could reasonably be exempted on the ground that part of these emissions would likely occur without the project.[11]

Debates on the assessment of indirect climate impacts have often unfolded in relation to downstream scope 3 emissions from fossil-fuel projects, including coal

---

[8] See Amit Garg and others, 'Introduction', in Simon Eggleston and others (eds), *2006 IPCC Guidelines for National Greenhouse Gas Inventories* (IGES 2006) vol 2, ch 1, 20 (noting that '[t]he carbon content may vary considerably both among and within primary fuel types on a per mass or per volume basis', due to different chemical composition).

[9] *An Taisce—National Trust for Ireland v An Bord Pleanála* [2022] IESC 8, [2022] 2 IR 173 [105].

[10] See *Sylvester v US Army Corps of Engineers* (9th Cir 1989) 884 F 2d 394, 400 ('Environmental impacts are in some respects like ripples following the casting of a stone in a pool. The simile is beguiling but useless as a standard. So employed it suggests that the entire pool must be considered each time a substance heavier than a hair lands upon its surface. This is not a practical guide.').

[11] At the moment, magnitude-based assessments of significance do not generally give different weight to direct and indirect climate impacts at the scoping stage. See Chapter III, subsection B.2.

mines,[12] oil and gas wells,[13] as well as pipelines[14] and railways[15] aimed at transporting these fuels. Occasionally, debates have extended to various other indirect climate impacts, including the scope 2 emissions from the electricity consumption of warehouse[16] and a desalinization plant,[17] the upstream scope 3 emissions from power plants in relation to a transmission line project[18] and from milk production in relation to a cheese-making factory,[19] or other scope 3 emissions related from transportation infrastructure projects such as airports[20] and roads.[21] On the other

[12] See eg *Gray v Minister for Planning* [2006] 152 LGERA 258 (NSWLEC); *Xstrata Coal Queensland Pty Ltd v Friends of the Earth-Brisbane* [2012] QLC 13; *Hunter Environment Lobby Inc v Minister for Planning (No 2)* [2012] NSWLEC 40; *West Coast ENT Inc v Buller Coal Ltd* [2013] NZSC 87, [2014] 1 NZLR 32; *High Country Conservation Advocates v US Forest Service* (D Colorado 2014) 52 F Supp 3d 1174; *Hancock Coal Pty Ltd v Kelly (No 4)* [2014] QLC 12; *Dine Citizens against Ruining Our Environment v Office of Surface Mining, Reclamation and Enforcement* (D Colorado 2015) 82 F Supp 3d 1201; *Gloucester Resources Ltd v Minister for Planning* [2019] NSWLEC 7; *Waratah Coal Pty Ltd v Youth Verdict Ltd (No 6)* [2022] QLC 21; *R (Friends of the Earth) v South Lakeland Action on Climate Change* [2024] EWHC 2349 (admin).

[13] See eg *Greenpeace Nordic Association v Ministry of Petroleum and Energy* (2020) Case No 20-051052SIV-HRET (Supreme Court); *Center for Biological Diversity v Bernhardt* (9th Cir 2020) 982 F 3d 723; *Greenpeace Ltd v Advocate General* [2021] CSIH 53, 2021 Scot (D) 9/10; *Mullaley Gas and Pipeline Accord Inc v Santos NSW (Eastern) Pty Ltd* [2021] NSWLEC 110; *Sierra Club Canada Foundation v Minister of Environment and Climate Change* [2023] FCJ 840, 2003 FC 849 (Federal Court); *Friends of the Earth v Haaland* (D DC 2022) 583 F Supp 3d 113; *R (Finch) v Surrey County Council* [2024] UKSC 20, [2024] All ER (D) 71 (Jun); *Greenpeace Nordic Association v Ministry of Energy* (2024) Case No 23-099330TVI-TOSL/05 (District Court of Oslo). See also Peter Splinter, 'Slow Down Mr. Sunak: The UK Has an Obligation to Consult its Neighbours before You Can Authorise Millions of Tonnes of CO_2 Emissions' *OpinioJuris* (29 March 2024) <https://opiniojuris.org/2024/03/29/slow-down-mr-sunak-the-uk-has-an-obligation-to-consult-its-neighbours-before-you-can-author ise-millions-of-tonnes-of-CO2-emissions/>.

[14] See eg *Sierra Club v Federal Energy Regulatory Commission* (DC Cir 2017) 867 F 3d 1357; *Food and Water Watch v Federal Energy Regulatory Commission* (DC Cir 2022) 28 F 4th 277.

[15] See eg *Mid States Coalition for Progress v Surface Transportation Board* (8th Cir 2003) 345 F 3d 520.

[16] Tribunal administratif de Cergy-Pointoise (Administrative Tribunal of Cergy-Pontoise), 17 February 2023, 2005602, paras 5, 7; Tribunal administratif de Strasbourg (Administrative Tribunal of Strasbourg), 25 July 2023, 2102476, para 7.

[17] See eg *North Coast Rivers Alliance v Marin Municipal Water District Board of Directors* (2013) 216 Cal App 4th 614, 651.

[18] See eg *Border Power Plant Working Group v Department of Energy* (SD California 2003) 260 F Supp 2d 997.

[19] *An Taisce* (n 9).

[20] *Barnes v US Department of Transportation* (9th Cir 2011) 655 F 3d 1124; Verfassungsgerichtshof (Constitutional Court) VfGH E 875/2017 (1 June 2017); *R (Friends of the Earth Ltd) v Heathrow Airport Ltd* [2020] UKSC 52, [2021] 2 All ER 967; *R (Goesa Ltd) v Eastleigh Borough Council* [2022] EWHC 1221 (Admin), [2022] PTSR 1473; Tribunal administratif de Lille (Administrative Tribunal of Lille), 5 December 2022, 2208424; *Bristol Airport Action Network Co-ordinating Committee v Secretary of State for Levelling Up, Housing and Communities* [2023] EWHC 171 (Admin), [2023] PTSR 853; Cour administrative d'appel de Nantes (Administrative Court of Appeal of Nantes), 14 December 2023, 22MA02967, para 8.

[21] *North Carolina Alliance for Transportation Reform Inc v US Department of Transportation* (MD North Carolina 2010) 713 F Supp 2d 491; *Coalition for Advancement of Regional Transportation v Federal Highway Administration* (6th Cir 2014) 576 F Appx 477; Conseil d'État (State Council) (6ème–5ème chambres réunies), 19 November 2020, 417362, ECLI:FR:CECHR:2020:417362.20201119; Cour administrative d'appel de Paris (Administrative Court of Appeal of Paris), 23 June 2021, 20PA02347; Conseil d'État (State Council) (10ème–9ème chambres réunies), 30 December 2021, 438686, ECLI:FR:CECHR:2021:438686.20211230; *R (Boswell) v Secretary of State for Transport* [2024] EWCA Civ 145, [2024] All ER (D) 161 (Feb).

hand, one of this chapter's main findings is that the assessment of indirect climate impacts has almost exclusively been decided at the scoping stage of the EA, and that potential indirect climate impacts have not generally been taken into account, at the screening stage, in preliminary decisions on whether an EA is required. As such, projects whose significant environmental impacts consist exclusively in indirect climate impacts, such as data centres aimed at 'mining' cryptocurrency, are not ordinarily subjected to any EA requirement.

The next section provides an overview of the debates that have taken place in multiple jurisdictions on the assessment of indirect climate impacts. The following sections address the main objections that have been made to the assessment of such impacts. While market substitution is an important consideration in the assessment of indirect climate impacts, Section B shows that it is unlikely to be 'perfect': opening a new coal mine, for instance, can be expected to lead to price adjustments resulting in an increase in global coal consumption. Section C refutes objections to the assessment of indirect climate impacts involving extraterritorial sources of GHG emissions, showing that states have the right—perhaps even the obligation—to take certain measures aimed at limiting and reducing GHG emissions beyond their territory. Lastly, Section D outlines practical ways to address the uncertainties associated with indirect climate impacts.

A. The Debate on Indirect Climate Impacts

This section takes stock of the various views on the opportunity of assessing indirect climate impacts. These views were reflected not only in deliberations on statutory, regulatory, and policy documents, but also in judicial decisions, in particular with regard to fossil-fuel production and transportation projects. The first two subsections observe that, while some lawmakers, agencies, and courts have included indirect climate impacts within the scope of CAs, others have decided not to. The third subsection identifies the most influential arguments in this debate, including the main objections to the assessment of indirect climate impacts, some of which are further discussed in the following sections.

1. Assessing indirect climate impacts

Indirect climate impacts have been included in EA frameworks in some jurisdictions. This is sometimes done by clear and specific provisions in EA statutory or regulatory instruments, or by guidance documents on the implementation of these instruments.[22] In most cases, however, decisions have been made by national

---

[22] See eg Directive 2011/92/EU of the European Parliament and of the Council of 13 December 2011 on the assessment of the effects of certain public and private projects on the environment, [2012]

agencies and courts. The following documents three examples of EA frameworks that explicitly require, or have been interpreted as requiring, the assessment of indirect climate impacts, respectively under US federal law, New South Wales law, and EU law.

The National Environmental Policy Act (NEPA) procedure, under US federal law, has long been interpreted as requiring an assessment of indirect environmental impacts, including climate impacts.[23] NEPA Regulations require these reviews to account for all 'reasonably foreseeable' effects, whether 'direct' or 'indirect'.[24] Among such indirect effects, courts have held that agencies should consider the 'growth-inducing effect' of infrastructure projects such as airport extensions.[25] With regard specifically to climate impacts, courts have required the assessment of scope 2 emissions from power plants in relation to a project to build transmission lines that would import electricity from Mexico,[26] as well as scope 3 emissions from downstream consumption of fossil fuels in relation to fossil-fuel production[27] and transportation projects.[28] Taking stock of this case law, the Council for Environmental Quality has guided agencies to 'quantify all reasonably foreseeable emissions associated with a proposed action', including 'indirect GHG emissions'.[29] The Council for Environmental Quality has further highlighted that,

OJ L 26/1, consolidated as of 15 May 2014, art 3(1) ('direct and indirect effects'); Federal-Provincial-Territorial Committee on Climate Change and Environmental Assessment, 'Incorporating Climate Change Considerations in Environmental Assessment: General Guidance for Practitioners' (2003) <https://publications.gc.ca/collections/collection_2012/acee-ceaa/En106-50-2003-eng.pdf> 7 ('direct and indirect emissions'); New York Comp. Codes R & Regis (2024) title 6, § 617-7(c)(2) ('direct [and] indirect ... impacts').

23 See Chapter II, subsection B.3(a).
24 40 CFR (2024) § 1508.1(g)(2). See also CEQ, 'National Environmental Policy Act Regulations' (29 November 1978) 43 Federal Register 55978, 56004; CEQ, 'Update to the Regulations Implementing the Procedural Provisions of the National Environmental Policy Act' (16 July 2020) 85 Federal Register 43304, 43343–4 (no longer including an explicit requirement to assess indirect effects); CEQ, 'National Environmental Policy Act Implementing Regulations Revisions' (20 April 2022) 87 Federal Register 23453, 23469 (reinstating a requirement to assess indirect effects).
25 Barnes (n 20) 1136–39.
26 Border Power Plant Working Group (n 18) 1017, 1029.
27 See San Juan Citizens Alliance v US Bureau of Land Management (D New Mexico 2018) 326 F Supp 3d 1227 (sale of oil and gas leases); Montana Environmental Information Center v US Office of Surface Mining (D Montana 2017) 274 F Supp 3d 1074, 1081 (coal mine).
28 See Mid States Coalition for Progress (n 15) 532 (railway project aimed at transporting coal); Sierra Club v Federal Energy Regulatory Commission (n 14) 1372 (natural gas pipeline); Food and Water Watch (n 14) 281 (natural gas pipeline).
29 CEQ, 'National Environmental Policy Act Guidance on Consideration of Greenhouse Gas Emissions and Climate Change' (9 January 2023) 88 Federal Register 1196, 1204. See also CEQ, 'Draft NEPA Guidance on Consideration of the Effects of Climate Change and Greenhouse Gas Emissions' (18 February 2010) <https://perma.cc/4P7U-RFYW>, 2–3; CEQ, 'Revised Draft Guidance for Federal Departments and Agencies on Consideration of Greenhouse Gas Emissions and the Effects of Climate Change in NEPA Reviews' (24 December 2014) 69 Federal Register 77802, 77825–26; CEQ, 'Final Guidance for Federal Departments and Agencies on Consideration of Greenhouse Gas Emissions and the Effects of Climate Change in National Environmental Policy Act Reviews' (1 August 2016) <https://perma.cc/2A9H-GPNH> 4; CEQ, 'Draft National Environmental Policy Act Guidance on Consideration of Greenhouse Gas Emissions' (26 June 2019) 84 Federal Register 30097, 30098.

'where the proposed action involves fossil fuel extraction', the review should assess 'effects associated with the processing, refining, transporting, and end-use of the fossil fuel being extracted, including combustion of the resource to produce energy'.[30]

Similar developments have taken place in some Australian states, in particular in New South Wales.[31] In 2006, the Land and Environment Court decided in *Gray v Minister for Planning* that the EA of a coal mine project had to take into account the downstream GHG emissions from the combustion of the coal.[32] The following year, the government reflected this case law in an environmental planning policy requiring a consent authority to 'consider an assessment of the greenhouse gas emissions (including downstream emissions) of the development'.[33] The New South Wales Land and Environment Court further clarified this requirement in subsequent decisions. In 2018, it observed that, in the government's policy, '"downstream emissions" ... denote the greenhouse gas emissions relating to sold goods and services and thus caused by end users' use of the product (eg coal) produced by a project'.[34] In *Gloucester Resources Ltd v Minister for Planning*, the Court further specified that downstream emissions include 'Scope 3 emissions from the transportation and combustion of coal product from the mine'.[35] Following several administrative and judicial decisions unfavourable to the coal mining industry,[36] the state government introduced a bill to 'prevent consent authorities from imposing conditions seeking to control ... downstream greenhouse gas emissions ... occurring outside Australia'.[37] However, this bill was not adopted, and the Legislative Council expressed support for the existing policy, affirming that 'considering downstream greenhouse gas emissions supports international agreements aimed at reducing emissions and combating climate change'.[38]

---

[30] CEQ, 2023 Guidance (n 29) 1204.

[31] On developments in other Australian states, see, eg *Australian Conservation Foundation v Latrobe City Council* (2004) 140 LGERA 100 (Vic) [49]; Environmental Protection Authority, 'Environmental Factor Guideline: Greenhouse Gas Emissions' (April 2023) (Western Australia) 4; Climate Change Act 2017 (Vic), s 17(4)(b) (requiring consideration for 'direct and indirect greenhouse gas emissions').

[32] *Gray* (n 12) [100].

[33] State Environmental Planning Policy (Mining, Petroleum Production and Extractive Industries) 2007, s 14(2). See also State Environmental Planning Policy (Resources and Energy) 2021 (NSW), s 2.20(2) (identical provision).

[34] *Wollar Property Progress Association v Wilpinjong Coal Pty Ltd* [2018] NSWLEC 92 [126].

[35] *Gloucester Resources* (n 12) [492].

[36] See eg *Rocky Hill Coal Project*, SSD 5156, NSW Independent Planning Commission, Determination Report (14 December 2017) <https://perma.cc/XD97-VP6V>; *Gloucester Resources* (n 12); *Bylong Coal Project*, SSD 6367, NSW Independent Planning Commission, Statement of Reasons for Decision (18 September 2019) <https://perma.cc/4K5C-HBGH>; *United Wambo Open Cut Coal Mine Project*, SSD 7142, NSW Independent Planning Commission, Statement of Reasons for Decision (29 August 2019) <https://perma.cc/7VLT-BT4Y>. See generally Brian J Preston, Contemporary Issues in Environmental Impact Assessment (2020) 37 Environmental & Planning Law Journal 423, 433.

[37] New South Wales, *Parliamentary Debates*, Legislative Assembly, 24 October 2019, 1576 (Rob Stokes, Minister for Planning and Public Spaces).

[38] Environmental Planning and Assessment Amendment (Territorial Limits) Bill 2019 (NSW) (introduced 24 October 2019, lapsed 27 February 2023), s 2.100.

The EU Directive on Environmental Impact Assessment (EIA) also requires the assessment of 'the direct and indirect significant effects of a project', including on the climate.[39] In 2017, the EU Commission recommended that an 'EIA should include an assessment of the direct and indirect greenhouse gas emissions of the Project, where these impacts have been deemed significant'.[40] The Commission specified that the indirect climate impacts include 'greenhouse gas emissions generated or avoided as a result of other activities encouraged by the Project'.[41] These could include scope 2 emissions, such as 'emissions associated with energy use for the operation' of a transportation infrastructure project, as well as scope 3 emissions, such as 'emissions due to consumer trips to the commercial zone' where a proposed commercial development project would be located.[42] However, the Commission further acknowledges that the assessment of indirect climate impact may not be as straightforward as that of direct impact.[43] Consistently with the Commission's guidance, the District Court of Oslo and the UK Supreme Court are among the domestic courts who interpreted instruments transposing the EU EIA directive as requiring the assessment of all significant sources of GHG emissions associated with a project, including the downstream emissions from the combustion of fossil fuels produced by the project.[44]

2. Excluding indirect climate impacts

Other jurisdictions have confined CA to direct climate impacts. Some decisions not to consider indirect climate impacts were based on the understanding that, in the case at issue, such impact was unlikely to be significant.[45] More relevantly for the present purpose, other decisions sought to exclude any consideration of indirect climate impacts as a matter of principle, notwithstanding the potential magnitude of these impacts.

Some of these developments have taken place in Australia. In particular, the Land Court of Queensland has held in several cases that the CA for coal mines did not need to consider downstream scope 3 emissions, mainly based on the Court's assumption that these emissions would occur anyway as coal consumers would be able to find alternative sources of coal.[46] Further, the Federal Court of Australia

[39] Directive 2011/92/EU (n 22), consolidated as of 15 May 2014, art 3(1).
[40] European Commission, 'Environmental Impact Assessment of Projects: Guidance on the Preparation of the Environmental Impact Assessment Report' (2017) 39.
[41] Ibid.
[42] Ibid.
[43] Ibid 40.
[44] *Greenpeace Nordic* (2024) (n 13) s 3.5.4; *Finch* (SC) (n 13).
[45] This argument was frequently made with regard to road infrastructure projects. See eg *North Carolina Alliance for Transportation Reform Inc* (n 21) 520–21 (highway); *Coalition for Advancement of Regional Transportation* (n 21) 491 (bridges).
[46] *Xstrata Coal* (n 12) [559]; *Adani Mining Pty Ltd v Land Services of Coast* [2015] QLC 48 [456]; *Hancock Coal* (n 12); *New Acland Coal v Ashman (No 4)* [2017] QLC 24. In a more recent case, *Waratah*

found in *Australian Conservation Foundation Inc v Minister for the Environment* that the Minister could validly approve a coal mine without considering its downstream emissions overseas on the ground that such emissions were 'subject to a range of variables' and, thus, could not be predicted reliably.[47]

Similar developments have taken place in the British Isles. The Scottish Court of Session once held in *Greenpeace Ltd v Advocate General* that an assessment of downstream emissions was not legally required as part of the EA for the approval of oil and gas production wells, noting that this assessment 'would not be practicable' and that the impact would likely not be 'material' anyway.[48] In *An Taisce v An Bord Pleanála*, the Supreme Court of Ireland rejected the claim that the EIA for a cheese-making factory should have included an assessment of the upstream GHG emissions from milk production, based on a restrictive interpretation of what constitutes 'indirect' effects for the purpose of the EU EIA directive.[49]

Some other jurisdictions excluded some indirect climate impacts while including others. For instance, the Government of Canada determined that an EA should include 'an upstream GHG emission assessment', when relevant, but that '[a]n estimate of downstream emission [would] not [be] required'.[50] Observers have criticized this decision as being 'unreasonably generous to greenhouse gas emitters'.[51] The Federal Court in *Sierra Club Canada Foundation v Minister of Environment and Climate Change* sided with the government when upholding the approval of an oil production project despite the absence of assessment of its

---

Coal, the Queensland Land Court distinguished the exclusion of indirect emissions in those cases as 'factual conclusions based on evidence about the market for coal considering propositions about substitution and displacement of the coal and what that would mean for emissions'. See *Waratah Coal* (n 12) [670]. See also *Coast and Country Association of Queensland Inc v Smith* [2016] QCA 242 [45].

[47] *Adani Mining Pty Ltd*, Department of the Environment (Cth), 2010/5736, Statement of Reasons (13 October 2015) <http://epbcnotices.environment.gov.au/_entity/annotation/45c02035-e672-e511-b93f-005056ba00a7/a71d58ad-4cba-48b6-8dab-f3091fc31cd5?t=1712258955789> [140], cited in *Australian Conservation Foundation Inc v Minister for the Environment* (2016) 251 FCR 308 [58], and approved at [173]–[174]. See also *Wildlife Preservation Society of Queensland Proserpine v Minister for the Environment and Heritage* 232 ALR 510 [72]; *Australian Conservation Foundation Inc v Minister for the Environment and Energy* (2017) 251 FCR 359 [60].

[48] *Greenpeace Ltd v Advocate General* (n 13) [68]. This precedent was overruled in *Finch* (SC) (n 13).

[49] *An Taisce* (n 9).

[50] Government of Canada, 'Strategic Assessment of Climate Change' (October 2020) <www.canada.ca/en/services/environment/conservation/assessments/strategic-assessments/climate-change.html> 5. See also Government of Canada, 'Draft Technical Guide Related to the Strategic Assessment of Climate Change: Guidance on Quantification of Net GHG Emissions, Impact on Carbon Sinks, Mitigation Measures, Net-Zero Plan and Upstream GHG Assessment' (August 2021) <www.canada.ca/en/environment-climate-change/corporate/transparency/consultations/draft-technical-guide-strategic-assessment-climate-change.html> 37–41.

[51] Julia Levin, 'Comments on the Government of Canada's Draft Strategic Assessment of Climate Change' (Environmental Defence Canada 2019) <https://environmentaldefence.ca/wp-content/uploads/2019/09/Environmental-Defence-Draft-SACC-Submission.pdf> 3. See also Meinhard Doelle and Adebayo Majekolagbe, 'Meaningful Public Engagement and the Integration of Climate Considerations into Impact Assessment' (2023) 101 Environmental Impact Assessment Review (article# 107103) 4; Flavia Vieira De Castro, 'Canada's Climate Change Mitigation Commitments and the Role of the Federal Impact Assessment Act' (2020) 33 Journal of Environmental Law & Practice 211, 248.

downstream emissions.[52] In a similar fashion, the European Investment Bank's CA procedure focuses on scope 1 and scope 2 emissions on the ground that, '[f]or the majority of projects financed by the Bank, these are the most significant emissions associated with projects'.[53] Accordingly, scope 3 emissions are only to be taken into account when they 'are significant and can be estimated',[54] as would be the case of 'a power plant that exists solely to supply the project ... or a waste disposal site for the exclusive use of the project'.[55]

Even in jurisdictions that broadly consider indirect climate impacts at the scoping stage (such as the US, New South Wales, and most EU Member States), these impacts are seldom considered at the screening stage—that is, when deciding whether a proposed activity must be the object of an EA procedure at all. Screening decisions tend to focus on the risk of direct environmental impacts. As such, EA procedures may not be undertaken in the first place for some activities that lack any direct impact, even if they are a significant indirect climate impact.

Data centres are a case in point, in particular those developed to mine cryptocurrencies. These facilities use large quantities of electricity,[56] which, depending on the source of the energy, can result in large amounts of scope 2 emissions.[57] The construction and operation of such facilities are not usually subjected to any EA requirement because they have limited direct environmental impact[58] (although the operation of some of these facilities is reported to cause significant noise pollution).[59] While many jurisdictions routinely consider energy consumption as part of an EA,[60] they do not consider energy consumption as a ground to conduct an EA at the first place.

Some jurisdictions have sought to address this issue by extending EA requirements to activities with an indirect climate impact. For instance, a 2019 revision of the Rules on the implementation of the Hawaii Environmental Policy Act directs

---

[52] *Sierra Club Canada Foundation* (n 13).

[53] European Investment Bank, 'Project Carbon Footprint Methodologies: Methodologies for the Assessment of Project Greenhouse Gas Emissions and Emission Variations' (version 11.3, 2023) 8. See also Sébastien Godinot and others, 'Reviewing the European Investment Bank's Carbon Footprint Methodology' (Bankwatch Network, December 2016) <http://caneurope.org/content/uploads/2016/12/EIB-carbon-footprint-methodology.pdf>, 2.

[54] European Investment Bank, 'Project Carbon Footprint Methodologies' (n 53) 8.

[55] Ibid, 9.

[56] Felix Creutzig and others, 'Demand, Services and Social Aspects of Mitigation', in Shukla and others (eds) (n 5) 503, 540 (estimating that 'consumer devices, data centres, and data networks account for anywhere from 6% to 12% of global electricity use').

[57] Camilo Mora and others, 'Bitcoin Emissions Alone Could Push Global Warming above 2°C' (2018) 8 Nature Climate Change 931.

[58] Martin Roeck and Thomas Drennen, 'Life Cycle Assessment of Behind-the-Meter Bitcoin Mining at US Power Plant' (2022) 27 International Journal of Life Cycle Assessment 355, 355 (noting that 'not a single environmental impact assessment has been conducted on this type of operation').

[59] Gabriel JX Dance, 'Anxiety, Mood Swings and Sleepless Nights: Life near a Bitcoin Mine', *The New York Times* (3 Feb 2024).

[60] See eg Umweltverträglichkeitsprüfungsgesetz 2000 (Environmental Impact Assessment Law of 2000) BGBl I No 89/2000, as amended <www.ris.bka.gv.at/GeltendeFassung.wxe?Abfrage=Bundesnormen&Gesetzesnummer=10010767> (Austria), s 6(1)(1)(e).

agencies to consider whether the proposed action may '[r]equire substantial energy consumption or emit substantial greenhouse gases', among other factors, when determining whether the action may have a significant effect on the environment requiring further assessment.[61] Similarly, New York City's 2021 Technical Manual requires agencies to evaluate 'on a case-by-case basis … whether an assessment of consistency with the City's GHG reduction goals should be conducted' for projects that do not fall within general criteria for inclusion.[62] The Manual specifies that, 'if a project would result in the construction of a building that is particularly energy-intense, such as a data processing center or health care facility, a GHG emissions assessment may be warranted'.[63]

The absence of EA requirements in most jurisdictions has permitted the rapid, unchecked development of crypto-mining facilities. It is only years after these facilities entered into operations that some jurisdictions started to take measures to address their indirect climate impact, for instance by seeking better information on the energy use of these cryptocurrency mining facilities,[64] imposing a temporary ban on the purchase of power plants for the purpose of mining cryptocurrency,[65] or banning any mining of cryptocurrency within their territory.[66] Absent proper consideration for these indirect climate impacts at the screening stage, as Md Abu Bakar Siddik and colleagues noted, 'the environmental implications of data centers have been obscured from public view', despite the significance of these impacts and possible ways of reducing or avoiding them.[67]

[61] Hawaii Code R (2024) § 11-200.1-13. See also Minnesota Environmental Quality Board, 'Environmental Assessment Worksheet' (December 2022) <www.eqb.state.mn.us/environmental-review/overview/environmental-assessment-worksheet-eaw-process>, s 18 (requiring information on scope 2 emissions from off-site electricity generation, as part of the screening process under the Minnesota Environmental Policy act).

[62] Mayor's Office of Environmental Coordination, CEQR Technical Manual (December 2021) <www.nyc.gov/site/oec/environmental-quality-review/technical-manual.page>, ch 18, 7.

[63] Ibid.

[64] Regulation (EU) 2023/1114 of the European Parliament and of the Council of 31 May 2023 on markets in crypto-assets, and amending Regulations (EU) No 1093/2010 and (EU) No 1095/2010 and Directives 2013/36/EU and (EU) 2019/1937, [2023] OJ L 150/40, preamble para 7. See also S 661, 118th Congress (2023) §3(a)(1)(C) (US).

[65] New York Env't Conserv Law (McKinney 2024) § 19-0331. See also 2022 New York Sess Laws ch 628 (A 7389-C) (McKinney), §3 (requiring the department of environmental conservation to prepare a 'generic environmental impact statement on cryptocurrency mining operations' that would address the indicated climate impact of these operations); New York State Department of Environmental Conservation, SEQR Draft Public Scoping Document for General Environmental Impact Statement for Cryptocurrency Mining Operations that Use Proof-of-Work Authentication Methods to Validate Blockchain Transactions (January 2024) <https://dec.ny.gov/sites/default/files/2024-01/draftpublicscopingdocumentcrypto.pdf>.

[66] See National Development and Reform Commission Order No 29 (中华人民共和国国家发展和改革委员会令 第29号) of 6 November 2019 (China).

[67] Md Abu Bakar Siddik and others, 'The Environmental Footprint of Data Centers in the United States' (2021) 16 Environmental Research Letters 1, 9.

3. The arguments at play

The various political and judicial decisions regarding the assessment of indirect climate impacts tend to refer to a finite pool of arguments. On the one hand, the proponents of an assessment of these impacts highlight the need for EA to provide a comprehensive picture of all significant environmental impacts, direct or indirect. For instance, the US Council for Environmental Quality points out that including indirect environmental effects is 'critical to ensuring that agency decision-makers have a complete view of reasonably foreseeable effects of their proposed actions'.[68] Likewise, the European Court of Justice has long interpreted EA requirements as having 'a wide scope and a broad purpose'.[69] And the New South Wales Land and Environment Court justified that all GHG emissions are to be assessed as part of an EIA by observing that '[a]ll of the direct and indirect GHG emissions' of a project would 'impact on the environment' and 'contribute to climate change'.[70]

On the other hand, those opposed to the assessment of indirect climate impacts object to this, mainly on three grounds that are discussed more thoroughly in the following sections. First, one of the most controversial arguments in the debate on the scope of CA relies on the assumption of a perfect market substitution. If a national authority rejects a coal mine project in order to avoid downstream emissions from the combustion of coal—the argument goes—the downstream emissions would nonetheless occur because the consumers would purchase the same amount of coal from another mine. This reasoning has been influential in some jurisdictions. For instance, the Land Court of Queensland concluded that approving a coal mine project would cause 'no increase of greenhouse gas emissions' as 'alternative supply will be sourced elsewhere to meet global demand if the mine is not approved'.[71] Section B critically assesses this argument. It shows that, while market substitution is likely to occur with regard to many indirect GHG emissions, it is unlikely ever to be perfect. By increasing the supply of coal, a new coal mine would likely reduce the global price of coal, thus boosting global coal consumption. A CA could assess the extent to which the coal produced by a new mine would be additional to coal production in a baseline scenario without the mine.

Second, objections have been made to the assessment of indirect GHG emissions that occur outside of the territory of the state in which the CA is conducted. These arguments suggest that a state has no right, no obligation, or else no capacity to reduce GHG emissions from activities that take place beyond its territory.

---

[68] CEQ, 'National Environmental Policy Act Implementing Regulations Revisions' (20 April 2022) 87 Federal Register 23453, 23467. See also *Robertson v Methow Valley Citizens Council* (1989) 490 US 332, 349.

[69] C-72/95, *Aannemersbedrijf PK Kraaijeveld BV v Gedeputeerde Staten van Zuid-Holland* [1996] ECR I-5431 [31].

[70] *Gloucester Resources* (n 12) [415].

[71] *Adani Mining* (QLC) (n 46) [449].

For instance, the High Court of New Zealand considered that a national decision-maker could not 'form an accurate view as to whether the overseas discharges are adverse', as this may depend on the context in which these discharges occur.[72] Section C looks further into these arguments. It suggests that, while states' mitigation obligations are primarily territorial, they may extend to extraterritorial emissions under the state's jurisdiction or control. Thus, even though a state does not have the right to implement mitigation policies on other states, the territorial implementation of measures and policies may have intended implications for overseas activities.

Third, the most valid concern with the assessment of indirect climate impacts is a practical one. Indirect climate impacts are difficult to assess because they depend on extraneous factors that are difficult to predict. The Supreme Court of Norway, for instance, observed that 'the net impact of Norwegian exports of oil and gas on global emissions is complex and debated'.[73] Similarly, the Scottish Court of Session considered that assessing downstream emissions 'would not be practicable ... in an assessment of the environmental effects of a project for the extraction of fossil fuels'.[74] And the European Investment Bank's decision not to account for scope 3 emissions relates primarily to practical concerns relating to a lack of data to accurately predict these emissions.[75] Section D argues that decision-makers are better informed by EA reports that acknowledge indirect climate impacts and the uncertainty associated with these impacts, than by EA reports concealing these impacts on the ground that they are difficult to predict. Further, some indirect climate impacts can be predicted more accurately than others, and the distinction between direct and indirect climate impacts does not always coincide with a distinction between emissions that can and cannot be assessed. While there must be a limit to the assessment of indirect emissions, the key, it is suggested, is a test of reasonableness: a CA must account for all indirect impacts that can reasonably be anticipated to be potentially significant.

Three other objections are occasionally made to the assessment of indirect climate impacts but are not thoroughly considered in the present chapter, either because they are less prominent in existing debates and appear less convincing, or because they are discussed elsewhere in this book. One of these objections concerns the potential overlap between the assessment of a proposed activity's indirect climate impact and the CA of another proposed activity.[76] Guy Dwyer, for

---

[72] *Royal Forest and Bird Protection Society of New Zealand Inc v Buller Coal Ltd* [2012] NZHC 2156 [53]. See also *West Coast ENT* (n 12) [120] (approving the reasoning).

[73] *Greenpeace Nordic* (2020) (n 13) [234].

[74] *Greenpeace Ltd v Advocate General* (n 13) [68].

[75] See European Investment Bank, 'EIB Climate Action Public Consultation: Issues Matrix' (22 September 2015) <www.eib.org/attachments/consultations/eib_car_final_issues_matrix-_20150922_en.pdf>, 63 (Table 10); European Investment Bank, 'Project Carbon Footprint Methodologies' (n 53) note to readers.

[76] For a general discussion on potential policy overlaps, see Chapter II, subsection C.2.

instance, argued that accounting for scope 2 and scope 3 emissions would result in 'double-counting of emissions' because, he assumes, these emissions would also be considered as scope 1 emissions in relation to other projects.[77] Double counting is an important issue in international emission accounting,[78] but it is less of a concern with regard to CA, the primary objective of which is not to produce emission data that can be aggregated to one another. At any rate, overlaps among EAs are likely to be limited, first, because many activities are not subjected to any EA requirements and, second, because overlapping CA would likely look at the same emissions from different angles. For instance, while the CA of a power plant and that of a data centre may predict the same GHG emissions, the former would consider measures to reduce the power plant's emission intensity, while the latter would consider steps to reduce electricity consumption from the data centre.

A second occasional objection is that the assessment of indirect climate impact is unlikely to lead to effective mitigation measures. Several courts rejected this objection on the ground that, at the very least, the national agency could decide not to approve the proposed activity.[79] To the extent that the argument on ineffectiveness relies on the assumptions of perfect substitution or of a 'division of responsibilities'[80] in global efforts on climate change mitigation, it is addressed in Sections B and C respectively. The broader objection to the effectiveness of CA procedures in regulating climate impacts, including indirect ones, is further discussed in Chapter V along with other questions concerning the effectiveness of CA.

A last objection, in the context of judicial proceedings, regards the competence of courts to impose an assessment of indirect climate impacts on national agencies in the absence of a clear and specific statutory requirement. In the United Kingdom, courts recognized a 'substantial margin of appreciation' to decision-makers in deciding whether and how to assess scope 3 emissions,[81] considering that 'issues as to whether an effect is significant and the adequacy of any assessment of significant effects are matters of judgment for the decision-maker [that] are only open to challenge in the courts applying the conventional "Wednesbury" standard'.[82] Similarly, the Supreme Court of Norway held that it was for the government to decide

---

[77] Guy Dwyer, '"Market Substitution" in the Context of Climate Litigation' (2022) 12 Climate Law 1, 7. A similar concern was voiced in Godinot and others (n 53) 2.

[78] See generally Lambert Schneider, 'Robust Accounting of International Transfers under Article 6 of the Paris Agreement: Discussion Paper' (German Emissions Trading Authority 2017); Christina Hood, Gregory Briner, and Marcelo Rocha, 'GHG or not GHG: Accounting for Diverse Mitigation Contributions in the Post-2020 Climate Framework' (Climate Change Expert Group Paper No 2014–2, OECD Environment Directorate 2014).

[79] See eg Dine Citizens against Ruining Our Environment (n 12) 1217; Waratah Coal (n 12) [137].

[80] Greenpeace Nordic (2020) (n 13) [159].

[81] R (Friends of the Earth) v Secretary of State for International Trade [2023] EWCA Civ 14, [2023] 1 WLR 2011 [63]. See also Greenpeace Ltd v Advocate General (n 13) [68]; R (Finch) v Surrey County Council [2022] EWCA Civ 187, [2022] All ER (D) 93 (Feb) [61]; Bristol Airport Action Network (n 20) [163]. See also R (Greenpeace Ltd) v Secretary of State for Energy Security and Net Zero [2023] EWHC 2608 (Admin), [2024] PTSR 345 [113]–[115].

[82] Goesa Ltd (n 20) [100]. See also R (Friends of the Earth Ltd) v Heathrow Airport Ltd (n 20) [119]; Finch (SC) (n 13) [56]–[57]. See generally Associated Provincial Picture Houses Ltd v Wednesbury

whether to assess downstream emissions from oil-and-gas projects as part of a CA or rather at an overarching level, 'as a part of the Norwegian climate policy'.[83] Ultimately, legal systems have different approaches to a court's power to control the scope of a CA, and some of these questions are addressed in Chapter V. At any rate, observing that the question is a political one merely displaces it from judicial to political institutions—national agencies implementing EA, or even law-making institutions[84]—where similar substantive arguments can be considered.

B. The Market Substitution Objections

A frequent argument against the assessment of indirect climate impacts relies on the assumption of a perfect market substitution. The argument suggests that, in a competitive market, a proposed activity would cause no additional indirect climate impacts as it would only substitute for similar activities. This section reviews this argument. The first subsection recounts the making of the market substitution argument and its positive reception, in particular in Australia, New Zealand, and the United Kingdom. The second subsection considers moral issues raised by the market substitution argument. The third subsection questions the validity of the premise of perfect market substitution.

1. The theory of perfect market substitution

Proponents of the assessment of indirect climate impacts, such as the District Court for the District of Colorado, have argued that, if an agency is able to estimate the amount of coal that a mine would produce, it 'can likewise attempt to predict the environmental effects of its combustion'.[85] The Court added: '[j]ust because [the agency] does not possess perfect foresight as to the timing or rate of combustion or as to the state of future emissions technology does not mean that it can ignore the effects completely'.[86] The Oslo District Court made a similar point more

---

Corporation [1948] 1 KB 223 (Court of Appeal), 234 (setting out that an authority must not make a decision 'so unreasonable that no reasonable authority could ever have come to it').

   [83] *Greenpeace Nordic* (2020) (n 13) (unofficial translation by the Court) [234].
   [84] *Environment Council of Central Queensland Inc v Minister for the Environment and Water (No 2)* (2023) 413 ALR 318 [7].
   [85] *WildEarth Guardians v US Office of Surface Mining, Reclamation and Enforcement* (D Colorado 2015) 104 F Supp 3d 1208, 1231. See also *High Country Conservation Advocates* (n 12) 1196–97.
   [86] *WildEarth Guardians v US Office of Surface Mining, Reclamation and Enforcement* (n 85) 1231. See also *Dine Citizens against Ruining Our Environment* (n 12) 1213 (noting that, in the case at issue, there was 'virtually no uncertainty regarding when, where, and how the coal' from the proposed mine would be used; the respondents themselves conceded that the project's downstream GHG emissions were 'reasonably foreseeable').

strongly when noting that downstream emissions from oil and gas production facilities might actually be easier to predict than emissions from the operation of these facilities.[87]

However, other stakeholders have pointed out that some scope 2 and scope 3 emissions would occur notwithstanding whether a proposed activity is approved. As the activity would take place in a competitive market, they suggested, a competitor would take on the same business if the project was rejected. The US Bureau of Land Management, for instance, contended that, if a coal lease was not approved, 'it is likely that alternative sources would maintain the use of coal at national levels'.[88] Consequently, the Bureau of Land Management considered that climate change should not be considered 'beyond the scope' of the EA of fossil-fuel extraction projects.[89]

Similar arguments have frequently been made in relation to downstream GHG emissions in the context of fossil-fuel extraction or transportation activities. In particular, the Queensland Land Court has repeatedly invoked market substitution to justify that the assessment of downstream emissions from coal mines was not necessary.[90] In *Xstrata Coal Queensland Pty Ltd v Friends of the Earth*, the Court was persuaded that 'stopping the project will have no impact on climate change because it will have no impact on the global demand for coal and therefore no impact on global GHG emissions'.[91] In subsequent cases, the Land Court reiterated that refusing new coal mine projects would achieve 'no reduction of GHGs'[92] as 'other coal will be obtained from elsewhere'.[93] In one case, the Court came to the conclusion that a proposed coal mine would actually be beneficial to climate change, on the ground that the coal it would provide would substitute for more carbon-intensive coal.[94] However, the Court's position seems to have shifted in recent years, as evidenced in *Waratah Coal Pty Ltd v Youth Verdict Ltd*, where the Court noted 'evidence from the market experts' according to which 'perfect substitution is not likely, but some substitution is possible'.[95]

---

[87] *Greenpeace Nordic* (2024) (n 13) s 3.5.3. See also *Finch* (SC) (n 13) [2] ('It can ... be said with virtual certainty that, once oil has been extracted from the ground, the carbon contained within it will sooner or later be released into the atmosphere as carbon dioxide and so will contribute to global warming'). The UK Supreme Court ruled out perfect market substitution, but it did not further assess the relevance of imperfect market substitution.

[88] *East Lynn Lake Coal Lease*, EIS-ES-030-2008-0004, Bureau of Land Management, Final Land Use Analysis and Final Environmental Impact Statement (2009), 266, cited in Michael Burger and Jessica Wentz, 'Downstream and Upstream Greenhouse Gas Emissions: The Proper Scope of NEPA Review' (2017) 41 Harvard Environmental Law Review 110, 134.

[89] *Flat Canyon Federal Coal Lease Tract*, UTU-77114, Bureau of Land Management, Final Environmental Impact Statement, (January 2002), ch 2, 11.

[90] See generally Justine Bell-James and Sean Ryan, 'Climate Change Litigation in Queensland: A Case Study in Incrementalism' (2016) 33 Environmental & Planning Law Journal 515.

[91] *Xstrata Coal* (n 12) [559], [563].

[92] *Hancock Coal* (n 12) [232].

[93] *Adani Mining* (QLC) (n 46) [456].

[94] *Hancock Coal* (n 12) [227], [232]. Similar arguments were made, but rejected, in *Xstrata Coal* (n 12) [559], [581]; *Waratah Coal* (n 12) [32], [1393].

[95] *Waratah Coal* (n 12) [1005].

Similarly, the Federal Court of Australia found 'nothing ... which bespeaks of legal irrationality'[96] in the Minister's assessment that a proposed coal mine would cause 'no net increase to GHG emissions'[97] because, 'should the proposed action not proceed, the market would respond through an increase in supply elsewhere'.[98] An obiter dictum of the Supreme Court of New Zealand in *West Coast ENT Inc v Buller Coal Ltd* suggested that carbon dioxide emissions from coal burning might not need 'to be seen as a consequence ... of the mining of coal in New Zealand' because 'steel manufacturers will, whatever happens in New Zealand, burn whatever coal is required for their purposes'.[99] In *Greenpeace Nordic Association v Ministry of Petroleum and Energy*, the Supreme Court of Norway also seemed to accept that new oil and gas wells would merely substitute for other fossil-fuel projects.[100] And, in *Greenpeace Ltd v Advocate General*, the Scottish Court of Session upheld the Secretary of State's submission that '[t]he production of oil from the Vorlich field [would] not increase the use of oil',[101] accordingly that it was 'difficult to argue that [the project] would have any material effect on climate change'.[102]

While market substitution has generally been invoked in relation to scope 2 and scope 3 emissions, it could be also made in relation to direct climate impacts.[103] A possible argument is that, assuming a fixed global demand for fossil fuels, even the direct emissions of a fossil-fuel production project (eg fugitive emissions) could substitute to similar emissions from competing projects.[104] Likewise, the direct emissions from the operation of a new fossil-fuel fired power plant could substitute to the emissions of existing power plants.[105] Beyond fossil fuels, the Supreme Court of Ireland suggested that milk consumption by a cheese-making factory 'would not have any significant indirect environmental effect ... because ... this milk [would] be produced in any event'.[106] And a new highway or airport would likely capture traffic from other transportation infrastructure.[107]

---

[96] *Environment Council of Central Queensland* (n 84) [161].

[97] Ibid [127]. But see *Environment Council of Central Queensland Inc v Minister for the Environment and Water* [2024] FCAFC 56 [104] (suggesting that 'the Minister did not engage in substitution reasoning').

[98] Ibid. See also, International Energy Agency, 'Coal 2022: Analysis and Forecasts to 2025' (2022) 8, 11.

[99] *West Coast ENT* (n 12) [122].

[100] *Greenpeace Nordic* (2020) (n 13) [234].

[101] *Greenpeace Ltd v Advocate General* (n 13) [40].

[102] Ibid [68].

[103] Dwyer (n 77) 8. See also Godinot and others (n 53) 6. See also The Greenhouse Gas Protocol, 'The GHG Protocol for Project Accounting' (World Business Council for Sustainable Development and World Resources Institute 2005) <https://ghgprotocol.org/project-protocol> 32.

[104] Justine Bell-James and Briana Collins, '"If We Don't Mine Coal, Someone Else Will": Debunking the "Market Substitution Assumption" in Queensland Climate Change Litigation' (2020) 37 Environmental & Planning Law Journal 167, 183; *Hancock Coal* (n 12) [249].

[105] See eg *Greenpeace Nordic* (2024) (n 13) s 3.6.3.

[106] *An Taisce* (n 9) [108] (emphasis removed). The Court nuanced this argument in the following paragraph, suggesting that the market substitution effect may be imperfect, but that the causal link would be difficult to establish.

[107] Alexander Zahar, 'Environmental Impact Assessment for Greenhouse Gas Emissions Is Pie in the Sky', in Benoit Mayer and Alexander Zahar (eds), *Debating Climate Law* (CUP 2021) 297, 306–07.

However, to justify the exclusion of indirect (or even direct) climate impacts, the market substitution argument needs not only to show that market substitution occurs, but also that it is 'perfect'. In other words, the argument needs to be that '[t]here will ... be ... *the same amount* of GHG emissions caused' with or without the proposed activity.[108] As Brian Preston put it, assuming imperfect market substitution does not justify excluding the assessment of indirect climate impacts (although it may justify higher thresholds of significance).[109] Even if only a small proportion of a coal mine's downstream emissions are additional to the emissions that would occur without the project, this indirect climate impact could well be significant and, therefore, relevant to decision-makers. At times, however, the debate shows some confusion between evidence of 'a significant degree of substitution' and assertions that a proposed activity would have no additional indirect climate impact at all.[110]

2. Normative counter-arguments

Judges and scholars have sometimes responded to the market substitution objections on moral grounds. Saul Holt and Chris McGrath, for instance, denounced the 'drug dealers' defence' as '[m]ental gymnastics that ignores reality'.[111] Justine Bell-James and Briana Collins have observed that '[t]here are few other contexts where a harmful behaviour or action is excused purely because another entity would have otherwise caused the harm'.[112] The New South Wales Land and Environment Court held in *Gloucester* that a coal mine's downstream emissions do 'not become acceptable because a hypothetical and uncertain alternative development might also cause the same unacceptable environmental impact'.[113] Rather, the Court submitted, '[t]he environmental impact remains unacceptable regardless of where it is caused'.[114]

These moral arguments against the market substitution objections generally rely on non-consequentialist moral theories: authors consider that, 'even when the consequences of two acts ... are the same, one might be wrong and the other

---

[108] Brian Preston, 'Contemporary Issues in Environmental Impact Assessment' (2020) 37 Environmental & Planning Law Journal 423, 438 (emphasis added).

[109] See text above at n 11.

[110] Friends of the Earth, 'Whitehaven Coal Mine Legal Challenge Briefing' (April 2023) <https://friendsoftheearth.uk/climate/whitehaven-coalmine-legal-challenge-briefing-1>. See also *Former Marchon Site*, APP/H0900/V/21/3271069, Department for Levelling Up, Housing and Communities, Application Ref 4/17/9007, Decision (7 December 2022) <https://perma.cc/ZN3N-6LQU> [21].

[111] Saul Holt and Chris McGrath, 'Climate Change: Is the Common Law up to the Task?' (2018) 24 Auckland University Law Review 10, 29.

[112] Bell-James and Collins (n 104) 185.

[113] *Gloucester Resources* (n 12) [545].

[114] Ibid.

right'.[115] Thus, Augustin Fragnière suggested that, 'even if refraining from per-forming GHG intensive actions makes no difference, we ought to do it all the same', either 'because it says something about our integrity as moral agents',[116] or because individuals (and by extension states) 'have a duty to reduce *their* carbon footprint irrespective of what others do'.[117] In the same line, Matthew Rendall argued that, 'if the industrialised countries go on emitting, then they will profit from an injustice, and they will act wrongly', notwithstanding the consequences.[118]

Yet, one could question the relevance of non-consequentialist moral theories to EA procedures. EA has often been approached as a tool to assess the consequences of a proposed activity rather than its inherent moral value or its conformity with posited moral duties. Even the proponents of non-consequentialist moral argu-ments acknowledge that these are 'unlikely to be the strongest arguments before a court' when market substitution has been invoked as an objection to the assess-ment of indirect climate impacts.[119]

By contrast, market substitution arguments are relevant from a consequentialist perspective, as they help in assessing the potential consequences of the proposed activity.[120] Guy Dwyer sought to argue otherwise when he submitted that the con-sequences of approving a proposed activity differed from the consequences of re-jecting this activity: while the two decisions would generate '*similar* impacts', he submitted, they would not generate the exact '*same*' impacts.[121] Dwyer thus refers to the fact that, in the case of a coal mine, different atoms of carbon from different coal fields would be discharged into the atmosphere at slightly different times and places. Yet this distinction is not relevant if one approaches climate impacts in ab-stract terms, as an incremental exacerbation of global risks. The Supreme Court of Queensland rejected an argument similar to Dwyer's when it held that the market substitution argument 'was not that there was "replacement harm", but that there would be the "same or greater harm" if the mine did not proceed than if it did proceed'.[122]

---

[115] Frances M Kamm, 'Nonconsequentialism', in Frances M Kamm, *Intricate Ethics: Rights, Responsibilities, and Permissible Harm* (OUP 2006) 10, 11. See also Jonathan Glover and M Scott-Taggart, 'It Makes no Difference whether or Not I Do It' (1975) 49 Proceedings of the Aristotelian Society (Supplementary Volumes) 171.

[116] Augustin Fragnière, 'Climate Change and Individual Duties' (2016) 7 WIREs Climate Change 798, 803.

[117] Ibid 804 (emphasis added).

[118] Matthew Rendall, 'Carbon Leakage and the Argument from no Difference' (2015) 24 Environmental Values 535, 541.

[119] Bell-James and Collins (n 104) 185. See also Dwyer (n 77) 30; Fleur Kingham, President of the Land Court of Queensland, 'Climate Change Litigation' (Speech Delivered at the Environmental Defenders Office Climate Law Update, Banco Court, 18 September 2018) <https://archive.sclqld.org.au/judgepub/2018/kingham20180918.pdf>, 9.

[120] See Zahar (n 107) 303.

[121] Dwyer (n 77) 14 (emphasis in the original).

[122] *Coast and Country Association of Queensland* (n 46) [45]. See also *Hancock Coal* (n 12) [232].

3. Factual counter-arguments

A more compelling response to the market substitution objections engage with the objections' economic rather than moral foundations. Basic economic theory confirms that some market substitution is likely to happen, but it also shows that the 'perfect' nature of this market substitution is implausible.[123] Perfect market substitution would be 'an economic anomaly in which supply and demand do not interact to determine the level of consumption'.[124] Thus, despite a market substitution effect, measures aimed at reducing the supply of fossil fuels can play a role in supporting the mitigation of climate change.[125]

Classic economic theory suggests that, when supply in a commodity is further constrained, the price of this commodity increases, and the quantity sold decreases.[126] Perfect market substitution would assume either that the price elasticity of supply is infinite (ie fossil-fuel producers can extend production without increasing the price at all), or that the price elasticity of demand is null (ie fossil-fuel consumers are willing to pay more without decreasing their consumption at all). Yet the former assumption would go against what Paul Samuelson and William Nordhaus describe as the 'law of downward-sloping demand': '[w]hen the price of a commodity is raised ... buyers tend to buy less of the commodity'.[127] And the latter assumption would violate the 'law of diminishing return': 'the marginal product of each unit of input will decline as the amount of that input increases'.[128]

As far as the energy sector is concerned, neither assumption is supported by empirical economic research on price elasticities.[129] A fossil-fuel extraction project increases the supply of this fossil fuel, which leads to a reduction in the price of this fuel, incentivizing consumers to increase their consumption, for instance by consuming more energy or by relying less on alternative sources of energy. Other existing or planned projects may offer viable alternatives sources, but this would

---

[123] See eg Holt and McGrath (n 111) 21; Bell-James and Collins (n 104) 175; Daria Shapovalova, 'Arctic Petroleum and the 2°C Goal: A Case for Accountability for Fossil-Fuel Supply' (2020) 10 Climate Law 282, 292–93.

[124] Thomas Michael, Donovan S Power, and Joel M Brown, 'Comments on the Greenhouse Gas Impacts and the Modeling of Coal Flows in the Millennium Bulk Terminals Longview SEPA Draft Environmental Impact Statement' (Earth Justice and Sierra Club 2016) 36.

[125] See eg GB Asheim and others, 'The Case for a Supply-Side Climate Treaty' (2019) 365 Science 325, 325; Fergus Green and Richard Denniss, 'Cutting with both Arms of the Scissors: The Economic and Political Case for Restrictive Supply-Side Climate Policies' (2018) 150 Climatic Change 73, 77; Michael Lazarus and Harro van Asselt, 'Fossil Fuel Supply and Climate Policy: Exploring the Road Less Taken' (2018) 150 Climatic Change 1, 4–5.

[126] Paul A Samuelson and William D Nordhaus, *Economics* (19th edn, McGraw-Hill Irwin 2010) 57.

[127] Ibid 47.

[128] Ibid 108–09.

[129] See eg Bureau of Ocean Energy Management, 'Consumer Surplus and Energy Substitutes for OCS Oil and Gas Production: The 2021 Revised Market Simulation Model (MarketSim)' (OCS Study BOEM 2021-072, November 2021) <www.boem.gov/marketsim-model-documentation>, 17–20 (providing default demand and supply elasticities based mainly on 'peer-reviewed studies in the empirical economics literature').

likely be at a slightly higher price, as one can assume that a profit-seeking project proponent selected what appeared to be the most profitable project.

US courts have relied on such economic arguments to reject suggestions of a perfect market substitution. In *Mid States Coalition for Progress v Surface Transportation Board*, the Court of Appeal for the Eighth Circuit found it 'illogical' to assume that 'the demand for coal will be unaffected by an increase in availability and a decrease in price' resulting from the project.[130] In *High Country Conservation Advocates v US Forest Service*, the District Court for the District of Colorado 'could not make sense' of arguments invoking a perfect market substitution[131]: a new coal mine would inevitably 'increase the supply of … coal' and thus 'impact the demand for coal relative to other fuel sources'.[132] In *WildEarth Guardians v US Bureau of Land Management*, the Court of Appeal for the Tenth Circuit found that 'the blanket assertion … that coal would be substituted from other sources' was 'arbitrary and capricious because … irrational (ie contrary to basic supply and demand principles)'.[133] In particular, the Court rejected the defendant's presumption 'that either the reduced supply [of coal] will have no impact on price, or that any increase in price will not make other forms of energy more attractive and decrease coal's share of the energy mix, even slightly'.[134] Consistently, the Council on Environmental Quality (CEQ) recommended that agencies 'should not simply assume that if the federal action does not take place, another action will perfectly substitute for it and generate identical emissions'.[135]

When responding to these arguments, those opposing the assessment of indirect climate impacts have often confused perfect and imperfect market substitution. Guy Dwyer, for instance, noted that, '[w]hen the Chinese government moved to place restrictions on the importation of Australian coal as part of a dispute between the two countries, it resorted to other markets for substitute coal'.[136] But while there is no doubt that *some* market substitution can occur, Dwyer fails to demonstrate that this substitution was *perfect*. China's ban on Australian coal, in fact, has been reported to have 'hurt China with higher prices on domestic markets and the need to pay premiums for relatively low-quality imports' from alternative sources,[137] leading to supply shortages.[138] Contrary to Dwyer's assertion,

[130] *Mid States Coalition for Progress* (n 15) 549.

[131] *High Country Conservation Advocates* (n 12) 1197.

[132] Ibid 1198.

[133] *WildEarth Guardians v US Bureau of Land Management* (10th Cir 2017) 870 F 3d 1222, 1235–36.

[134] Ibid 1229.

[135] CEQ, 2023 Guidance (n 29) 1205.

[136] Dwyer (n 77) 21 (references omitted).

[137] Ron Wickes, Mike Adams, and Nicolas Brown, *Economic Coercion by China: The Impact on Australia's Merchandise Exports* (Institute of International Trade, Working Paper 4, July 2021) 18.

[138] Amanda Lee, 'China Faces Coal Shortage as Import Restrictions, Tighter Environmental Checks Begin to Bite', *South China Morning Post* (18 October 2020) <www.scmp.com/economy/china-economy/article/3105854/china-faces-coal-shortage-import-restrictions-tighter>; Muyu Xu, 'Analysis: Quantity Over Quality – China Faces Power Supply Risk Despite Coal Output Surge', *Reuters* (21 June 2022).

it is not necessary for supply-side fossil-fuel regulation to be 'effected universally across jurisdictions [for] prices [to] go up and demand [to] go down':[139] a (relatively) small reduction in the supply of fossil fuels can be expected, under basic economic laws of supply and demand, to have a proportional impact on price and consumption.[140]

The CA of a proposed activity can attempt to assess market substitution effects as a way to better assess indirect climate impacts. In broad terms, these effects depend on the market conditions that determine the price elasticity of demand and supply.[141] Thus, market substitution is more likely to occur when supply can easily be increased. In the energy factor, this may depend, among other things, on international energy markets (eg the amount of stock available) and the geopolitical factors that constrain these markets.

The market effects of a new activity also depend on the possibility for consumers to turn to alternative products. Thus, scholars have noted that, in the fossil-fuel sector, substitution of fossil fuel with alternative sources of energy is more likely 'where competitive clean energy sources' are available.[142] But substitution can also occur between different fossil fuels.[143] Thus, in *Allegheny Defense Project v Federal Energy Regulatory Commission*, the US Court of Appeal for the District of Columbia Circuit noted that downstream emissions from a natural gas extraction project were predicted to be 'partially offset by reductions in higher carbon-emitting fuel that the Project's natural gas would replace'.[144] This observation alluded to a finding of the Environmental Impact Statement according to which, in the circumstances of the project, 'increased production and distribution of natural gas would likely displace some use of' fuel oil and coal.[145] Along similar lines, the Supreme Court of Norway noted the possibility that increasing oil-and-gas production could turn consumers away from coal, thus reducing global GHG emissions.[146]

Another important consideration is that the alternative supply of the same product may be provided in a different way, for instance by companies operating under a laxer regulatory framework. For instance, natural gas production from a

[139] Dwyer (n 77) 25.

[140] Green and Denniss (n 125) 77.

[141] Samuelson and Nordhaus (n 126) 66, 73. See also GHG Protocol, 'The GHG Protocol for Project Accounting' (n 103) 32.

[142] Bell-James and Collins (n 104) 175. See also Shapovalova (n 123) 292–93.

[143] See Apostolos Serletis, Govinda R Timilsina, and Olexandr Vasetsky, 'Interfuel Substitution in the United States' (2010) 32 Energy Economics 737.

[144] *Allegheny Defense Project v Federal Energy Regulatory Commission* (DC Cir 2019) 932 F 3d 940, 946. See also references above at n 94.

[145] *Atlantic Sunrise Project*, Docket No CP15-138-000, Federal Energy Regulatory Commission, Final Environmental Impact Statement (2016) <www.ferc.gov/final-environmental-impact-statement-atlantic-sunrise-project>, vol 1 ch 4, 318.

[146] *Greenpeace Nordic* (2020) (n 13) [234]. See also *R (Friends of the Earth) v South Lakeland Action on Climate Change* (n 12).

well could substitute to more polluting wells, such as wells using more energy to exploit unconventional reservoirs or causing larger amounts of fugitive emissions. Thus, assuming a partial market substitution, it is possible (albeit highly unlikely) that authorizing a new fossil-fuel supply project could occasionally achieve a reduction in global GHG emissions in the short-term. However, a new fossil-fuel supply project would probably be running longer than pre-existing ones, thus impeding the phasing out of fossil fuels on the longer-term.

Lastly, the magnitude of market substitution depends on the timeframe under consideration. Given more time, alternative suppliers may find new ways to increase their production, whereas consumers are more likely to find alternative products.[147] Thus, the US Court of Appeal for the Eighth Circuit noted that, while additional coal supply may 'not affect the short-term demand for coal ... it will most assuredly affect the nation's long-term demand' by disincentivizing clean energy investments.[148]

In these circumstances, the Supreme Court of Norway rightly noted that assessing market effects is 'complicated'.[149] Yet clarifying such complicated matters is precisely one of the functions of EA procedures. As such, the CEQ rightly encourages agencies implementing NEPA reviews on relevant projects, including those related to the extraction and transportation of fossil fuels, 'to conduct substitution analysis to provide more information on how a proposed action and its alternatives are projected to affect the resulting resource or energy mix, including resulting GHG emissions'.[150] The CEQ also notes that several economic modelling tools have already been developed to facilitate such substitution analysis.[151]

C. The Extraterritoriality Objections

This section considers the application of CA to indirect climate impacts that involve extraterritorial sources of GHG emissions. The first subsection defines the problem at issue. It shows that, while there is no question that national EA frameworks apply only to activities that occur within the state's jurisdiction, EAs may consider those activities' scope 2 and scope 3 emissions that occur extraterritorially. The second subsection considers an objection to the right of states to seek to reduce extraterritorial GHG emissions. It is argued that a state is entitled—and under certain circumstances might even be required—to take measures that aim at reducing

---

[147] Samuelson and Nordhaus (n 126) 66, 73.
[148] *Mid States Coalition for Progress* (n 15) 549. See also *High Country Conservation Advocates* (n 12) 1197–98.
[149] *Greenpeace Nordic* (2020) (n 13) [234].
[150] CEQ, 2023 Guidance (n 29) 1205.
[151] Ibid (referring to Bureau of Ocean Energy Management's Revised Market Simulation Model, the US Energy Information Administration's National Energy Modelling System, and ICF's Integrated Planning Model).

GHG emissions abroad, provided that these measures are respectful of the sovereignty of other states. The third subsection assesses the practicality of predicting overseas emissions. It contends that, while there are real difficulties, these do not justify the exclusion of indirect climate impacts from the scope of CAs.

1. Conceptualizing extraterritoriality

The extraterritoriality of indirect climate impacts has often been invoked as a justification for excluding these impacts from the scope of CA procedures. To properly understand these arguments, a distinction needs to be drawn between three geographical dimensions of EA procedures: (1) the place where the activity is implemented, (2) the place where GHGs are emitted, and (3) the place where the environment is impacted. For instance, a coal mine may be located in Australia; its fuel may be consumed in India; and its environmental impacts could unfold in Australia (eg water pollution from the operation of mine), India (eg local air pollution from the combustion of the coal), and globally (eg climate impact of the combustion of the coal).

The first dimension—the place where the activity is implemented—is seldom problematic, as it is well accepted that EA procedures apply to a state's jurisdiction: mainly its territory[152] or, at times, other areas under its jurisdiction (eg continental shelf or exclusive economic zone)[153] or activities under its control (eg activities implemented in areas beyond national jurisdictions or overseas activities by national agencies).[154] Issues in this regard are not specific to the assessment of climate impacts and, as such, are not extensively discussed in this book. The third dimension—the place where the environment is impacted—was discussed in Chapter II, where it was established that EA has often been used to consider extraterritorial and global impacts, including climate impacts.[155] The discussion in the present section focuses on the second dimension—the location of the source of emissions. It is rare, beyond the context of indirect climate impacts, for an activity taking place within a state's jurisdiction to cause significant environmental impacts from extraterritorial sources.

---

[152] See eg Convention on Environmental Impact Assessment in a Transboundary Context (adopted on 1 March 1991, entered into force 10 September 1997) 1989 UNTS 309 (Espoo Convention) art 1(ii); Directive 2011/92/EU (n 22), consolidated as of 15 May 2014, art 7(1) (by implication); 42 USC (2024) § 4336e (10)(vi).

[153] See eg Lov om petroleumsvirksomhet (Petroleum Act), 72/1996 (Norway), ss 1.4, 3.1. See also *Greenpeace Nordic* (2020) (n 13); *Center for Biological Diversity v Bernhardt* (n 13) 731; *Natural Resources Defense Council Inc v US Department of Navy* (CD California 17 September 2002) CV-01-07781 CAS(RZX), 2002 WL 32095131.

[154] See Impact Assessment Act, SC 2019, c 28, s 6(1)(l); *Iron Rhine Railway (Belgium v Netherlands)* (2005) 27 RIAA 35; *Environmental Defense Fund Inc v Massey* (DC Cir 1993) 986 F 2d 528.

[155] See Chapter II, subsection A.3(a).

Some confusion can occasionally be observed between these three dimensions of extraterritoriality. For instance, the Federal Court of Canada suggested that the downstream emissions of an oil project may not fall 'within the legislative authority of Parliament' and, therefore, did not have to be assessed.[156] Similarly, the Constitutional Court of Austria applied a constitutional provision defining the state's territory as implying that the assessment of an airport extension project could not consider the GHG emissions from international flights arriving to or departing from national airports.[157] These two courts thus confused the place where the activity would be implemented (first dimension) and the place where the emissions would occur (second dimension).[158] While the Canadian and Austrian governments have no authority to regulate GHG-emitting activities carried out beyond their jurisdiction, they do have the authority to regulate activities within their territory—and, in doing so, they could consider consequences unfolding overseas.[159]

Most of the prominent cases concerned with indirect climate impacts overseas relate to projects that extract or transport fossil fuels for exportation. Yet scope 2 and scope 3 emissions may also unfold extraterritorially under a number of other scenarios. For instance, a new airport may increase international civil aviation, thus increasing GHG emissions beyond the state's territory.[160] Developing a large-scale photovoltaic power station could lead to significant overseas emissions for the production of the solar panels, thus reducing the climate benefit of such projects.[161] And a policy promoting the use of biofuels could cause overseas emissions from the land sector (eg deforestation).[162]

---

[156] *Sierra Club Canada Foundation* (n 13) [67].

[157] Verfassungsgerichtshof (n 20).

[158] For another instance of a similar confusion, see *West Coast ENT* (n 12) [75].

[159] See Case C-366/10, *Air Transport Association of America v Secretary of State for Energy and Climate Change* [2011] 2011 ECR I-13755 [125].

[160] See eg Xueli Xiong and others, 'Aviation and Carbon Emissions: Evidence from Airport Operations' (2023) 109 Journal of Air Transport Management (article# 102383) 1–9, at 5–6. See *R (Friends of the Earth Ltd) v Heathrow Airport Ltd* (n 20) [161]; *Bristol Airport Action Network* (n 20) [81].

[161] Da Liu and others, 'Contribution of International Photovoltaic Trade to Global Greenhouse Gas Emission Reduction: The Example of China' (2019) 143 Resources, Conservation & Recycling 114. There is anecdotal evidence that these indirect GHG emissions are not consistently considered in EA reports. See eg Longfield Solar Farm Ltd, Environmental Assessment: Scoping Report, 'Longfield Solar Farm' (October 2020) <https://infrastructure.planninginspectorate.gov.uk/wp-content/ipc/uploads/projects/EN010118/EN010118-000009-LSF%20-%20Scoping%20Report.pdf> (UK), 44 (acknowledging that '[r]aw material extraction and manufacturing of products required to build the equipment [are] expected to make a significant contribution to overall GHG emissions'); but see KMH Environmental and Pitt & Sherry, Environmental Impact Statement, 'Gunnedah Solar Farm' (17 April 2018) <https://gunnedahsolar.com.au/working/wp-content/uploads/2019/10/Gunnedah-Solar-Farm-EIS-download.pdf> (NSW) 48 (merely asserting that the project will 'emit ... negligible GHG emissions'), 231 (quantifying the GHG emission reductions without accounting for upstream emissions).

[162] See Directive (EU) 2015/1513 of the European Parliament and of the Council of 9 September 2015 amending Directive 98/70/EC relating to the quality of petrol and diesel fuels and amending Directive 2009/28/EC on the promotion of the use of energy from renewable sources, [2015] OJ L 239/1, preamble para 4–5.

The location where GHGs are emitted does not meaningfully affect their effect on the climate system.[163] As such, the downstream emissions from the combustion of coal will have the same climate effect notwithstanding where it occurs. Such consideration led several US courts to find that national agencies should assess a project's downstream emissions, notwithstanding whether these are located within or beyond the state's territory, as part of NEPA reviews concerned with the sale of oil and gas leases.[164] Similarly, the Norwegian Supreme Court in *Greenpeace Nordic Association* admitted the 'need to consider all emissions from Norwegian petroleum production as a whole', even though the Court suggested that this could be done at an overarching level, 'as part of the Norwegian climate policy', rather than 'in the individual environmental assessment' of an oil-and-gas lease project.[165]

Yet two further objections were made to the assessment of extraterritorial scope 2 and scope 3 emissions. The first objection suggests that a state does not have the right to take into account extraterritorial sources of GHG emissions in the assessment of a proposed activity within its jurisdiction. The second objection relates to the practical difficulty of assessing GHG emissions from extraterritorial sources. The following subsections address these two objections.

2. The right of states to reduce overseas emissions

A frequent objection to the assessment of indirect climate impacts from extraterritorial sources of GHG emissions relies on the assumption that a state should only seek to reduce GHG emissions within its own jurisdiction. For instance, the Supreme Court of Norway in *Greenpeace Nordic Association* identified what it presented as a 'division of responsibilities between states in accordance with international agreements', including a 'clear principle' according to which 'each state is responsible for [fossil-fuel] combustion on its own territory'.[166] Similarly, in *Australian Conservation Foundation Inc v Minister for the Environment and Energy*, the Federal Court of Australia upheld the exclusion of indirect GHG emissions overseas on the ground that 'the provisions of [climate treaties] place

---

[163] See eg Mahbod Rouhany and Hugh Montgomery, 'Global Biodiesel Production: The State of the Art and Impact on Climate Change', in Meisam Tabatabaei and Mortaza Aghbashlo (eds), *Biofuel* (Springer 2019) 1.

[164] See *Center for Biological Diversity v Bernhardt* (n 13) 740; *Friends of the Earth v Haaland* (n 13) 141. See also *Border Power Plant Working Group* (n 18) 1017.

[165] *Greenpeace Nordic* (2020) (n 13) (unofficial translation by the Court) [234]. On the false dichotomy between EA and 'climate policy', see Chapter II, subsection C.2.

[166] *Greenpeace Nordic* (2020) (n 13) (unofficial translation by the Court) [159]. See also *Greenpeace v Netherlands*, ECLI:NL:RBDHA:2020:12440 (District Court of The Hague, 9 December 2020) s 4.4 (reflecting a similar assumption about the territorial nature of mitigation obligations under climate treaties).

responsibility for dealing with the overseas emissions upon the countries consuming the coal'.[167]

However, closer analysis shows that there is no support to the assumption that states cannot take extraterritorial sources of GHG emissions into account when adopting measures and policies on climate change mitigation. While treaties apply in principle to the territory of their parties,[168] this only means that states must implement measures within their territory and does not exclude the territorial application of measures aimed partly or wholly at reducing extraterritorial sources of emissions. The seminal decision of the Permanent Court of International Justice in *SS Lotus* noted that, while a state 'may not exercise its power ... in the territory of another State', it may nonetheless 'exercis[e] jurisdiction in its own territory ... in respect of any case which relates to acts which have taken place abroad'.[169] This principle seems to allow a state to adopt and implement measures within its territory with the aim of reducing extraterritorial sources of GHG emissions. Thus, according to the US Court of Appeal for the Ninth Circuit, an agency may have the 'authority to act on the emissions resulting from foreign oil consumption' of nationally produced oil.[170]

Joanne Scott has suggested that climate treaties have imposed 'a system boundary that is principally territorial in nature'.[171] This, however, is based on a confusion between reporting obligations and substantive mitigation obligations.[172] When states prepare national inventories of GHG emissions, they are generally directed to focus on those emissions that occur within their territory,[173] with a few exceptions justified by practical considerations (eg emissions from road vehicle fuels are accounted in the country where the fuel is purchased[174] and emissions from international transportation are reported separately).[175] This territorial focus seeks to ensure the completeness of national inventories and to avoid double

---

[167] *Australian Conservation Foundation* (2017) (n 47) [51].

[168] Vienna Convention on the law of Treaties (adopted 23 May 1969, entered into force 27 January 1980) 1155 UNTS 331, art 29.

[169] *SS Lotus (France v Turkey)* [1927] PCIJ Series A No 10, 18–19. See also Menno T Kamminga, 'Extraterritoriality', in Rüdiger Wolfrum (ed), *Max Planck Encyclopedias of Public International Law* (OUP 2020) para 6 (noting that '[d]ue to the scarcity of relevant international case law, the general principles governing the conflict are still derived from' the *Lotus* judgment).

[170] *Center for Biological Diversity v Bernhardt* (n 13) 740. See also *Gloucester Resources* (n 12) [556]; but see *Sierra Club Canada Foundation* (n 13) [67].

[171] Joanne Scott, 'Unilateralism, Extraterritoriality and Climate Change', in Michael Faure (ed), *Elgar Encyclopedia of Environmental Law* (Edward Elgar 2023) vol 1, 167, 168.

[172] See ibid 172.

[173] Eg Decision 4/CP.1, 'Methodological Issues', FCCC/CP/1995/7/Add 1 (6 June 1995) 15, para 1(a)–(b). See also Kristin Rypdal and others, 'Introduction to the 2006 Guidelines', in Eggleston and others (eds) (n 8) vol 1, ch 1, 4.

[174] Maria José Sánchez, 'Reporting Guidance and Tables', in Eggleston and others (eds) (n 8) vol 1, ch 8, 4–5.

[175] Decision 18/CMA.1, 'Modalities, Procedures and Guidelines for the Transparency Framework for Action and Support Referred to in Article 13 of the Paris Agreement', FCCC/PA/CMA/2018/3/Add 2 (19 March 2019) 18, Annex para 53.

counting,[176] with the goal, among other things, that national inventories could be compiled into accurate global emissions inventories.

However, there is no reason to assume that the scope of states' mitigation obligations coincides with the scope of their GHG inventories.[177] With the historical exception of the Kyoto Protocol,[178] the scope of treaty obligations on the mitigation of climate change are not explicitly limited to the emissions that take place within the state's own territory.[179] While climate treaties do not require states to take measures aimed at reducing extraterritorial sources of GHG emissions, they also do not prevent them from doing so.[180] Overlaps and redundancies among national mitigation policies do not raise the same issues as double counting in the context of emission accounting. To the contrary, some studies suggest for instance that, in the fossil-fuel sector, demand- and supply-side mitigation policies may be mutually reinforcing: measures aimed at reducing reliance on fossil fuels are more attractive when fossil fuels are less abundant and more expensive.[181]

In practice, states have sought to limit and reduce GHG emissions notwithstanding whether these emissions fell within the scope of their national emission inventories. For instance, they have sought to prevent carbon leakage and to promote technology spillovers beyond their borders,[182] and they have taken measures to reduce emissions from road vehicles and from international transportation.[183] Several parties have committed to taking measures aimed at reducing emissions from international transportation[184] and emissions embedded in international trade.[185] Some authorities even suggested that states or corporations have not only

---

[176] See eg ibid Annex para 3.

[177] Thus, art 4(13) of the Paris Agreement establishes an accounting mechanism for the implementation of NDCs, whose scope, sector and gas coverage, and methodological assumptions may differ from those of the Enhanced Transparency Mechanism under art 13.

[178] Kyoto Protocol to the United Nations Framework Convention on Climate Change (adopted 11 December 1997, entered into force 16 February 2005) 2303 UNTS 162, art 3(1).

[179] United Nations Framework Convention on Climate Change (adopted 9 May 1992, entered into force 21 March 1994) 1771 UNTS 107 (UNFCCC) art 4(1)(b), 4(2)(a); Paris Agreement (adopted 12 December 2015, entered into force 4 November 2016) 3156 UNTS 79, art 4(2).

[180] See Benoit Mayer and Zhuoqi Ding, 'Climate Change Mitigation in the Aviation Sector: A Critical Overview of National and International Initiatives' (2023) 12 Transnational Environmental Law 14, 20.

[181] Asheim and others (n 125) 325 (noting that supply-side measures could enhance the efficacy of demand-side ones by avoiding a drop in the price of fossil fuels that would incentivize increased consumption by free-riders).

[182] Navroz K Dubash and others, 'National and Sub-national Policies and Institutions', in Shukla and others (n 5) 1355, 1393.

[183] See eg UNFCCC Subsidiary Body for Implementation, Report by the Secretariat, 'Compilation and synthesis of fifth biennial reports of Parties included in Annex I to the Convention', FCCC/SBI/2023/INF 7 (17 October 2023) para 31.

[184] EU, Updated NDC (16 October 2023) 12. See generally Air Transport Association of America (n 159) [134]–[135]; Joanne Scott, 'The Geographical Scope of the EU's Climate Responsibilities' (2015) 17 Cambridge Yearbook of European Legal Studies 92; Natalie L Dobson, Extraterritoriality and Climate Change Jurisdiction: Exploring EU Climate Protection under International Law (Bloomsbury 2021) 39–40.

[185] Switzerland, First NDC (9 December 2020) 1 ('compensation of imported "grey" emissions through additional emission reductions abroad'), 4 (support for 'the inclusion of international aviation

a right, but also a duty, to take measures aimed at reducing GHG emissions over-seas. Thus, the District Court of the Hague in *Milieudefensie v Royal Dutch Shell* interpreted the duty of care of the defendants towards Dutch citizens as requiring a reduction in all global carbon dioxide emissions under the corporation's control, including downstream emissions, notwithstanding their location.[186] Similarly, human rights treaty bodies have interpreted the obligation of states to protect human rights as an obligation to mitigate climate change, including by limiting their supply of fossil fuel to other countries.[187] And the European Court of Human Rights accepted that an applicant could invoke a state's control over emissions em-bedded in international trade when interpreting that state's obligation to mitigate climate change.[188]

3. Practical issues

Another common objection to consideration of GHG emissions by extrater-ritorial sources is that these emissions are difficult to assess. More specifically, it has been argued that extraterritorial emissions cannot be reliably predicted, that their significance cannot be properly determined, or that they cannot effectively be avoided or minimized. The following refutes these objections. While extrater-ritorial sources of GHG emissions may at times be more difficult to assess and to address, this does not justify the complete exclusion of these emissions from the scope of CA procedures.

a) Predicting overseas emissions
Practical concerns have been raised about the possibility of predicting indirect climate impacts from extraterritorial sources of emissions. In *Sierra Club Canada Foundation*, for instance, the Federal Court of Canada noted that the oil from the project 'may be used all over the world and for numerous purposes', thus 'elicit[ing] different GHG emissions'.[189] This observation led the Court to suggest that '[t]he

---

and navigation on the basis of existing and future internationally agreed rules applicable to all Parties'). See however Switzerland, NDC update (17 December 2021) (no longer including this language).

[186] *Milieudefensie v Royal Dutch Shell*, ECLI:NL:RBDHA:2021:5337 (District Court of The Hague, 26 May 2021) s 4.1.4.
[187] See eg UN Committee on Economic, Social and Cultural Rights, Concluding Observations, Fourth Periodic Report of Ecuador, E/C.12/ECU/CO/4 (14 November 2019) para 12; UN Committee on Economic, Social and Cultural Rights, Concluding Observations, Fourth Periodic Report of Argentina, E/C.12/ARG/CO/4 (1 November 2018) [13]–[14]; UN Committee on the Elimination of Discrimination against Women, Concluding Observations, Eighth Periodic Report of Australia, CEDAW/C/AUS/CO/8 (25 July 2018) para 30(c).
[188] *Verein Klimaseniorinnen Schweiz v Switzerland*, Application No 53600/20 (9 April 2024) <https://hudoc.echr.coe.int/?i=001-233206> [287].
[189] *Sierra Club Canada Foundation* (n 13) [67].

Agency would merely be speculating in considering the environmental effects of downstream GHG emissions'.[190] Similarly, the High Court of South Africa in *South Durban Community Environmental Alliance v Minister of Forestry, Fisheries and the Environment* found that the EA of a gas-fired power plant did not need to assess upstream emissions relating to the extraction and transportation of natural gas. As '[t]he specific source of the gas ha[d] not been identified' and as the project 'even may source fuel from outside the Republic', the Court reasoned, requiring an assessment of upstream emissions would 'create an almost impossible situation'.[191]

The fact that indirect climate impacts could occur overseas may be a complicating factor, but it is certainly not an absolute obstacle to the prediction of these emissions. As Michael Burger and Jessica Wentz noted, downstream GHG emissions from fossil-fuel extraction projects 'can be meaningfully evaluated even when there is considerable uncertainty about [their] exact timing and location'.[192] The end-use of fossil fuels is relatively predictable—the 'numerous purposes' of purchasing oil, for instance, often involve its combustion. The most accurate GHG emission predictions would build on more thorough information, for instance on the amount of methane leaked during the transportation, refining, and combustion of the fossil fuel. In practice, however, such information may be missing even for domestic sources of GHG emissions, and it may depend on circumstances that are expected to evolve during the implementation of the project. In such circumstances, GHG emission accounting methodologies provide ways to deal with incomplete information, for instance by relying on default assumptions.[193]

These observations lend support to recent US court decisions against agencies that found that they were unable to assess overseas GHG emissions. In particular, in *Center for Biological Diversity v Bernhardt*, the US Court of Appeal for the Ninth Circuit held that the Bureau of Ocean Energy Management 'acted arbitrarily and capriciously by failing to quantify' the emissions resulting from foreign oil consumption' or at least by failing to 'explain … thoroughly why it [could] not do so'.[194] To come to this conclusion, the Court rejected the Bureau's contention that it lacked 'reliable information on foreign emissions factors and consumption patterns'.[195] 'Various studies' had estimated the foreign oil consumption of the project at issue and, while the Bureau claimed that these studies relied on 'simplistic assumptions', the Court appeared to suggest that an approximate estimate of overseas emissions would be preferable to no estimate at all.[196] District Courts reached

[190] Ibid.

[191] *South Durban Community Environmental Alliance v Minister of Forestry, Fisheries and the Environment* [2022] ZAGPPHC 741 [29].

[192] Burger and Wentz (n 88) 114. See also *Finch* (SC) (n 13) [103] (observing that 'the effect of the combustion emissions on climate does not depend on when or where the combustion takes place').

[193] Jim Penman and others, 'Overview' in Eggleston and others (eds) (n 8) preliminary section, 8.

[194] *Center for Biological Diversity v Bernhardt* (n 13) 751.

[195] Ibid 737.

[196] Ibid 739.

similar conclusions in two subsequent cases, even when the agency had provided 'lengthier explanation'[197] or submitted that overseas emissions were 'too remote and speculative'.[198]

b) Assessing the significance of overseas emissions

Other practical concerns with an assessment of overseas GHG emissions relate to the determination of the significance and acceptability of the impact. While there are several ways of defining significance in relation to GHG emissions, it is generally accepted that the magnitude of the impact (eg the volume of emissions) must be assessed in the light of other factors.[199] Whether a given level of GHG emissions is significant and acceptable depends in part on the place where it takes place and the benefits drawn from the GHG-emitting activity. Exporting fossil fuels to one country could contribute to unjustifiable GHG emissions; exporting it to another country, however, could be instrumental to alleviating extreme poverty.

In this regard, climate treaties recognize that national mitigation action should reflect states' common but differentiated responsibilities and respective capabilities in the light of national circumstances,[200] with developed states 'taking the lead',[201] so as to respect the right of all states to sustainable development.[202] A likely implication is that the same amount of overseas emissions could be assessed differently depending on where it would occur, what country or population it would benefit, and how it would contribute to sustainable development. Concretely, one could argue that a fossil-fuel extraction project is more easily justified if it is aimed at consumption in a developing rather than developed country.

The Supreme Court of New Zealand raised such concerns in an obiter dictum in *West Coast ENT*. The Court noted that seeking to regulate overseas emissions 'poses ... practical and legal problems, for instance, in assessing the regulatory context in which the discharges will occur and the extent to which they will be balanced by mitigation or compensation measures'.[203] The Supreme Court thus echoed the High Court's assessment of 'the implausibility of applying sustainable management principles to overseas jurisdictions'[204]—a reference to principles that the Resource Management Act imposes on decision-makers as a touchstone to determine whether to approve a project.[205] The High Court noted that, 'in order to form an accurate view as to whether the overseas discharges are adverse and

---

[197] *Sovereign Inupiat for a Living Arctic v Bureau of Land Management* (D Alaska 2021) 555 F Supp 3d 739, 764.
[198] *Friends of the Earth v Haaland* (n 13) 139.
[199] See Chapter III, subsection B.2.
[200] See eg UNFCCC (n 179) arts 3(1), 4(1); Paris Agreement (n 179) arts 2(2), 4(3).
[201] Paris Agreement (n 179) art 4(4). See also UNFCCC (n 179) arts 3(1), 4(2)(a).
[202] UNFCCC (n 179) art 3(4); Paris Agreement (n 179) preamble para 12, arts 2(1), 4(1).
[203] *West Coast ENT* (n 12) [175].
[204] *Royal Forest and Bird Protection Society of New Zealand* (n 72) [53], cited with approval in *West Coast ENT* (n 12) [111].
[205] *Royal Forest and Bird Protection Society of New Zealand* (n 72) [24].

contrary to the sustainable management purpose, an authority would need to assess the management of those effects in those overseas jurisdictions', including by considering 'whether an end use of coal ... is subject to sustainable environmental policy, regulatory control, mitigation or compensation in [various] foreign jurisdictions'.[206]

The main concern in West Coast ENT, however, was with the role of the judge. As the lower court put it, the matter may not be 'properly justiciable ... with acceptable judicial method'.[207] An entirely different question is whether the assessment of the significance and acceptability of extraterritorial emissions can be made through political deliberation during the CA process itself, with the role of judges potentially limited to a procedural control. Put this way, there appears to be no obvious reason to exclude relevant information from public deliberations. Political decisions on whether to allow an activity could certainly benefit from whatever limited information is available on the extraterritorial effects of this activity, even if this information could support a range of views on the significance and acceptability of the proposed activity.

c) Avoiding or minimizing overseas emissions

A last potential concern about the assessment of extraterritorial emissions relates to the effectiveness of the assessment of extraterritorial sources of GHG emissions. By hypothesis, these extraterritorial emissions are not within the exclusive control of the state where the CA is carried out. Nonetheless, an agency's decision can often, if not avoid entirely, at least reduce sources of emissions overseas.

At least, it is always possible for the agency to reject the proposed activity entirely.[208] In Gloucester, for instance, the New South Wales Land and Environment Court upheld the government's rejection of a coal mine project on the ground that 'the negative impacts of the Project, including ... climate change impacts, outweigh the economic and other public benefits of the Project'.[209] Subsequently, the New South Wales Independent Planning Commission refused the Bylong coal mine project, partly out of concern for the significance of its direct and indirect GHG emissions.[210] Such decisions may not entirely avoid a source of GHG emissions, as power plants, for instance, may be able to find alternative sources of coal; yet, for the reasons explained above, this market substitution cannot be assumed to be perfect.[211] Blocking a project aimed at producing fossil fuels for exportation is generally an effective way to reduce extraterritorial sources of GHG emissions.

---

[206] Ibid [53], cited with approval in West Coast ENT (n 12) [111]. See also Greenpeace New Zealand Inc v Genesis Power Ltd [2008] NZSC 112, [2009] 1 NZLR 730 [12].

[207] Royal Forest and Bird Protection Society of New Zealand (n 72) [53], cited with approval in West Coast ENT (n 12) [111].

[208] See eg Waratah Coal (n 12) [1370]; Dine Citizens against Ruining Our Environment (n 12) 1217.

[209] Gloucester Resources (n 12) [688].

[210] Bylong Coal Project (n 36) [687]–[697]. See also ibid [817] (noting that the project proponent had 'not minimised Scope 1, 2 and 3 GHG emissions to the greatest extent practicable').

[211] See this chapter, subsection B.3.

Conditions imposed on the project proponent may sometimes be able to re-duce GHG emissions by extraterritorial sources. The New South Wales Land and Environment Court found in *Hunter Environment Lobby Inc v Minister for Planning* and again in *Mullaley Gas and Pipeline Accord Inc v Santos NSW (Eastern) Pty Ltd* that the agency's decision not to impose conditions to reduce scope 3 emissions was reasonable because the project proponent did not have suf-ficient control over these emissions.[212] However, the Court also noted in *Mullaley Gas* that there are 'circumstances where the proponent of the development might have a sufficient degree of control over the indirect emissions' for conditions to be imposed on the project to minimize these emissions.[213] For instance, the pro-ponent could be required to export coal only to power plants with particular char-acteristics and processes in place.[214]

Discussing national measures intended to reduce extraterritorial emissions, Joanne Scott introduced a distinction between transaction-specific measures and country-specific measures.[215] Transaction-specific measures could include condi-tions concerning the companies with which the project proponent could trade. For instance, a coal mine could be required to sell coal exclusively to highly efficient power plants, thus ensuring that its coal is used to produce as much electricity as possible, which would reduce the need for other sources of electricity. Country-specific measures, on the other hand, could require the project proponent to trade only with specified countries. These measures could seek to ensure that the pro-posed activity does not support activities in countries that do not cooperate in global efforts on the mitigation of climate change. Further, the approval of new coal mines could be subjected to a condition preventing the exportation of coal for power generation to high-income countries, on the ground that such countries should shift to other sources of energy.

At present, the rare measures that are imposed on proposed activities with the aim of reducing extraterritorial sources of GHG emissions tend to be rather sym-bolic. For instance, the New South Wales Independent Planning Commission required that the proponent of the United Wambo coal mine adopts an 'Export Management Plan ... to ensure that all reasonable and feasible measures are adopted ... to minimise [scope 3 emissions] to the greatest extent practicable.[216] The core condition that this plan must implement is that the coal should only be exported to countries that are Parties to the Paris Agreement or are considered by the Planning Secretary as having similar mitigation policies.[217] These conditions

---

[212] See *Hunter* (n 12) [94]; *Mullaley Gas* (n 13) [110]–[111], [118].

[213] See *Mullaley Gas* (n 13) [107]. See also *Waratah Coal* (n 12) [93].

[214] *Mullaley Gas* (n 13) [105]–[106].

[215] Scott (n 171) 170. See also Christina Voigt, 'The First Climate Judgment before the Norwegian Supreme Court: Aligning Law with Politics' (2021) 33 Journal of Environmental Law 697, 706.

[216] *United Wambo* (n 36) 18 (condition B32).

[217] Ibid.

do not appear particularly onerous given that the Paris Agreement has achieved quasi-universal membership. Nonetheless, national practice in this regard could develop in the years to come in ways that may reflect an emerging agreement on the legality and acceptability of particular conducts. For instance, the exportation of fossil fuels could be limited to countries that have achieved their past nationally determined contribution (NDC) under the Paris Agreement.

D. Epistemological Objections

This section considers a third type of objections to the assessment of indirect climate impacts, namely, objections that are based on the practical difficulty of assessing these impacts. As noted above, courts have found that assessing scope 2 and scope 3 emissions 'would not be practicable',[218] and that it 'was not possible to draw robust conclusions',[219] because these emissions depend on 'a range of variables'.[220] Yet, excluding potentially major effects from the scope of the assessment could mislead decision-makers about the merits of the proposed activity. It is argued that the goal of informing decisions is better served with an assessment of all potentially significant effects, whether direct or indirect, even when this assessment can only be qualitative and uncertain: an imprecise prediction of indirect climate impacts, with a prominent acknowledgement of uncertainties, is preferable to no assessment at all. The first subsection identifies some of the main sources of uncertainty affecting the prediction of indirect climate impacts, whereas the second subsection surveys the methodologies that can be used to deal with these sources of uncertainty.

1. Sources of uncertainty

An activity's indirect climate impacts depend, in part, on decisions by third actors. For instance, an activity's scope 2 emissions depend not only on the level of electricity consumption, but also on the way public utility companies generate electricity at the relevant place and time.[221] Similarly, the downstream (scope 3) emissions of

---

[218] *Greenpeace Ltd v Advocate General* (n 13) [68].
[219] *Australian Conservation Foundation* (2016) (n 47) [140].
[220] *Adani Mining*, Statement of Reasons (n 47) [140]. See also *An Taisce* (n 9) [110].
[221] Given the heterogeneity of energy sources, the marginal GHG intensity of the last unit of electricity generated may differ significantly from the average GHG intensity. Further, the average GHG intensity of electricity depends on the time of the day, with less efficient plants being used typically for peak power generation. The GHG Protocol recommends the calculation of scope 2 emissions based on annual average emission intensity. See GHG Protocol, 'GHG Protocol Scope 2 Guidance' (n 3) 53. While this may be justified by practical concerns of data availability, this approach fails to account for certain measures that could be taken to reduce a project's GHG emissions, such as the timing electricity consumption in ways that may reduce GHG emissions.

a natural gas production project depend in part on the level of fugitive emissions during the transportation of the fuel.[222] Overall, the existence of a market substitution effect implies that not all of the direct and indirect emissions associated with a project are necessarily additional to emissions that would occur without the project.

In turn, relevant decisions by third actors are influenced by various factors, such as regulation, technology development, and market conditions. For instance, the net climate effect of the combustion of fossil fuels may be increased by measures aimed at limiting local air pollution by aerosols, if these measures do not also reduce GHG emissions, because aerosols tend to have a climate cooling effect.[223] Further, new technologies may make some activities more or less efficient. For instance, more efficient power plants would generate more electricity and, possibly, substitute to other sources of GHG emissions.[224] Lastly, market conditions may affect the way activities substitute to one another. For instance, whether a new coal plant would substitute to coal- or natural gas-fired power plants may depend on the relative price of the two fuels.[225]

Courts have often struggled with these sources of uncertainties. In *High Country Conservation Advocates*, the District Court for the District of Colorado rejected the agencies' 'contention that new technology might reduce carbon emissions from future coal combustion', noting that the agencies could not 'rely on unsupported assumptions that future mitigation technologies will be adopted'.[226] Yet, the contrary assumption, namely that new technologies will not be deployed over a period of several decades, appears just as questionable. More generally, it is unclear how an agency could rely on well-supported assumptions in the face of deep uncertainty.

2. Dealing with uncertainty

The concerns with an assessment of indirect climate impacts often relates to a misunderstanding of the nature of EA. As a tool for the prediction of potential future impacts, EA constantly deals with uncertainty.[227] There would hardly be a need for thorough scientific study if all environmental impacts were self-evident. EA studies deal with uncertainty in two ways: by reducing it as far as possible, and

---

[222] See n 6 above. But see *Finch* (SC) (n 13) [2], asserting 'virtual certainty that, once oil has been extracted from the ground, the carbon contained within it will sooner or later be released into the atmosphere as carbon dioxide', thus ignoring the possibility that carbon be released in other forms (eg as methane) or stored in solid forms (eg landfilled plastics).

[223] See Johannes Quaas and others, 'Robust Evidence for Reversal of the Trend in Aerosol Effective Climate Forcing' (2022) 22 Atmospheric Chemistry & Physics 1221; Paola A Arias and others, 'Technical Summary', in Valerie Masson-Delmotte and others (eds) (n 6) 35, 42, 92.

[224] *Australian Conservation Foundation* (2016) (n 47) [138].

[225] See text above at n 141.

[226] *High Country Conservation Advocates* (n 12) 1197.

[227] Glasson and Therivel (n 1) 122.

then by disclosing remaining sources of uncertainty.[228] EA aims at supplying the best information possible under existing constraints, but it cannot provide perfect information.

Various methodological tools can guide the prediction of climate impacts in this regard. These include, for instance, the Guidelines for National Greenhouse Gas Inventories of the Intergovernmental Panel on Climate Change (IPCC)[229] and several guidelines and guidance documents developed by the GHG Protocol, for instance on project accounting[230] and on the calculation of scope 2 and scope 3 emissions.[231] These tools define principles conducive to relatively complete, precise, and accurate prediction of climate impacts.[232] For instance, the GHG Protocol for Project Accounting recommends that GHG projects '[c]onsider all relevant information that may affect the accounting and quantification' of GHG emissions[233] and '[r]educe uncertainties as much as is practical'.[234] Further, these methodologies suggest multiple ways to overcome practical issues, such as a lack of data, and to deal with uncertainty in GHG emission assessments.[235]

The principles contained in these GHG emission accounting and prediction tools are often similar to those emerging from EA law and practice. For instance, EA frameworks generally reflect the principle that '[e]nvironmental effects are not to be disregarded merely because they are difficult to identify or quantify'.[236] As the New South Wales Land and Environment Court noted in *Gray*, '[t]he fact that it is difficult to quantify an impact with precision does not mean it should not be done'.[237] Decision-makers are better informed with an approximative assessment of indirect climate impacts than with no assessment at all.

US law and practice illustrates how these principles can guide the assessment of indirect climate impacts. Interpreting the requirement for agencies to assess those '[i]ndirect effects' that 'are ... reasonably foreseeable',[238] the Court of Appeal for the District of Columbia Circuit observed that agencies 'need not foresee the unforeseeable', but may need to engage in 'some degree of forecasting'.[239] In another

---

[228] Ibid 122–24.

[229] See eg Eggleston and others (eds) (n 8).

[230] GHG Protocol, 'The GHG Protocol for Project Accounting' (n 103).

[231] GHG Protocol, 'GHG Protocol Scope 2 Guidance' (n 3); The Greenhouse Gas Protocol, 'Technical Guidance for Calculating Scope 3 Emissions' (World Resources Institute and World Business Council for Sustainable Development 2013) <https://ghgprotocol.org/scope-3-calculation-guidance-2>.

[232] See eg Decision 18/CMA.1 (n 175) Annex para 3(d); Justin Goodwin and others, 'Approaches to Data Collection' in Eggleston and others (eds) (n 8) vol 1, ch 2, 11, 37.

[233] GHG Protocol, 'The GHG Protocol for Project Accounting' (n 103) 23.

[234] Ibid 24.

[235] See eg William Irving, 'Time Series Consistency' in Eggleston and others (eds) (n 8) vol 1, ch 5, 8–14.

[236] Murray Raff, 'Ten Principles of Quality in Environmental Impact Assessment' (1997) 14 Environmental and Planning Law Journal 207, 209.

[237] *Gray* (n 12) [138].

[238] 40 CFR (2024) § 1508.8.

[239] *Scientists' Institute for Public Information Inc v Atomic Energy Commission* (DC Cir 1973) 481 F 2d 1079, 1092.

case, the Court defined 'reasonably foreseeable' effects as those that are 'suffi-ciently likely to occur that a person of ordinary prudence would take [them] into account in reaching a decision'.[240] The Supreme Court added that agencies must be guided by a 'rule of reason' when deciding 'whether and to what extent' to as-sess specific effects 'based on the usefulness of any new potential information to the decisionmaking process'.[241] US courts have also acknowledged that, while a quantitative assessment of indirect GHG emissions is desirable, it is not always feasible, and that agencies could justify a purely qualitative analysis under some circumstances.[242]

GHG accounting and prediction methodologies suggest a pragmatic approach whereby more attention is given to potentially larger sources of GHG emissions. For instance, the IPCC recommends that states use more sophisticated ('higher tier') assessment methodologies for 'key' categories of emissions and sinks, while possibly relying on more rudimentary methodologies to estimate other sources of emissions.[243] Similarly, the GHG Protocol notes that a GHG assessment should as-sess '*significant* secondary effects', particularly 'those that would negate the project activity's primary effect'.[244] Consistently, NEPA requires that environmental im-pact statements 'discuss impacts in proportion to their significance',[245] and courts have held that an agency is 'only required to discuss the combustion-related ef-fects to the extent necessary under the circumstances for the evaluation of the pro-ject'.[246] On the one hand, a succinct analysis is sufficient when the indirect effects 'are so remote as to be indiscernible' or 'are expected to be of such small scale that they would have little bearing on climate change'.[247] On the other hand, 'sparse discussions of downstream GHG emissions are insufficient' in relation to projects whose 'entire purpose ... is to generate a greater supply of oil and gas for down-stream use'.[248]

[240] *Sierra Club v Federal Energy Regulatory Commission* (n 14) 1371.

[241] *Department of Transportation v Public Citizen* (2004) 541 US 752, 767.

[242] See *Sierra Club v US Department of Energy* (DC Cir 2017) 867 F 3d 189, 202; *Sierra Club v Federal Energy Regulatory Commission* (n 14) 1374; *WildEarth Guardians v Zinke* (D DC 2019) 368 F Supp 3d 41, 74–75; *Center for Biological Diversity v Bernhardt* (n 13) 740; *Utah Physicians for a Healthy Environment v US Bureau of Land Management* (D Utah 2021) 528 F Supp 3d 1222.

[243] Anke Herold and others, 'Methodological Choice and Identification of Key Categories' in Eggleston and others (eds) (n 8) vol 1, ch 4, 4. Each 'tier' represents 'a level of methodological com-plexity'. See Rypdal and others (n 173) 6. See also Decision 18/CMA.1 (n 175) Annex para 21.

[244] GHG Protocol, 'The GHG Protocol for Project Accounting' (n 103) 34 (emphasis added). See also GHG Protocol, 'Technical Guidance for Calculating Scope 3 Emissions' (n 231) 11–12.

[245] 40 CFR (2024) § 1502.2(b).

[246] *Dine Citizens against Ruining Our Environment* (n 12) 1214 (references omitted).

[247] *Swomley v Schroyer* (D Colorado 2020) 484 F Supp 3d 970, 976, affirmed (10th Cir 2021) 2021 WL 4810161. See also *Hapner v Tidwell* (9th Cir 2010) 621 F 3d 1239, 1245; *Earth Island Institute v Gibson* (ED California 2011) 834 F Supp 2d 979, 990; *League of Wilderness Defenders v Connaughton* (D Oregon 17 July 2013) 3:12–cv–02271–HZ, 2013 WL 3776305, at *14. But see *Center for Biological Diversity v US Forest Service* (D Montana 2023) 687 F Supp 3d 1053, 1065 (concerning a larger project).

[248] *WildEarth Guardians v Zinke* (n 242) 75.

GHG emission accounting and prediction tools further highlight the importance of transparency, including for what concerns the uncertainty associated with any impact assessment.[249] Thus, the GHG Protocol emphasizes the need for GHG assessments to '[p]rovide clear and sufficient information ... to assess ... credibility and reliability', including the identification of '[s]pecific exclusions or inclusions' and the explanation of 'assumptions' informing the assessment.[250] Likewise, the IPCC notes that '[u]ncertainty estimates are an essential element of a complete inventory of greenhouse gas emissions and removals'.[251] Consistently, EA practitioners and scholars have long recognized that 'the fundamental challenge' in developing useful EA procedures 'is not simply to better mobilize known information', but rather 'to cope with the uncertain and the unexpected' or 'to plan in the face of the unknown'.[252] They have also called for 'improving the communication of uncertainty in EIA predictions' in order to 'improve EIA as a decision-aiding tool'.[253]

These considerations are also reflected in some EA practice. For instance, the Court of Appeal for the District of Columbia observed that 'one of the functions of a NEPA statement is to indicate the extent to which environmental effects are essentially unknown'.[254] In another case, the Court of Appeal for the District of Columbia noted that 'the effects of assumptions on estimates can be checked by disclosing those assumptions so that readers can take the resulting estimates with the appropriate amount of salt'.[255] Consistently, the CEQ recommends that CAs 'disclose ... any assumptions used in the analysis and explain any uncertainty' when quantifying direct and indirect GHG effects.[256] In *Gray*, the New South Wales Land and Environment Court observed more hesitantly that the limitations of the methodologies used to predict scope 3 emissions 'can be ... taken into account in the environmental assessment process'.[257]

---

[249] See eg Raff (n 236) 217.

[250] GHG Protocol, 'The GHG Protocol for Project Accounting' (n 103) 23.

[251] Christopher Frey and others, 'Uncertainties' in Eggleston and others (eds) (n 8) vol 1, ch 3, 6. See also Decision 18/CMA.1 (n 175) Annex para 29 (requiring each Party to 'quantitatively estimate and qualitatively discuss the uncertainty of' their estimates).

[252] CS Holling, *Adaptive Environmental Assessment and Management* (Wiley 1978) 7.

[253] Aud Tenney, Jens Kværner, and Karl Idar Gjerstad, 'Uncertainty in Environmental Impact Assessment Predictions: The Need for Better Communication and More Transparency' (2006) 24 Impact Assessment & Project Appraisal 45, 45.

[254] *Scientists' Institute for Public Information* (n 239) 1092.

[255] *Sierra Club v Federal Energy Regulatory Commission* (n 14) 1374.

[256] CEQ, 2023 Guidance (n 29) 1204.

[257] *Gray* (n 12) [138].

Conclusion

This chapter has observed a variety of approaches to the assessment of indirect climate impacts. The debate tends to focus on three main issues: the potential for market substitution, the relevance of overseas emissions, and the practical difficulty of ascertaining these indirect climate impacts. While some market substitution is likely to take place, the economic laws of supply and demand suggest that increased supply in fossil fuel will generally result in increased consumption of fossil fuels, thus also in increased GHG emissions. States have both the capacity and the right—perhaps even the obligation—to seek to reduce indirect GHG emissions from extraterritorial sources. And there are practical ways, found in GHG emission account and prediction tools and also in EA law and practice, to address challenges to our capacity to measures or predict indirect climate impacts, in particular through a prominent acknowledgment of uncertainty in CA studies.

The chapter thus argued that significant indirect climate impacts should be assessed before a decision can be made on a proposed activity. Practical challenges in assessing these indirect climate impacts are not a proper justification for a complete exclusion of these impacts from the scope of the EA: even when these impacts cannot be quantified, they should be acknowledged in order to inform decisionmakers. For what concerns fossil-fuel projects, in particular, the regulation of supply can be a useful complement to the regulation of demand. On the other hand, while the debates tend to focus on the scoping of CA, attention also needs to be given to decisions at the screening stage. Activities with little or no direct environmental impacts, such as data centres, tend to be excluded from the scope of EA in spite of their significant indirect climate impact, resulting in a missed opportunity for public scrutiny on activities responsible for large amounts of GHG emissions.

V

Achieving Better Decisions

Introduction

This chapter turns back to the fundamental question of whether environmental assessment (EA) should be used as a tool for climate change mitigation. Climate assessment (CA) could be expected to lead to more 'pluralistic' and legitimate decision-making processes[1] and ensure that decisions are better informed, hence more likely to be in line with a community's policy objectives, including the mitigation of climate change.[2] On the other hand, CA could impose additional cost and burden on project proponents, thus potentially hindering activities that could bring about various economic, social, and environmental benefits. Whether the benefits of CA justify its cost for society largely depends on CA's ability to improve decisions.

The present chapter approaches this question by exploring three related issues: the design of the decision-making process, the content of the decisions adopted as part of a CA, and the effect of these decisions. Focusing on the decision-making process, Section A explores the respective roles of the public, decision-makers, and courts. It shows that the public can play a useful and legitimate role despite the fact that no individual or population would be directly affected by the climate impact of a proposed activity. Further, the section observes a wide variety of arrangements on the respective roles of administrative, political, and judicial authorities, reflecting differences in legal and political culture, with implications for the nature of the decision. An extended role of political institutions is justified by CA decisions' extensive reliance on value-based judgements, but, it is argued, this should not exclude a role for the courts.

Section B reviews the different substantive decisions to which CAs can lead. It shows that CAs rarely result in the rejection of proposed activities, but that the proposed activities that are approved are often subjected to conditions that aim at reducing climate impacts. For instance, these conditions may impose a limit on a project's greenhouse gas (GHG) emissions or promote the use of cleaner technologies or processes, or the adoption of a GHG-emission reduction plan. These

---

[1] Neil Craik, 'The Assessment of Environmental Impact', in Emma Lees and Jorge Vinuales (eds), *The Oxford Handbook of Comparative Environmental Law* (OUP 2019) 876, 882.
[2] Ibid 881; Jane Holder, *Environmental Assessment: The Regulation of Decision Making* (OUP 2005) 22.

Environmental Assessment as a Tool for Climate Change Mitigation. Benoit Mayer, Oxford University Press.
© Benoit Mayer 2024. DOI: 10.1093/oso/9780198939184.003.0005

conditions need to be designed with the view of facilitating the monitoring of their implementation.

Lastly, Section C assesses the real-world effects of these decisions. Counterintuitively, this section argues that the rejection of a proposed GHG-emitting activity may not achieve as much climate benefit as a decision to approve the activity with conditions aimed at reducing its climate impact. For one, most rejection decisions are based on procedural flaws that can be addressed in a subsequent CA procedure. Overall, any decision that does prevent the implementation of the proposed activity is prone to some degree of carbon leakage or political push-back. By contrast, an approval decision with conditions may contribute to changing the way activities are implemented, for instance incentivizing the development and deployment of new technologies or by testing innovative legal standards that could then be imposed on other projects through regulation.

A. Decision-Making Process

This section considers how substantive CA decisions are arrived at. The first subsection explores the role of public participation, the second subsection identifies the primary decision-making authorities, and the third subsection discusses the involvement of courts in these decision-making processes.

1. Public participation

Reflecting on the recognition of public participation as a central component of EA, this section identifies the particularities of CA, namely the absence of a public that would be directly affected by the climate impact of a proposed activity. In spite of this, it is argued that the public can play a legitimate and useful role in expressing a community's concerns for climate impacts and suggesting ways to reduce these impacts.

a) The soul of EA
With the adoption of the Rio Declaration, states have recognized that '[e]nvironmental issues are best handled with the participation of all concerned citizens'.[3] A similar principle is reflected in the Aarhus Convention[4] and the Escazu

---

[3] Rio Declaration on Environment and Development, A/CONF.151/26 (Vol I) (12 August 1992), principle 10.

[4] Convention on Access to Information, Public Participation in Decision-Making and Access to Justice in Environmental Matters (adopted 25 June 1998, entered into force 30 October 2001) 2161 UNTS 447 (Aarhus Convention), preamble para 10. See also Aarhus Convention Compliance Committee, Findings and recommendations with regard to Communication ACCC/C/2008/24 concerning compliance by Spain, ECE/MP.PP/C.1/2009/8/Add.1 (8 February 2011) [82]

Agreement.[5] As conceptualized by the Aarhus Convention, public participation in environmental matters includes not only the possibility to submit views on a proposed activity,[6] but also access to relevant information[7] and to judicial review procedures 'to challenge the substantive and procedural legality' of relevant decisions.[8] In other instruments, states have also recognized public participation as an essential feature of efforts to address climate change.[9]

Neil Craik has characterized public participation is the 'soul' of EA.[10] At least, as Anne Glucker put it, public participation is largely 'considered as an integral part of the assessment procedure'.[11] Statutory EA instruments require public participation at various stages of the EA procedure, including screening, scoping, and appraisal of the EA report.[12] Consistently, the Espoo Convention requires the establishment of an environmental impact assessment (EIA) procedure 'that permits public participation',[13] while the Kiev Protocol mandates 'early, timely and effective opportunities for public participation' in strategic environmental assessment (SEA) procedures.[14]

Public participation in environmental matters may serve a large variety of intrinsic and instrumental goals. First, public participation has been presented as having 'value in its own right'[15] and being justified 'quite independently of whether

(noting that '[t]he Convention does not make the EIA a mandatory part of public participation; it only requires that when public participation is provided for under an EIA procedure in accordance with national legislation ... such public participation must apply the provisions of its article 6').

[5] Regional Agreement on Access to Information, Public Participation and Justice in Environmental Matters in Latin America and the Caribbean (adopted 4 March 2018, entered into force 22 April 2021) <www.cepal.org/en/escazuagreement> (Escazu Agreement).

[6] Aarhus Convention (n 4) art 6(7).

[7] Ibid, art 4(1).

[8] Ibid, art 9(2).

[9] See eg United Nations Framework Convention on Climate Change (adopted 9 May 1992, entered into force 21 March 1994) 1771 UNTS 107 (UNFCCC) arts 4(1)(i), 6(a)(iii); Paris Agreement (adopted 12 December 2015, entered into force 4 November 2016) 3156 UNTS 79, preamble para 15, art 12.

[10] Neil Craik, *The International Law of Environmental Impact Assessment: Process, Substance and Integration* (CUP 2008) 31.

[11] Anne N Glucker and others, 'Public Participation in Environmental Impact Assessment: Why, Who and How?' (2013) 43 Environmental Impact Assessment Review 104, 104.

[12] See eg Directive 2011/92/EU of the European Parliament and of the Council of 13 December 2011 on the assessment of the effects of certain public and private projects on the environment, [2012] OJ L 26/1, consolidated as of 15 May 2014, arts 4(5), 6–7; 42 USC (2024) § 4332(C); 40 CFR (2024) § 6.203; Environmental Assessment Act (环境影响评价法) of 28 October 2002 (China) arts 5, 11, 21; World Bank, 'Environmental and Social Framework' (2016) 100.

[13] Convention on Environmental Impact Assessment in a Transboundary Context (adopted on 1 March 1991, entered into force 10 September 1997) 1989 UNTS 309 (Espoo Convention) art 2(2).

[14] Protocol on Strategic Environmental Assessment to the Convention on Environmental Impact Assessment in a Transboundary Context (adopted on 21 May 2003, entered into force 11 July 2010) 2685 UNTS 140 (Kiev Protocol) art 8(1).

[15] Ciaran O'Faircheallaigh, 'Public Participation and Environmental Impact Assessment: Purposes, Implications, and Lessons for Public Policy Making' (2010) 30 Environmental Impact Assessment Review 19, 22.

the decision-maker believes that the citizen will be able to enhance the process or add anything of value'.[16] Thus, public participation can aim to buttress the 'transparency'[17] and 'legitimacy'[18] of the decision-making processes as well as their 'democratic'[19] and 'inclusive'[20] nature, or to grant a 'social licence to operate' to an activity.[21]

Second, scholars have suggested that EIA may 'empower ... local people' and 'enhance ... the position of ... disadvantaged or marginalised members of society'.[22] However, there is also a well-known risk that public participation may 'represent the views of the most vocal interest groups rather than of the general public'.[23] As Ciaran O'Faircheallaigh noted, '[t]he powerless in society are in fact the least likely to participate in EIA, both because they lack the resources to do so and often find the processes involved alien and intimidating'.[24] To mitigate this risk, EA instruments and agencies can strive to encourage participation by, or even to actively reach out to, 'groups that are hard to reach'[25] or have historically been disadvantaged,[26] thus seeking to 'alter the power relationships in existing policy-making processes'.[27]

Third, public participation has also been justified as a way 'to improve the quality ... of environmental assessments'[28] and to 'render EIA more effective'.[29] In particular, public participation has been presented as a tool 'to ensure that all

---

[16] Jenny Steele, 'Participation and Deliberation in Environmental Law: A Problem-Solving Approach' (2001) 21 Oxford Journal of Legal Studies 415, 420. See also *R (Finch) v Surrey County Council* [2024] UKSC 20, [2024] All ER (D) 71 (Jun) [105].

[17] Directive 2011/92 (n 12) preamble para 16.

[18] Paul C Stern and others, *Public Participation in Environmental Assessment and Decision Making* (National Academies Press 2008) 1.

[19] Glucker and others (n 11) 104; *Concerned Citizens of Costa Mesa Inc v 32nd District Agricultural Association* (1986) 42 Cal 3d 929. See also UNECE, 'Good Practice Recommendations on Public Participation in Strategic Environmental Assessment' (2016) 5 (suggesting that public participation contributes 'towards greater environmental democracy'); Chiara Armeni and Maria Lee, 'Participation in a Time of Climate Crisis' (2021) 48 Journal of Law & Society 549, 560 ('Public participation is necessary ... because we have a right to be involved in decisions about our world').

[20] *Berkeley v Secretary of State for the Environment* [2001] 2 AC 603 (HL).

[21] Anna-Sofie Hurup Olsen and Anne Merrild Hansen, 'Perceptions of Public Participation in Impact Assessment: A Study of Offshore Oil Exploration in Greenland' (2014) 32 Impact Assessment & Project Appraisal 72, 75.

[22] Frank Vanclay, 'International Principles for Social Impact Assessment' (2003) 21 Impact Assessment & Project Appraisal 5, 7. See also Sherry R Arnstein, 'A Ladder of Citizen Participation' (1969) 35 Journal of the American Planning Association 216, 216.

[23] John Glasson and Riki Therivel, *Introduction to Environmental Impact Assessment* (5th edn, Routledge 2019) 150.

[24] See O'Faircheallaigh (n 15) 23.

[25] UNECE, 'Good Practice Recommendations' (n 19) 6 (box 1). The Recommendations were endorsed in Meeting of the Parties to the Protocol on Strategic Environmental Assessment, Decision II/8, 'Good Practice Recommendations on Public Participation in Strategic Environmental Assessment', ECE/MP.EIA/20/Add.2–ECE/MP.EIA/SEA/4/Add.2 (15 July 2014) 12.

[26] Environment Protection Act 2019 (NT) s 43(b).

[27] Richard K Morgan, 'Environmental Impact Assessment: The State of the Art' (2012) 30 Impact Assessment & Project Appraisal 5, 10.

[28] Stern and others (n 18) 1.

[29] Glucker and others (n 11) 104.

relevant information, including input from those affected, is available so that the decision-maker can make the most informed and well-considered decision'.[30] This argument is premised 'on a belief that citizens can make important contributions to environmental protection'.[31] This belief, in turn, may be justified in the light of 'the dispersal and fragmentation of knowledge and information'[32] and of the fact that the information gathered by project proponents and national agencies 'may well be deficient'.[33] More pragmatically, it has been argued that (some forms of) participation reduces risks of conflicts and protests that may 'delay or hinder project implementation',[34] and that it may even 'generate a sense of "environmental citizenship" that, in turn, might catalyse behavioural change'.[35]

In practice, the influence of public participation on decisions varies,[36] for reasons relating to the design and implementation of EA instruments and possibly also to the political culture of the community where these instruments are implemented.[37] In some contexts, studies have shown that the public had little direct influence on final decisions.[38] Some authorities certainly approach public consultation as a mere formality that is unlikely to have any influence on the final decision-making.[39] Alternatively, efforts to reach the most vulnerable populations may fail.[40] Thus, Meinhard Doelle and Adebayo Majekolagbe noted that, while EA

[30] O'Faircheallaigh (n 15) 21.

[31] Daniel P Selmi, 'The Judicial Development of the California Environmental Quality Act' (1984) 18 UC Davis Law Review 197, 215.

[32] Armeni and Lee (n 19) 560.

[33] R (Blewett) v Derbyshire County Council [2003] EWHC 2775 (Admin), [2003] All ER (D) 332 (Nov) [41].

[34] Glucker and others (n 11) 108. See also Stephen Connelly and Tim Richardson, 'Value-Driven SEA: Time for an Environmental Justice Perspective?' (2005) 25 Environmental Impact Assessment Review 391, 398; Anne Shepherd and Christi Bowler, 'Beyond the Requirements: Improving Public Participation in EIA' (1997) 40 Journal of Environmental Planning & Management 725, 729–30. Participation through judicial proceedings, on the other hand, can significantly delay project implementation.

[35] Armeni and Lee (n 19) 560.

[36] Arnstein (n 22).

[37] See eg Thabang Maphanga and others, 'The State of Public Participation in the EIA Process and its Role in South Africa: A case of Xolobeni' (2023) 105 South African Geographical Journal 277, 294–96; Cristiane B Dias, Getting Heard but not Listened to: An Analysis of Public Participation in Environmental Impact Assessment (EIA) in Brazil (Lexington 2021) 51; Akeem M Lawal, Stefan Bouzarovski, and Julian Clark, 'Public Participation in EIA: The Case of West African Gas Pipeline and Tank Farm Projects in Nigeria' (2013) 31 Impact Assessment & Project Appraisal 226, 230.

[38] See eg Maria Lee and others, 'Public Participation and Climate Change Infrastructure' (2013) 25 Journal of Environmental Law 33; Yvonne Rydin, Maria Lee, and Simon J Lock, 'Public Engagement in Decision-Making on Major Wind Energy Projects' (2015) 27 Journal of Environmental Law 139; Yvonne Rydin and others, 'Local Voices on Renewable Energy Projects: The Performative Role of the Regulatory Process for Major Offshore Infrastructure in England and Wales' (2018) 23 Local Environment 565.

[39] See eg Barbara Pozzoni and Nalini Kumar, 'A Review of the Literature on Participatory Approaches to Local Development for an Evaluation of the Effectiveness of World Bank Support for Community Based and Driven Development Approaches' (World Bank Operations Evaluation Department 2005); Azizan Marzuki, 'A Review on Public Participation in Environmental Impact Assessment in Malaysia' (2009) 3 Theoretical & Empirical Researches in Urban Management 126, 134.

[40] See Brendan FD Barrett and Riki Therivel, Environmental Policy and Impact Assessment in Japan (2nd edn, Routledge 2019) 118–19.

may give a 'say' to powerless populations, often, 'powerful interests will have their way'.[41] Practical barriers to participation range from the language and medium of consultation (eg written form inaccessible to illiterate populations) to the sheer technicality of the documentation.[42] Participation can further be limited by a poor public knowledge of, and interest in, EA processes.[43]

b) Grounds for participation in CA

Across the various justifications for public participation in EA, a certain emphasis tends to be attached to the populations affected by the proposed activity. One goal of public participation, according to the European Court of Justice, is to ensure that 'the environmental concerns of … the persons affected … are genuinely taken into account'.[44] Likewise, Paul Burton has argued that 'everyone affected by a decision' should be able to participate in the making of that decision.[45] Doelle and Majekolagbe added that a purely expert-led process 'becomes less tenable when impacts with more direct implications for people and communities are being considered'.[46]

This emphasis on affected individuals and communities is widely reflected in EA law and practice. The US National Environmental Policy Act (NEPA), for instance, directs agencies to request comments from 'State, Tribal, or local governments that may be affected by the proposed action'.[47] India's 2006 EIA Notification

[41] Meinhard Doelle and Adebayo Majekolagbe, 'Meaningful Public Engagement and the Integration of Climate Considerations into Impact Assessment' (2023) 101 Environmental Impact Assessment Review 7.

[42] See Juan Palerm and Carla Aceves, 'Environmental Impact Assessment in Mexico: An Analysis from a "A Consolidating Democracy" Perspective' (2004) 22 Impact Assessment & Project Appraisal 99, 103; Philip M Omenge, Stanley M Makindi, and Gilbert O Obwoyere, 'Public Participation in Environmental Impact Assessment and its Substantive Contribution to Environmental Risk Management: Insights from EIA Practitioners and Other Stakeholders in Kenya's Renewable Energy Sub-Sector' (2019) 8 Energy & Sustainability 133, 141; Geoffrey V Hurley, 'Environmental Assessment Process Needs Overhaul', Saltwire (21 January 2023) < www.saltwire.com/atlantic-canada/opinion/commentary-environmental-assessment-process-needs-overhaul-100816550/>; Mohammad A Alomari and Raphael J Heffron, 'Environmental Impact Assessment: A Middle Eastern Experience' (2021) 33 Journal of Environmental Law 309, 327.

[43] See Hans Wiklund, 'Why High Participatory Ideals Fail in Practice: A Bottom-Up Approach to Public Nonparticipation in EIA' (2011) 13 Journal of Environmental Assessment Policy & Management 159, 167; Nicola Hartley and Christopher Wood, 'Public Participation in Environmental Impact Assessment: Implementing the Aarhus Convention' (2005) 25 Environmental Impact Assessment Review 319, 333; Carla Lostarnau and others, 'Stakeholder Participation within the Public Environmental System in Chile: Major gaps between Theory and Practice' (2011) 92 Journal of Environmental Management 2470, 2472; Mariska Wouters, Ned Hardie-Boys, and Carla Wilson, Evaluating Public Input in National Park Management Plan Reviews: Facilitators and Barriers to Meaningful Participation in Statutory Processes (New Zealand Department of Conservation 2011) 17.

[44] C-474/10, Department of the Environment for Northern Ireland v Seaport [2011] ECR I-10227, Opinion of AG Bot [29]. See also Stern and others (n 18) 15.

[45] Paul Burton, 'Power to the People? How to Judge Public Participation' (2004) 19 Local Economy 193, 194.

[46] Doelle and Majekolagbe (n 41) 2.

[47] 40 CFR (2024) § 1503.1(2)(i)–(v).

contains specific provisions 'for ascertaining concerns of local affected persons'.[48] In Indonesia, a 2020 law confines participation in EA procedures to 'communities directly affected' by the proposed activity.[49] The Asian Infrastructure Investment Bank requires project proponents to consult '[p]roject-affected people'.[50] And the Espoo Convention requires its parties to permit participation by 'the public in the areas likely to be affected'.[51]

In a CA, however, affected populations cannot readily be identified. Climate impacts are global and diffuse, contributing very incrementally to a very wide range of risks.[52] Even if one were to consider that populations can be directly affected by climate change,[53] they are not in any proximate way affected by the climate impact of a particular project. Thus, while some courts have suggested that plaintiffs were affected by climate change,[54] these courts did not hold that plaintiffs could be meaningfully affected *by a particular decision* leading to an increase in GHG emissions that contributed marginally to exacerbating climate change.

Practice has been divided as to whether individuals, groups, or even countries could justify having a particular interest to participate in CA procedures on the ground that they are affected by climate impacts. On the one hand, the government of Czechia allowed Micronesia to present its views in an EIA procedure concerning the modernization of the Prunéřov coal-fired power plant, based on a Czech statute allowing a state to participate if its territory 'may be affected by significant effects' of the project.[55] However, the theory that Micronesia was directly

[48] Ministry of Environment and Forests, 'Environmental Impact Assessment Notification' (14 September 2006) s 7III(ii)(a).

[49] Law No 32/2009, Protection and management of the environment (Perlindungan dan pengelolaan lingkungan hidup) (3 October 2009) art 26(2), as modified by Law No 11.2020, Omnibus Law on Job Creation (UU Cipta Kerja) (5 October 2020) art 22(5). See also Hans N Jong, 'Indonesia's Omnibus Law A "Major Problem" for Environmental Protection', *Mongabay* (4 November 2020) <https://news.mongabay.com/2020/11/indonesia-omnibus-law-global-investor-letter/>.

[50] Asian Infrastructure Investment Bank, 'Environmental and Social Framework' (2019) 30.

[51] Espoo Convention (n 13) art 2(6).

[52] See Chapter II, subsection C.3(a).

[53] Identifying 'climate victims' would be difficult. How individuals or even society is affected by climate change depends not only on physical phenomena that can be caused or exacerbated by climate change, but also on how society faces these physical hazards—on policies such as adaptation to climate change, disaster-risk reduction, and development more generally. See eg Anthony Oliver-Smith, 'Peru's Five-Hundred-Year Earthquake: Vulnerability in Historical Context', in Anthony Oliver-Smith and Susanna M Hoffman (eds), *The Angry Earth: Disaster in Anthropological Perspective* (Routledge 1999) 74; Hans-Otto Pörtner and others, 'Technical Summary', in Hans-Otto Pörtner and others (eds), *Climate Change 2022: Impacts, Adaptation and Vulnerability. Working Group II Contribution to the Sixth Assessment Report of the Intergovernmental Panel on Climate Change* (CUP 2022) 37, at 42.

[54] See eg *Juliana v United States* (9th Cir 2020) 947 F 3d 1159, 1168. See also UN Committee on the Rights of the Child, Decision on a communication submitted by Chiara Sacchi against Argentina (22 September 2021) CRC/C/88/D/104/2019, paras 10.13–14 (finding that the authors had 'prima facie' the status of victim of a state's failure to take sufficient action on climate change mitigation). But see eg C-565/19, *Carvalho v European Parliament and Council* (General Court, 25 March 2021) ECLI:EU:C:2021:252; *Verein Klimaseniorinnen Schweiz v Switzerland* App no 53600/20 (ECHR, 9 April 2024) paras 485–88

[55] Law on environmental impact assessment, No 100/2001 (20 February 2001) <https://perma.cc/5AQN-SH9U>, s 11. See Office of Environment & Emergency Management, 'Request for a Transboundary Environmental Impact Assessment (EIA) proceeding from the plan for the

affected as a result of this particular project is questionable, given the limited contribution of the project to climate change.

On the other hand, the Oslo District Court rejected a comparable theory in
Greenpeace Nordic Association v Ministry of Energy, where the applicants relied on
the duty to consult children under the UN convention on the Rights of the Child.[56]
While acknowledging that children are impacted by climate change, the Court did
not find that this justified a specific legal duty to consult children or to assess their
best interest as part of the CA for the exploitation of three petroleum fields.[57] The
Court, citing the UN Committee on the Rights of the Child, noted that virtually
every state action affects children one way or another,[58] yet the Court considered
that not every state action could be subjected to a duty to consult children.[59] In the
Court's view, the best interest of the child should be assessed at a more general level
than in relation to the exploitation of specific petroleum fields[60]—presumably, as
part of more strategic decisions on the state's strategy on climate change mitigation.

While public participation in CA may not be readily justified based on the
interest of representing affected populations, there may be other justifications.
Indeed, public participation in the context of EA procedures is not limited to a
bargaining between project proponents and affected populations.[61] The best evidence of this lies in the fact that EA procedures routinely assess impacts that do
not directly affect any individual or community, such as purely ecological harm or
harm to future generations.[62] Alternatively to the representation of affected populations, public participation can be justified as a way to tap on public expertise or to

---

modernisation of the Prunerov II power plant' (3 December 2009) <https://perma.cc/765B-S4L3>
(Micronesia); Micronesia, Fourth National Report on the Implementation of the Convention on
Biological Diversity (22 June 2021) <www.cbd.int/doc/world/fm/fm-nr-04-en.pdf>, 160; James
Kanter, 'A Pacific Island Challenge to European Air Pollution', *The New York Times* (19 January 2010).

[56] Convention on the Rights of the Child (adopted 20 November 1989, entered into force 2
September 1990) 1577 UNTS 3, art 12. See *Greenpeace Nordic Association v Ministry of Energy* (2024)
Case No 23-099330TVI-TOSL/05 (District Court of Oslo), s 3.8.
[57] *Greenpeace Nordic* (2024) (n 56) s 3.8.
[58] UN Committee on the Rights of the Child, General comment No 14 (2013) on the right of the
child to have his or her best interests taken as a primary consideration (art 3, para 1), CRC/C/GC/14
(29 May 2013) [20] ('all actions taken by a State affect children in one way or another').
[59] *Greenpeace Nordic* (2024) (n 56) s 3.8. The Court noted that the decision to exploit petroleum
fields would affect children not only by causing GHG emissions, but also in other ways, including
through revenues to the state, welfare services and employment. Indeed, these impacts are likely to
be more direct, especially if one only considers the enjoyment of the rights of the existing child during
their own childhood (given the protracted nature of climate impacts), a fortiori if one emphasizes
the rights of the children with the state's own territory or jurisdiction. See Benoit Mayer, 'Climate
Change Mitigation as an Obligation under Human Rights Treaties?' (2021) 115 American Journal of
International Law 409, 417–18.
[60] *Greenpeace Nordic* (2024) (n 56) s 3.8.
[61] For the exception of Indonesia, and critics thereof, see references listed at n 49.
[62] Benoit Mayer, 'Climate Assessment as an Emerging Obligation under Customary International
Law' (2019) 68 International & Comparative Law Quarterly 271, 304.

enable interested individuals and organizations to express views on the pursuance of common goals transcending social interests.[63]

Expertise is well recognized as a ground for participation in EA procedures.[64] A critical question, however, concerns the selection of the relevant expertise. In the light of the goal of EA procedures, the relevance of expertise should be assessed based on the ability to inform the decision with regard to the activity proposed. In the context of a CA, the most relevant expertise is that related to the prediction of the GHG emissions resulting from the activity under consideration and to ways to avoid or reduce these emissions. By contrast, expertise about the existence and severity of climate change is less directly relevant, as the urgent need to mitigate climate change is widely recognized. A fortiori, expertise on specific climate impacts, for instance on the way climate change affects the frequency and severity of tropical cyclones, would not help to determine whether an activity should be approved and, if so, under what conditions.

Public and altruistic interests in environmental protection are also largely recognized as a ground for participation in EA procedures.[65] As public institutions are largely expected to have such public interests in mind, EA procedures must often involve consultation with 'the authorities likely to be concerned by [a proposed activity] by reason of their specific environmental responsibilities'.[66] But individuals and private organizations may also be interested in the public good and concerned with environmental impacts without being directly affected. Thus, both NEPA and the EU's EIA Directive allow participation by any 'interested' members of the public,[67] notwithstanding the nature of this interest. Similarly, the Aarhus Convention recognizes a right to participation in decisions on specific activities to 'the public concerned',[68] which includes not only 'the public affected of likely to be affected by' the decision, but also the public 'having an interest' in it, including environmental non-government organizations (NGOs).[69] Arguably, the very

---

[63] By analogy, see *Walton v Scottish Ministers* [2012] UKSC 44, [2013] 1 CMLR 858 [153], where Lord Hope held that individuals may have standing for environmental protection if 'they have a genuine interest in the aspects of the environment that they seek to protect ... and ... have sufficient knowledge of the subject to qualify them to act in the public interest'. See generally Leah Sprain, 'Paradoxes of Public Participation in Climate Change Governance' (2016) 25 Good Society 62, 65 (pointing out the difference between thinking about participation by stakeholders and by the public at large).

[64] See eg 40 CFR (2024) §6.203(a)(4)–(5); Environmental Assessment Act (China) (n 12) art 5.

[65] Glucker and others (n 11) 109.

[66] Directive 2011/92 (n 12), consolidated as of 15 May 2014, art 6(1); Directive 2001/42/EC of the European Parliament and of the Council of 27 June 2001 on the assessment of the effects of certain plans and programmes on the environment, [2001] OJ L 197/30, art 6(3). See also C-474/10, *Department of the Environment for Northern Ireland v Seaport* [2011] ECR I-10227 [29] (implying that at least one body must be designated as being concerned by any activity).

[67] See eg 40 CFR (2024) § 1503.1(a)(v), 1506.6(a)–(b); Directive 2011/92 (n 12), consolidated as of 15 May 2014, art 1(2)(e). See also eg Aarhus Convention (n 4) art 2(5); World Bank (n 12) 98.

[68] Aarhus Convention (n 4) art 6.

[69] Ibid, art 2(5). See also Aarhus Convention Compliance Committee, Findings and recommendations with regard to communication ACCC/C/2010/50 concerning compliance by the Czech Republic, ECE/MP.PP/C.1/2012/11 (2 October 2012) [66]; Attila Panovics, 'The Aarhus Convention Model' [2016] Hungarian Yearbook of International Law & European Law 251, 259.

decision of an individual or organization to submit views evidences some sort of 'interest' in the proposed activity. Notably, Canada's Impact Assessment Act did away from any condition of having an 'interest' in the proposed activity and allowed participation by any member of the public.[70]

Notwithstanding whether participation is justified based on expertise or on interest, it is frequently limited by an implicit condition that the prospective participant is a national or resident of, or otherwise based in, the country. The preamble to Canada's Impact Assessment Act, for instance, expresses the Government's commitment 'to providing *Canadians* with the opportunity to participate' in the impact assessment process.[71] Similarly, the US Environmental Protection Agency interpreted NEPA as allowing participation by 'citizens',[72] thus suggesting that, in the silence of the statute and regulations, foreign nationals and institutions are not normally invited to participate. While the Espoo Convention and EU law enable participation by foreigners, this is only in areas that are likely to be affected by significant environmental effects of the proposed activity.[73] The Aarhus Convention and the Escazu Agreement prohibit discrimination based on nationality[74] but do not prevent a distinction based on residence.

Questions about the boundaries of participatory rights in CA remain open and might be the object of litigation in years to come.[75] On the one hand, one could argue that interested foreigners should be allowed to participate in CA procedures as a way to tap their expertise and to ensure proper consideration for the global nature of climate impacts. On the other hand, allowing essentially everyone anywhere in the world to participate in every CA simply based on their expressed 'interest' in the decision could have undesirable results. First, the sheer volume of the submissions could challenge agencies' limited human and financial resources.[76] From this perspective, it could appear a waste of time and money to involve every public actors in every environmental decision. Second, allowing global participation in CA could distract from efforts to apply EA as a democratic tool allowing a political community to decide on how *it* wants to reconcile different priorities, such as economic development and a fair contribution to global efforts on climate change mitigation. In particular, the participation of foreign organizations with potentially

---

[70] Impact Assessment Act, SC 2019, c 28, ss 11, 27, 99.

[71] Ibid, preamble para 3 (emphasis added).

[72] US Environmental Protection Agency, 'How Citizens can Comment and Participate in the National Environmental Policy Act Process' (3 October 2023) <www.epa.gov/nepa/how-citizens-can-comment-and-participate-national-environmental-policy-act-process>.

[73] Espoo Convention (n 13) art 2(6); Directive 2011/92 (n 12), consolidated as of 15 May 2014, art 7.

[74] Aarhus Convention (n 4) art 3(9); Escazu Agreement (n 5) art 3(a). See also Mateusz Slowik, 'An Introduction to the Aarhus Convention's Cousin, the Escazu Agreement' (*Landmark Chambers Aarhus Blog*, 14 June 2023) <www.landmarkchambers.co.uk/resource-post/9-an-introduction-to-the-aarhus-conventions-cousin-the-escazu-agreement/>.

[75] Jonas Ebbesson, 'Public participation', in Lavanya Rajamani and Jacqueline Peel (eds), *The Oxford Handbook of International Environmental Law* (OUP 2021) 351, 357.

[76] Glucker and others (n 11) 109. See also Stern and others (n 18) 15.

greater expertise could make it even more difficult for local voices to be heard,[77] thus further shifting EA from a democratic process towards an expert-driven one.

By contrast to public participation in CA processes, the right to judicial review of the decision to approve a project tends to be more narrowly defined. For instance, the EU's EIA Directive only protects the right to judicial review to members of the public with a '*sufficient* interest' in the decision.[78] Similarly, Japanese courts have reportedly rejected two legal challenges to the approval of coal-fired power plants on the ground that the plaintiffs did not suffer any personal damage as a result of these decisions.[79] In another context, the European Court of Justice found that the EU's alleged lack of ambition on climate change was not 'of direct and individual concern' to an applicant and, thus, did not fulfil the condition, under EU treaty law, for an individual to institute proceedings against a European regulation.[80]

Here also, however, some other courts have suggested a more liberal approach to standing in judicial reviews. For instance, the UK Supreme Court admitted that 'there may ... be cases in which any individual, simply as a citizen, will have sufficient interest to bring a public authority's violation of the law to the attention of the court, without having to demonstrate any greater impact upon himself than upon other members of the public.'[81] Similarly, US courts have found that individuals and NGOs could have an interest in ensuring compliance with NEPA's procedural requirements,[82] and possibly also in the substantive measures adopted to mitigate climate change.[83] In France, respondents have not generally objected to the standing of the applicants[84] because administrative courts have long adopted a

---

[77] See Glasson and Therivel (n 23) 152. See also Brian D Clark, 'Improving Public Participation in Environmental Impact Assessment' (1994) 20 Built Environment 294, 295 (questioning whether foreigners 'have a right to express their views on, and attempt to influence, a decision on a project which may be on the other side of the world').

[78] Directive 2011/92 (n 12), consolidated as of 15 May 2014, art 11(1)(a) (emphasis added). See also eg Senior Courts Act 1981 (UK), s 31(3).

[79] Tōkyō Chihō Saibansho (Tokyo District Court) (27 January 2023), 令和元年（行ウ）第275号, 令和元年（行ウ）第598号 (Yokosuka Climate Case). See also Maria A Tigre, 'Guest Post: Climate Litigation in Japan: Citizens' Attempts for the Coal Phase-Out', *Climate Law: A Sabin Center Blog* (1 June 2022) <https://blogs.law.columbia.edu/climatechange/2022/06/01/climate-litigation-in-japan-citizens-attempts-for-the-coal-phase-out/>.

[80] *Carvalho* (n 54) [48]–[52]. See also consolidated version of the Treaty on European Union, [2012] OJ C 326/13, art 263(4) ('Any natural or legal person may, under the same conditions, institute proceedings against a decision addressed to that person or against a decision which, although in the form of a regulation or a decision addressed to another person, is of direct and individual concern to the former').

[81] *Walton* (n 63) [94].

[82] *Center for Biological Diversity v US Department of Interior* (DC Cir 2009) 563 F 3d 466, 479.

[83] *Juliana* (2020) (n 54) 1168; *Juliana v United States* (D Oregon 1 June 2023) Case 6:15-CV-01517-AA, 2023 WL 3750334, *18–19. But see *Center for Biological Diversity v US Department of Interior* (n 82) 478. See also *Juliana* (2020) (n 54), 1171–72; *Juliana v United States* (9th Cir 1 May 2024) 24-684 <https://climatecasechart.com/wp-content/uploads/case-documents/2024/20240501_docket-24-684_order.pdf>, *4–5.

[84] For exceptions, see Tribunal administratif de Guyane (Administrative Tribunal of French Guiana), 18 July 2022, 2001348, paras 3–6; Cour administrative d'appel de Nantes (Administrative Court of Appeal of Nantes), 21 May 2019, 17NT03927.

liberal approach to standing[85] and a statutory provision has expressly recognized standing to environmental NGOs to challenge decisions with an adverse environmental impact.[86]

c) Types of public interventions

The public have made various types of contributions to deliberations during CA procedures. Perhaps most obviously, the public can present views about climate change, for instance by raising concerns about the impacts of climate change and calling for enhanced mitigation action. Thus, in *Waratah Coal Pty Ltd v Youth Verdict Ltd*, the Land Court of Queensland noted the extensive insights by First Nations witnesses about the impacts of climate change on their communities.[87] On the other hand, the public have also occasionally expressed doubts about the extent to which public authorities should seek to mitigate climate change, or even, more radically, about the existence or anthropogenic causes of climate change.[88] Notwithstanding whether these views are in favour or against climate action, they are not specific to the activity under consideration. The existence and anthropogenic nature of climate change as well as the urgency of action to mitigate climate change are widely recognized,[89] and considerable time and resources may be wasted if each CA procedure were to become a ritualistic rehearsing of unspecific arguments about the causes and impacts of climate change.

Public participation can play a more useful role by concentrating on the climate impact of the activity under consideration, whether to appraise the merits of the activity or to suggest ways to avoid or reduce this impact.[90] At times, public submissions may help to identify GHG emissions associated with a proposed activity or to assess the need for further research on these emissions, for instance with regard to scope 3 emissions that might not initially have been included within the ambit of the EA procedure.[91] In this regard, while participation can take place at various stages of the CA, studies on the effectiveness of EA have consistently shown that

---

[85] Conseil d'État (State Council), 29 March 1901, 94580 ('Casanova').

[86] Code de l'environnement (2024) art L 142-1.

[87] *Waratah Coal Pty Ltd v Youth Verdict Ltd (No 6)* [2022] QLC 21 [1557]–[1565].

[88] See eg *RES Southern Cross v Minister for Planning* [2008] NSWLEC 1333 [46] (noting objections to a renewable energy project that had been made 'on the basis of either climate change scepticism or denial').

[89] See eg Paris Agreement (n 9) art 2(1).

[90] See generally Massimo Cattino and Diana Reckien, 'Does Public Participation Lead to More Ambitious and Transformative Local Climate Change Planning?' (2021) 52 Current Opinion in Environmental Sustainability 100.

[91] *Narrabri Underground Mine Stage 3 Extension Project*, SSD 10269, NSW Independent Planning Commission, Statement of Reasons for Decision (1 April 2022) <https://perma.cc/G6SE-89MS> [169]. By contrast, see *McPhillamys Gold Project*, SSD 9505, NSW Independent Planning Commission, Statement of Reasons for Decision (30 March 2023) <https://perma.cc/R8Q4-YEUZ> [340] (noting that the EA Commission 'did not receive a large volume of submissions from the public raising GHG emissions as a key concern in relation to the Project'). See also eg *Gray v Minister for Planning* (2006) 152 LGERA 258 (NSWLEC) [28].

public participation is most impactful at the scoping stage, where the public can influence the design of the project or identify mitigation measures worth considering.[92] At other times, the public may suggest concrete ways to reduce an activity's climate impact, including for instance by commenting on the adequacy of mitigation measures under consideration[93] or on the ability of the project's proponent to implement these measures.[94] The public may also assist in ensuring the integrity of the CA procedure, for instance when independent experts scrutinize the procedure to identify subtle but significant inconsistencies or omissions.[95] Not unlike peer review in academic publishing, the process may enhance the quality of the information forming the basis of the final decision on the proposed activity.

2. Primary decision-making authority

The decision to approve a proposed activity at the end of a CA procedure is frequently made by the Executive branch of government, such as ministers or entities responding to them. For instance, in England and Wales, planning permissions are granted by the Secretary of State for Housing, Communities and Local Government[96] or by local (eg county or district) authorities.[97] A limitation of this approach is that it may create conflicts of interests. Such conflicts of interest are most obvious when governments are called upon to assess the environmental impact of an activity that they have themselves proposed or supported.[98] Yet, more

[92] Marina Nenasheva and others, 'Legal Tools of Public Participation in the Environmental Impact Assessment Process and their Application in the Countries of the Barents Euro-Arctic Region' (2015) 1 Barents Studies: Peoples, Economies and Politics 13, 22; Monika Suškevičs and others, 'Public Participation in Environmental Assessments in the EU: A Systematic Search and Qualitative Synthesis of Empirical Scientific Literature' (2023) 98 Environmental Impact Assessment Review 106944, 14.

[93] See eg *Former Marchon Site*, APP/H0900/V/21/3271069, Department for Levelling Up, Housing and Communities, Application Ref 4/17/9007, Decision (7 December 2022) <https://assets.publishing.service.gov.uk/government/uploads/system/uploads/attachment_data/file/1122625/22-12-07_Whitehaven_-_Decision_Letter_and_IR.pdf> [31] (reflecting opposition to reliance on carbon offsetting as a way to reduce an activity's climate impact).

[94] *Central Queensland Coal Project*, Department of Climate Change, Energy, the Environment and Water (Cth), 2016/7851, Statement of Reasons (20 August 2023) <https://epbcpublicportal.awe.gov.au/_entity/sharepointdocumentlocation/001cabd3-02a8-ed11-aad0-000d3ae0929c/2ab10dab-d681-4911-b881-cc99413f07b6?file=2016-7851-Statement-of-Reasons.pdf> [20] (noting comments according to which 'the proponent had a negative environmental history').

[95] See eg WildEarth Guardians, 'Protest of September 7, 2017 Competitive Oil and Gas Lease Sale' (6 July 2017) <https://perma.cc/K3RX-37LP>.

[96] See Town and Country Planning Act 1990, s 77; Town and Country Planning (Development Management Procedure) (England) Order 2015, art 17.

[97] See Town and Country Planning (Development Management Procedure) (England) Order 2015, art 34; Town and Country Planning Act 1990, s 1. See also, eg, Environment Protection and Biodiversity Conservation Act 1999 (Cth) (decision by the minister); Impact Assessment Act, SC 2019, c 28, s 60(1).

[98] This issue is most obvious when the environmental protection agency is in charge of deciding on the protect that it has itself proposed. See eg *Department of the Environment for Northern Ireland* (n 66); *Leung Hon Wai v Director of Environmental Protection* (2015) 18 HKCFAR 568. More ordinarily, however, the activity is proposed by another agency of the same government.

subtle conflicts of interests may also arise with regard to fully private projects, to the extent that a government may be more interested in the project's short-term and tangible economic benefits, which tend to unfold within their jurisdiction, than in the avoidance of long-term, diffuse, and mainly extraterritorial environmental impacts.

To avert such conflicts of interest, some EA frameworks defer the decision to approve certain activities to an institution with at least some degree of independence from the Executive government. Thus, in New South Wales, an Independent Planning Commission is in charge of deciding on some projects of state significance.[99] Similarly, the New Zealand Resource Management Act provides the possibility for the Minister to refer a resource consent application for a proposal of national significance to an ad hoc board of inquiry.[100] In both cases, these independent administrative authorities are constituted of individuals nominated by the Minister based on their expertise—in the case of New Zealand, their 'knowledge, skill, and experience relating to' the Resource Management Act, 'type of matter' to be considered, 'tikanga Māori' (Indigenous custom), and 'the local community', as well as their 'legal ... and technical expertise'.[101]

A limitation of this reliance on technical institutions is that it might concede the inherently political nature of the decision. In addition to information gathered through the EA procedure, the decision as to whether to approve a proposed activity involves value-based judgements, for instance when comparing the long-term effects of GHG emissions with the nearer-term benefits of the activity, or the ecological impact of the activity with its social and economic benefits.[102]

To reconcile the benefits of an independent assessment with an acknowledgment of the political nature of the ultimate decision, some EA frameworks require an independent institution to formulate recommendations without conferring it with full decision-making powers. Thus, in Canada, a Minister can decide to establish an independent panel of experts to conduct the impact assessment and make recommendations on the final decision.[103] India's EIA framework involves a comparable appraisal by an independent expert committee, whose recommendations have to be considered, but do not have to be followed, by the regulatory authority when considering whether to provide prior environmental clearance to a project.[104] And Queensland's EA framework provides for the possibility that a proposed activity be referred to the Land Court for recommendation.[105] A decision

---

[99] Environmental Planning and Assessment Act 1979 (NSW), s 4.5.

[100] Resource Management Act 1991 (New Zealand), s 147(1)(a).

[101] Ibid, s 149K(4). See also Environmental Planning and Assessment Act 1979 (NSW), s 2.8(3).

[102] See Chapter III, subsection D.2.

[103] Impact Assessment Act, SC 2019, c 28, arts 36(1), 51(1)(a), (d)(iv).

[104] Ministry of Environment and Forests (n 48), ss 7IV(i), 8(ii).

[105] Environmental Protection Act 1994 (Qld) (version consolidated as of 1 February 2024) ss 185. 190. See also Mineral Resources Act 1989 (Qld) (version consolidated as of 26 April 2024) ss 265, 269.

that is at odds with these recommendations, while lawful, would likely attract more public scrutiny.

Within the Executive branch, decisions on proposed activities can be made by different institutions. Under some EA frameworks, for instance in China, the decision belongs to environmental authorities at either the national or local level, depending on the scale of the proposed activity.[106] Reliance on environmental authorities, however, often reflects a narrow focus on the acceptability of environmental impacts, in particular their compatibility or consistency with environmental goals and standards,[107] as environmental authorities are unlikely to conduct a comprehensive analysis of all of the activity's economic and social costs and benefits. A more common approach is to refer decisions to authorities with competence that includes but is not limited to environmental matters, such as regional councils and territorial authorities under the New Zealand Resource Management Act.[108] In this alternative approach, environmental authorities may be consulted as a way to better manage potential conflicts of interest,[109] but they do not have the final say.

3. The role of courts

The role of courts in CA procedures varies significantly across jurisdictions, reflecting different legal cultures and conceptions of the separation of powers. At one extreme, courts may, in effect, substitute for primary decision-makers, in particular in the case of a merit review. For instance, the New Zealand Resources Management Act allows the Minister for the Environment to refer a decision to the Environment Court,[110] which must then make a discretionary decision on the merits of the application as if it were a consent authority.[111] Without going as far, several Australian states allow specialized courts to review the merit of the decisions, after this decision has been adopted by the competent authorities, thus 'exercis[ing] the same powers as those available to the original decision-maker'.[112]

---

[106] Environmental Assessment Act (China) (n 12) art 22(1); Regulations on Environmental Protection Management for Construction Projects (建设项目环境保护管理条例) of 29 November 1998, as revised by Decision of the State Council on Amending the Regulation on the Administration of Construction Project Environmental Protection (国务院关于修改《建设项目环境保护管理条例》的决定) of 16 July 2017, art 10. See also Environmental Impact Assessment Ordinance (1997) Cap 499, ss 8(3), 10.

[107] See eg Regulations on Environmental Protection Management for Construction Projects (n 106), as amended by Decision of the State Council (n 106) art 11.

[108] Resource Management Act 1991, ss 2(1), 104.

[109] Directive 2011/92 (n 12), consolidated as of 15 May 2014, art 6(1); Directive 2001/42 (n 66) art 6(3).

[110] Resource Management Act 1991, s 147(1)(b).

[111] Ibid, s 149U.

[112] Rosemary Lyster and others, *Environmental and Planning Law in New South Wales* (4th edn, Federation Press 2016) 39. See also Victoria McGinness and Murray Raff, 'Coal and Climate

More commonly, the role of courts is confined to assessing the legality of the decision that has been taken—that is, ensuring that the procedure has been followed and that the decision is not unlawfully unreasonable. In the context of such judicial reviews, judges have generally been inclined to treat CA decisions with great deference, given both the political and technical nature of these decisions. Policy goals are often in tension with one another, and whether a project hindering the realization of a goal on the mitigation of climate change is justifiable in light of other policy goals is often a question in relation to which there can be reasonable disagreement.[113] In a democratic society, such political decisions are more legitimately made by the elected branches of government rather than by the judiciary. Thus, English courts have insisted that they are 'only concerned with determining questions of law', and that they are not 'responsible for making political and socio-economic choices'.[114] US judges, likewise, have found that they 'may set aside an agency action only if it is "arbitrary, capricious, an abuse of discretion, or otherwise not in accordance with law"'.[115] And judges in New South Wales have noted that they must avoid 'the temptation to express [a] conclusion in terms of a recognised ground of review while in truth making a decision on the merits',[116] as doing so would 'undermine the basis for judicial independence and the fundamental role which judicial impartiality plays in the social stability of the nation and the maintenance of personal freedom of its citizens'.[117]

The technical nature of EA procedures is another reason that has often been invoked as a ground for judicial deference. As Emily Meazell noted, '[w]hen [US] courts review agencies' scientific and technical determinations, they often emphasize that the specialized subject matter requires them to be at their most deferential'.[118] Likewise, UK courts 'allow a substantial margin of appreciation to judgments based upon scientific, technical or predictive assessments by those with appropriate expertise'.[119] However, this 'technical' ground for judicial deference is

Change: A Study of Contemporary Climate Litigation in Australia' (2020) 37 Environmental & Planning Law Journal 87, 94–96.

[113] R (Boswell) v Secretary of State for Transport [2023] EWHC 1710 (Admin) [71]. See also R (Boswell) v Secretary of State for Transport [2024] EWCA Civ 145, [2024] All ER (D) 161 (Feb) [47]–[58].

[114] R (Greenpeace Ltd) v Secretary of State for Energy Security and Net Zero [2023] EWHC 2608 (Admin), [2024] PTSR 345 [21]. See also eg Boswell (2023) (n 113) [84], affirmed in Boswell (2024) (n 113).

[115] Barnes v US Department of Transportation (9th Cir 2011) 655 F 3d 1124, 1132, quoting 5 USC (2024) § 706(2)(A). See also eg WildEarth Guardians v Bernhardt (D New Mexico 2020) 501 F Supp 3d 1192, 1200.

[116] Australian Coal Alliance Inc v Wyong Coal Pty Ltd [2019] NSWLEC 31 [73], quoting Walsh v Parramatta City Council (2007) 161 LGERA 118 (NSWLEC) [56].

[117] Bruce v Cole (1998) 45 NSWLR 163, 184.

[118] See Emily H Meazell, 'Super Deference, the Science Obsession, and Judicial Review as Translation of Agency Science' (2011) 109 Michigan Law Review 733, 733. See also eg Marsh v Oregon Natural Resources Council (1989) 490 US 360, 377.

[119] R (Goesa Ltd) v Eastleigh Borough Council [2022] EWHC 1221 (Admin), [2022] PTSR 1473 [102]. See also R (Mott) v Environment Agency [2016] EWCA Civ 564, [2016] 1 WLR 4338 [78], upheld by

more questionable than the first, 'political' one. For one thing, the science under-pinning EAs, in particular CAs, is often much thinner, and the agency's expertise more limited, than what judges seem to assume,[120] and what agencies present as scientific finding may conceal political decisions.[121] Overall, the aim of EA is not to defer decisions to 'experts', but to inform the public, the agency, and judges by providing clearly intelligible scientific analysis. EA reports should provide scientific finding in a format that allows readers, including judges, to understand it and, thus, to assess the validity of objections.[122]

Nonetheless, courts have generally sought not to intrude in agencies' policy and technical decisions in EA procedures. English courts, in particular, have signalled that decision-makers have a 'substantial margin of appreciation',[123] applies, for instance, when holding that 'whether an effect is significant and the adequacy of any assessment of significant effects are matters of judgment for the decision-maker': accordingly, such decisions 'are only open to challenge in the courts applying the conventional "Wednesbury" standard'.[124] Various other courts in common and civil law countries have also rejected substantive challenges against CA decisions on the ground that these challenges raised policy issues that are 'the province of other branches of government'.[125]

In spite of such demanding tests, some courts have occasionally found EA decisions unlawful, either because they did not respect the prescribed procedure, or because they were unlawfully unreasonable. As detailed in Chapter II, a number of

---

the Supreme Court in *R (Mott) v Environment Agency* [2018] UKSC 10, [2018] 2 All ER 663; *R (Plan B Earth) v Secretary of State for Transport* [2020] EWCA Civ 214, [2020] PTSR 1446 [177].

[120] See eg Meazell (n 118) 746–47.

[121] See Wendy E Wagner, 'The Science Charade in Toxic Risk Regulation' (1995) 95 Columbia Law Review 1613, 1617; Shannon Roesler, 'Agency Reasons at the Intersection of Expertise and Presidential Preferences' (2019) 71 Administrative Law Review 491, 494.

[122] See Glasson and Therivel (n 23) 159–60.

[123] *R (Friends of the Earth) v Secretary of State for International Trade* [2023] EWCA Civ 14, [2023] 1 WLR 2011 [63].

[124] *Goesa Ltd* (n 119) [100]. See also *R (Friends of the Earth Ltd) v Heathrow Airport Ltd* [2020] UKSC 52, [2021] 2 All ER 967 [165]; *Bristol Airport Action Network Co-ordinating Committee v Secretary of State for Levelling Up, Housing and Communities* [2023] EWHC 171 (Admin), [2023] PTSR 853 [209]; *R (Friends of the Earth) v North Yorkshire County Council* [2016] EWHC 3303 (Admin), [2016] All ER (D) 104 (Dec) [57]; *R (Greenpeace Ltd) v Secretary of State for Energy Security and Net Zero* (n 114) [21]. Wednesbury unreasonableness 'applies to a decision which is so outrageous in its defiance of logic or of accepted moral standards that no sensible person who had applied his mind to the question to be decided could have arrived at it'. See *Council of Civil Service Unions v Minister for the Civil Service* [1985] AC 374 (HL) [410].

[125] *Hancock Coal Pty Ltd v Kelly (No 4)* [2014] QLC 12 [47]. See also eg *Environmental Defence Society Inc v Auckland Regional Council*, A183/2002 [2002] NZEnvC 315, [2002] NZRMA 492, (2003) 9 ELRNZ 1 [86]–[88]; *Greenpeace Nordic Association v Ministry of Petroleum and Energy* (2020) Case No 20-051052SIV-HRET (Supreme Court) [241]; *Lho'Imggi v Her Majesty the Queen* (2020) FCJ 1109, 2020 FC 1059 (Federal Court) [19]; *An Taisce—National Trust for Ireland v An Bord Pleanála* [2021] IEHC 254, [2021] 4 JIC 2003 [42]–[44], approved in *An Taisce—National Trust for Ireland v An Bord Pleanála* [2022] IESC 8, [2022] 2 IR 173 [107]; Verfassungsgerichtshof (Constitutional Court) VfGH E 875/2017 (1 June 2017) (Austria).

courts have found that EA instruments had to be interpreted as requiring the assessment of the climate impact associated with some GHG-intensive activities,[126] and, as documented in Chapters III and IV, decisions that did not properly justify the determination of significance[127] or the scope of the assessment[128] have occasionally been quashed.

In this context, courts have insisted that they were 'responsible for holding agencies to the standard the statute establishes'[129]—in particular, in the United States, by reference to the agency's obligation to take a 'hard look' at all relevant factors.[130] Courts have also noted that '[d]eference need not be afforded where [an EA instrument]'s basic requirements are not met'[131]—for instance, when the agency had not gathered the necessary information on which to base its decision,[132] had not provided a 'reasoned basis' for its decision,[133] or, more generally, had not complied with (the court's interpretation of) the EA instrument.[134] Even then, some of these decisions have received harsh criticisms, including by dissenting judges who expressed concern that courts were interfering in 'highly politicized scientific debates'[135] and making decisions on 'amorphous' grounds.[136]

---

[126] See Chapter II, subsection B.2(a).

[127] See eg *Diné Citizens against Ruining Our Environment v Haaland* (10th Cir 2023) 59 F 4th 1016; *Gloucester Resources Ltd v Minister for Planning* [2019] NSWLEC 7; *BT Goldsmith Planning Services Pty Ltd v Blacktown City Council* [2005] NSWLEC 210.

[128] See eg *Center for Biological Diversity v Bernhardt* (9th Cir 2020) 982 F 3d 723; *Gloucester* (n 127); C-404/09, *Commission v Spain* [2011] ECR I-11897.

[129] *Sierra Club v Federal Energy Regulatory Commission* (DC Cir 2017) 867 F 3d 1357, 1368. See also *WildEarth Guardians v Bernhardt* (D DC 2020) 502 F Supp 3d 237, 256; Corte Suprema de Justicia (Supreme Court), 19 April 2022, *Asociación de Prestadores de Servicios Turísticos de Mejillones v Director Regional del Servicio de Evaluación Ambiental*, Rol 71628-2021, civil (Chile).

[130] See eg *High Country Conservation Advocates v US Forest Service* (D Colorado 2014) 52 F Supp 3d 1174, 1193; *WildEarth Guardians v Zinke* (D DC 2019) 368 F Supp 3d 41; *California v Bernhardt* (ND California 2020) 472 F Supp 3d 573. See generally *Robertson v Methow Valley Citizens Council* (1989) 490 US 332, 350; *Kleppe v Sierra Club* (1976) 427 US 390, 410. See also, in state law, *Montana Environmental Information Center v Montana Department of Environmental Quality* (Montana District Court 6 April 2023) DV21-01307 <https://climatecasechart.com/wp-content/uploads/case-documents/2023/20230406_docket-DV21-01307_order.pdf>, *29.

[131] *California v Bernhardt* (n 130) 624.

[132] Ibid.

[133] *Pembina Institute for Appropriate Development v Canada (Attorney General)* [2008] FCJ 324, 2008 FC 302 (Federal Court) [79].

[134] See eg *Center for Biological Diversity v National Highway Traffic Safety Administration* (9th Cir 2008) 538 F 3d 1172 [1217]; *Earthlife Africa Johannesburg v Minister of Environmental Affairs* [2017] 2 All SA 519 (GP) [20]; *Save Lamu v National Environmental Management Authority*, NET 196/2016, judgment (26 June 2019) <https://climatecasechart.com/wp-content/uploads/non-us-case-documents/2019/20190626_Tribunal-Appeal-No.-Net-196-of-2016_decision.pdf> [138].

[135] *350 Montana v Haaland* (9th Cir 2022) 50 F 4th 1254, 1281.

[136] *Center for Biological Diversity v Department of Fish and Wildlife* (2015) 62 Cal 4th 204, 241–42. See also *League to Save Lake Tahoe Mountain Area Preservation Foundation v County of Placer* (2022) 75 Cal App 5th 63, 104.

B. The Contents of the Decisions

This section reviews the content of the decisions that have been reached through CA procedures, whether by political, administrative, or judicial authorities. The first subsection shows that many rejection decisions were based on procedural flaws that could subsequently be rectified, but that some others were occasionally motivated on more substantive grounds. The second subsection observes that the decisions approving the implementation of proposed activities often come with conditions aimed at avoiding, reducing, or offsetting climate impacts.

1. Rejection

CA procedures have sometimes led to the rejection of proposed activities, either by administrative decision or by a subsequent judicial decision. Most of these adverse decisions were based on the non-fulfilment of procedural requirements. In some cases, the procedural flaw was the failure of the EA to consider the activity's climate impact at all.[137] In other cases, a CA procedure was conducted but was found to be deficient, for instance because it failed to quantify emissions,[138] to account for indirect climate impacts,[139] to use relevant benchmarks to assess the significance of these impacts,[140] to justify a conclusion that these impacts were not significant,[141] or to facilitate public participation with regard to these findings.[142]

As Elena Aydos and others observed in relation to *Gray v Minister for Planning*, such rejections on procedural grounds do not necessarily 'have ... the practical implication of preventing the developments'.[143] This is because, as a general rule, an activity that has been rejected based on a procedural flaw can still be approved once this fall is rectified. Nonetheless, the rectification of the procedural flaw can

---

[137] Eg *Earthlife Africa Johannesburg* (2017) (n 134); *Save Lamu* (n 134) [151].

[138] Eg *WildEarth Guardians v Office of Surface Mining, Reclamation and Enforcement* (D Colorado 2015) 104 F Supp 3d 1208, 1230; *Sierra Club v Federal Energy Regulatory Commission* (n 129); *Sovereign Inupiat for a Living Arctic v Bureau of Land Management* (D Alaska 2021) 555 F Supp 3d 739; *Center for Biological Diversity v Bernhardt* (n 128) 751.

[139] Eg *Gray* (n 91); *Dine Citizens against Ruining Our Environment v Office of Surface Mining, Reclamation and Enforcement* (D Colorado, 2015) 82 F Supp 3d 1201; *Friends of the Earth v Haaland* (D DC 2022) 583 F Supp 3d 113. See also *An Taisce v An Bord Pleanála* [2015] IEHC 633, [2015] 10 JIC 0907.

[140] Eg *Sierra Club v Federal Energy Regulatory Commission* (n 129) 1374; *350 Montana v Haaland* (n 135) 1265–66; *Gray* (n 91) [98]; *Waratah Coal* (2022) (n 87) [1025].

[141] Eg *Pembina Institute for Appropriate Development* (n 133); *Center for Biological Diversity v Department of Fish and Wildlife* (n 136); *Diné Citizens against Ruining Our Environment v Haaland* (n 127) 1044; *WildEarth Guardians v Bernhardt* (n 129) 255; *350 Montana v Haaland* (n 135) 1269.

[142] See eg *Dartbrook Coal Mine*, DA 231-7-2000 MOD7, NSW Independent Planning Commission, Statement of Reasons for Decision (9 August 2019) <https://perma.cc/PG5H-W7Y9> [130], [243]; *Save Lamu* (n 134) [151]; *350 Montana v Haaland* (n 135) 1270.

[143] Elena Aydos and others, 'Rocky Hill: A Legal Breakthrough in the Consideration of Climate Change and Social Impacts of Coal Mines' (2020) 14 Carbon & Climate Law Review 98, 101.

delay the activity's approval significantly, sometimes by several years. To avoid such delays, the Montana legislature sought to allow the implementation of projects while procedural flaws were being addressed,[144] but the Montana District Court found that this was in breach of a constitutional guarantee of 'adequate remedies' for environmental protection.[145] Overall, as the new CA procedure provides more complete information, it could also lead to a different substantive decision.

Some other rejection decisions rely on more substantive grounds, such as an assessment of the merits of the project based on a comparison of its costs and benefits. As the South African Centre for Environmental Rights noted, 'where a project would have significant climate impacts that cannot be mitigated ... then the proposed activities should be refused'.[146] Thus, the Minister for Resources and Critical Minerals of Queensland followed the Land Court's recommendation in *Waratah Coal* and refused the development of a new coal mine and power plant project on the ground that the climate impacts of the project would exceed its economic benefits.[147] The Court found that '[t]he contribution of the combustion of the Project coal to the remaining carbon budget to meet the Paris Agreement goal is material', and, therefore, that '[a]pproving the Project would narrow the options for achieving that goal'.[148] The agency further highlighted that, based on its direct (scope 1) GHG emissions, the project would be Queensland's second highest emitter,[149] increase the state's emissions by close to 5 percent,[150] and, thus, impede the state's achievement of its climate change mitigation goals.[151] The agency also noted that over 90 percent of public representations related to the project expressed concerns about its climate impact and its effect on the achievement of mitigation goals.[152]

Most rejections on substantive grounds are based, not on an isolated assessment of the climate impacts of the proposed activity, but on a comprehensive assessment of the various environmental, social, and economic costs and benefits. Thus,

---

[144] Montana Code Ann (West 2024) § 75-1-201(6)(a)(ii), as amended by Montana Laws 2023, ch 703, effective 19 May 2023.

[145] *Held v State of Montana* (Montana District Court 14 August 2023) CDV-2020-307, 2023 WL 5229257, conclusions of law, [25]–[29]. See also *Park County Environmental Council v Montana Department of Environmental Quality* (Montana Sup Ct 2020) 477 P 3d 288, 306 [72], [89].

[146] Centre for Environmental Rights, 'Comments on the Intended Draft Guideline for Consideration of Climate Change Implications in Applications for Environmental Authorisations, Atmospheric Emissions Licenses and Waste Management Licences' (23 July 2021) <https://cer.org.za/wp-content/uploads/2021/07/LAC-Submission-National-Guideline-for-Climate-Change-Considerations_26-July-2021.pdf> para 5.

[147] *Waratah Galilee Coal Mine*, 101/0032704, C-EA-100147684, Department of Environment and Science, Decision (2 November 2023) <https://perma.cc/3LNL-JQA5>. See also *Waratah Coal* (2022) (n 87) [36]–[45]; Ben Smee, 'Queensland rejects Clive Palmer's bid to build "carbon neutral" coal-fired power station', *The Guardian* (23 November 2023).

[148] *Waratah Coal* (2022) (n 87) [1937].

[149] *Waratah Galilee Coal Mine* (2023) (n 147) s 3.3(f).

[150] Ibid, s 3.3(h).

[151] Ibid, s 3.3(i).

[152] Ibid, s 3.3(j).

in *Gloucester Resources Ltd v Minister for Planning*, the New South Wales Land and Environment Court upheld the Minister for Planning's decision to refuse consent to a coal mine on the ground that the project would have 'unacceptable planning, visual and social impacts'.[153] The Court, however, added that the project's climate impact 'adds a further reason for refusal'.[154] In a subsequent decision, the Independent Planning Commission of New South Wales rejected the Bylong Coal project, partly because it doubted that the applicant had minimized GHG emissions 'to the greatest extent practicable as required' under the applicable instrument, but also because of the project's other environmental, social and economic impacts.[155] Citing the Court, the Commission noted that 'it is rational to refuse fossil fuel developments with greater environmental, social and economic impacts … as this not only achieves the goal of not increasing GHG emissions by source, but also achieves the collateral benefit of preventing those greater environmental, social and economic impacts'.[156] By contrast, the Queensland government approved the Winchester South coal mine project shortly after it had rejected the Galilee coal mine project (at stake in *Waratah Coal*).[157] A majority of the output from the Winchester South project would be metallurgical rather than thermal coal[158] and, when explaining the decision, Queensland's premier observed the lack of alternatives to coal for steel making.[159] And when recommending against the approval of another coal project in *Waratah Coal*, the Queensland Land Court emphasized that the case was 'not about whether any new coal mines should be approved', but 'about whether this coal mine should be approved on its merits'.[160]

[153] *Gloucester* (n 127) [556]. See also *Rocky Hill Coal Project*, SSD 5156, NSW Independent Planning Commission, Determination Report (14 December 2017) <https://perma.cc/XD97-VP6V>, 19.

[154] Ibid. By contrast, climate change was not mentioned in the Minister's decision and was only a side concern in public deliberations. For one of a few exceptions, see James Whelan on behalf of Environmental Justice Australia, 'Air Pollution and the Proposed Rocky Hill Coal Mine' (10 November 2017) <www.ipcn.nsw.gov.au/cases/2017/10/rocky-hill-coal-project>.

[155] *Bylong Coal Project*, SSD 6367, NSW Independent Planning Commission, Statement of Reasons for Decision (18 September 2019) <https://perma.cc/4K5C-HBGH> [817].

[156] Ibid [692], [817]. The Independent Planning Commission's findings were accepted in *KEPCO Bylong Australia v Independent Planning Commission* [2021] NSWCA 216. See also *Gloucester* (n 127) [555].

[157] Department of Environment, Science and Innovation, 'Winchester South coal mine approved' (7 February 2024) <https://www.des.qld.gov.au/our-department/news-media/mediareleases/winchester-south-coal-mine-approved>.

[158] *Winchester South project*, Coordinator General's evaluation report on the environmental impact statement (November 2023) <https://eisdocs.dsdip.qld.gov.au/Winchester%20South/CGER/winchester-south-project-cger.PDF>, s 5.3.6.

[159] Eden Gillspie, 'Federal government approval the final hurdle for mega Queensland coalmine' *The Guardian* (7 February 2024).

[160] *Waratah Coal* (2022) (n 87) [1].

2. Approval

Most CA procedures conclude with an approval of the proposed activity, either because the activity's environmental impacts are not deemed significant, or because these significant impacts are deemed to be justified by the activity's expected benefits.[161] Often, however, these approvals are subjected to conditions that aim to mitigate the activity's environmental impacts, especially when these impacts are deemed significant. These conditions tend to follow a mitigation hierarchy: they aim at avoiding significant impact entirely; when this cannot be done, at minimising, rectifying, or compensating it; or, at least, at enhancing the activity's benefits.[162] This hierarchy can for instance be observed in the Asian Infrastructure Investment Bank's policy to '[a]void pollution or, when avoidance is not possible, minimize or control the intensity or load of pollutant emissions and discharges, including direct and indirect greenhouse gas emissions'.[163] While it is rarely possible to avoid all net GHG emissions from proposed activities, conditions can reduce or offset some or all of these emissions.[164]

Many EA instruments require decision-makers to consider ways to reduce the climate impacts of the projects they approve. The World Bank, for instance, is to contemplate '[o]ptions for reducing GHG emissions' such as 'alternative project locations; adoption of renewable or low carbon energy sources; alternatives to refrigerants with high global warming potential; more sustainable agricultural, forestry and livestock management practices; the reduction of fugitive emissions and gas flaring; carbon sequestration and storage; sustainable transport alternatives; and proper waste management practices'.[165] The UN Development Programme is committed to considering any potential options to 'reduce project-related GHG emissions and intensity' that are 'technically and financially feasible and cost-effective'.[166] And, in the NEPA review of relevant projects, the US Federal Aviation Administration must assess 'whether there are areas within the scope of a project where such emissions could be reduced', for instance through 'changes to more fuel efficient equipment, delay reductions, use of renewable fuels, and operational changes (eg performance-based navigation procedures)'.[167] Very technical measures are sometimes imposed: for instance, a French poultry farm was required to

---

[161] Angus Morrison-Saunders, *Advanced Introduction to Environmental Impact Assessment* (Edward Elgar 2018) 66 and references cited.

[162] Glasson and Therivel (n 23) 137.

[163] Asian Infrastructure Investment Bank (n 50) 33.

[164] But see *R (Finch) v Surrey County Council* [2024] UKSC 20, [2024] All ER (D) 71 (Jun) [105] (admitting that 'there are no measures within the control of the developer which, if the project proceeds, would avoid or reduce the combustion emissions and their impact on climate').

[165] World Bank (n 12) 41 fn 12.

[166] UN Development Programme, 'Social and Environmental Standards' (2021) 24.

[167] Federal Aviation Administration, 105.1 Desk Reference (version 3, October 2023) <www.faa.gov/media/71921>, 3–7.

regulate animal diet and managing manures in specified ways aimed at reducing GHG emissions.[168]

Consistently, decisions approving GHG-intensive activities have imposed conditions aimed at limiting and reducing these emissions. For instance, the New South Wales Independent Planning Commission approved the extension of the life of the Mount Pleasant coal mine on the condition that the project proponent adopted and implemented an 'Air Quality and Greenhouse Gas Management Plan' to be reviewed by state authorities.[169] This Plan should ensure that the project's on-site emissions remain below specified annual and project-life thresholds and that the project achieves a specified level of emission intensity.[170] The applicant must further 'minimise GHG emissions by using electricity generated by renewable or carbon neutral energy sources where reasonable and feasible'.[171] The project proponent must review and update this Plan every three years during the implementation of the project,[172] considering the possibility of using new technologies, for instance, to abate fugitive emissions[173] and emissions from non-road mobile diesel equipment.[174] The Commission imposed similar conditions on other coal mines that it approved,[175] as well as other large projects liable to causing significant levels of GHG emissions.[176]

Canadian federal and provincial authorities have sometimes imposed requirements for project proponents to use the 'best available technology economically achievable'.[177] In particular, the Minister of Environment and Climate Change has imposed a requirement for new oil and gas projects to comply with a 'best-in-class emissions performance' requirement—that is, to achieve 'the lowest level of GHG emissions intensity of oil and gas projects globally that are conducting the same activity as the proposed project'.[178] This standard can be implemented by

---

[168] See Cour administrative d'appel de Nantes (Administrative Court of Appeal of Nantes), 10 October 2023, 21NT02185, para 16.

[169] *Mount Pleasant Optimisation Project*, SSD 10418, NSW Independent Planning Commission, Statement of Reasons for Decision (6 September 2022) https://perma.cc/8NKD-L48A [124].

[170] Ibid [157], [158].

[171] Ibid [160].

[172] Ibid [158].

[173] Ibid [149].

[174] Ibid [160].

[175] See eg *Glendell Continued Operations*, SSD 9349, NSW Independent Planning Commission, Statement of Reasons for Decision (28 October 2022) <https://perma.cc/P23A-WFNV> [187]–[191]; *Narrabri Underground Mine* (n 91) [177], [182].

[176] See eg *Bowdens Silver*, SSD 5765, NSW Independent Planning Commission, Statement of Reasons for Decision (3 April 2023) <https://perma.cc/8PSC-AEHT> [255]; *McPhillamys Gold* (n 91) [343]–[344].

[177] Takafumi Ohsawa and Peter Duinker, 'Climate-Change Mitigation in Canadian Environmental Impact Assessments' (2014) 32 Impact Assessment & Project Appraisal 222, 226.

[178] Government of Canada, 'Draft Guidance for the Submission of Information Demonstrating Best-in-Class GHG Emissions Performance by Oil and Gas Projects Undergoing a Federal Impact Assessment' (2022) <www.canada.ca/en/services/environment/weather/climatechange/climate-plan/oil-gas-emissions-cap/best-class-draft-guidance.html>.

requiring the project proponent to develop a plan in which it demonstrates that it will achieve a best-in-class emission intensity.[179]

Other conditions seek to address indirect climate impacts. Thus, with regard to the downstream emissions of fossil-fuel production projects, the New South Wales Independent Planning Commission has required project proponents to prepare 'an Export Management Plan' setting out protocols to ensure that fossil fuels are only exported to parties to the Paris Agreement or countries with similar policies.[180] In another case, an urban development project in France was approved subject to measures aimed at reducing commuting needs and promoting cleaner modes of transportation.[181]

Decision-makers have also considered imposing offsetting conditions to make up for an activity's GHG emissions that could not be avoided.[182] Yet, in an early decision in *Environmental Defence Society Inc v Auckland Regional Council*, the Environment Court of New Zealand expressed 'considerable disquiet about the efficacy, appropriateness and reasonableness of a condition' that would require the proponent of a power plant to develop and implement a programme of forestry sequestration to offset the project's carbon dioxide emissions. The Court invoked, among other things, the 'doubtful efficacy of such a condition in the global context'[183] and the difficulty of monitoring and enforcing a condition if it were implemented extraterritorially.[184] In the two decades since *Environmental Defence Society*, various approaches and mechanisms have been developed to facilitate emission offsetting, including on the international plane, under the Paris Agreement.[185] Taking stock of these developments, the California Court of Appeal for the Fourth District found, in *Golden Door Properties LLC v County of San Diego*, that an agency could lawfully impose emission-offsetting conditions on a project proponent.[186] Such offsetting conditions are generally viewed as last-ditch

---

[179] See eg *Goldboro LNG—Natural Gas Liquefaction Plant and Marine Terminal Pieridae Energy (Canada) Ltd*, Environment and Climate Change (Nova Scotia), Environmental Assessment Approval (21 March 2014) <https://novascotia.ca/nse/ea/goldboro-lng/conditions.pdf> s 2.2 (requiring 'demonstration of how the facility achieves an overall carbon intensity in line with best-in-class'); *Teck Resources Ltd*, Alberta Energy Regulator, Report of the Joint Review Panel (25 July 2019) 2019 CarswellAlta 1562, 2019 ABAER 8 [916] (recommending that the proponent be quired to communicate 'a final detailed greenhouse gas management plan and an Energy Management System' that would 'demonstrate' how it 'will achieve emissions intensity "best-in-class" status').

[180] See eg *United Wambo Open Cut Coal Mine Project*, SSD 7142, NSW Independent Planning Commission, Statement of Reasons for Decision (29 August 2019) <https://perma.cc/7VLT-BT4Y> [309].

[181] Cour administrative d'appel de Paris (Administrative Court of Appeal of Paris), 23 June 2021, 20PA02347 [32]. See also the discussion of the Newhall Ranch project, below at n 195.

[182] Environmental Assessment Office, 'Effects Assessment Policy' (version 1.0, April 2020) 58–59.

[183] *Environmental Defence Society* (n 125) [88].

[184] Ibid [92].

[185] Paris Agreement (n 9) art 6(2), (4).

[186] *Golden Door Properties LLC v County of San Diego* (2020) 50 Cal App 5th 467, 562.

efforts to reduce climate impacts when other conditions cannot do so, rather than the main way to address the GHG emissions of a proposed activity.[187]

An important issue regards the implementation of the conditions imposed on approved activities. Ohsawa and Duinker suggested that many of the measures imposed by Canadian authorities, for instance speed limits for off-road vehicles operating within a project's site, 'may be difficult to enforce'.[188] In response to such concerns, agencies and courts have increasingly sought to ensure that the conditions imposed are crafted in a way that facilitates monitoring and enforcement.[189] Under California law, for instance, the mitigation measures imposed on project proponents 'must be fully enforceable'[190] and effective,[191] and these measures must normally be implemented before the proposed activity.[192] Emission-offsetting conditions, for instance, need to be defined by reference to specific offsetting protocols.[193] In California, like elsewhere, project proponents have been required to achieve specified goals before implementing particular components of their project—for instance, to develop an emission-reduction plan approved by relevant authorities,[194] to receive a 'zero net energy' certification from an independent agency,[195] or to submit building design plans for buildings equipped with electric vehicle charging stations.[196]

A more difficult issue regards the relation between the conditions imposed on proposed activities on a case-by-case basis and regulatory standards. One view is that it is both fairer and more effective if conditions such as energy standards for new buildings or emission intensity for power plants are defined in abstract and general terms through statutory law or regulation, applicable to all in a predictable way. Thus, the Environment Court of New Zealand noted in *Environmental Defence Society* that GHG emissions should in principle 'be considered at national level to ensure a consistency of approach to guarantee an efficiency compatible with achieving the best social, environmental and economic outcome'.[197] Similarly,

---

[187] *Former Marchon Site* (n 93) [31] (finding that the use of offsetting for 'some small amount of GHG ... is neither unusual nor inappropriate in the proposed development', a metallurgical coal mine). See also Asian Development Bank, 'Safeguard Policy Statement' (2009) 32 (providing that 'compensatory measures (offset) that aim to ensure that the project does not cause significant net degradation to the environment' can be imposed to address 'residual impacts [that] are likely to remain significant after mitigation').

[188] Ohsawa and Duinker (n 177) 229.

[189] See generally Glasson and Therivel (n 23) 178–82.

[190] California Code Regs (2024) title 14, § 15126.4(a)(2).

[191] *Federation of Hillside and Canyon Associations v City of Los Angeles* (2000) 83 Cal App 4th 1252, 1261.

[192] *POET LLC v California Air Resources Board* (2013) 217 Cal App 4th 681, 740. See also *AquAlliance v US Bureau of Reclamation* (ED California 2018) 287 F Supp 3d 969, 1040.

[193] *Golden Door Properties* (n 186) 562.

[194] *Mount Pleasant* (n 169) [124].

[195] *Newhall Ranch Resource Management and Development Plan and Spineflower Conservation Plan*, California Department of Fish and Wildlife, Final Actions and Supplemental Findings (14 June 2017) https://nrm.dfg.ca.gov/FileHandler.ashx?DocumentID=145821&inline 14–15.

[196] Ibid, 18.

[197] *Environmental Defence Society* (n 125) [88].

the Alberta Energy and Utilities Board rejected the provincial government's rec-ommendation to impose a 'best-in-class' condition on the GHG emissions of TrueNorth, a power plant project.[198] The Board noted that 'the issue of GHGs is best dealt with through initiatives and policies developed at the federal and pro-vincial levels', and it recommended that 'Alberta continue to implement measures that would achieve continuous improvement in' emission intensity.[199] With regard to another project to be implemented in Alberta, a Joint Review Panel refrained from imposing any conditions to limit GHG emissions, instead asserting 'that he onus is … on the Governments of Canada and Alberta to finalize and implement [a] regulatory framework for GHGs in a timely manner'.[200]

Yet such views fail to account for the various obstacles to the adoption and im-plementation of relevant regulatory standards. EA frameworks would not be useful devices for environmental protection if abstract regulatory standards were able to address every possible environmental impact of any activity and to impose the adoption of any reasonable step that could ever be taken to avoid or reduce such environmental impact. No abstract regulatory tool can define relevant stand-ards for every possible situation.[201] The best justification for the adoption of condi-tions on a case-by-case basis, in relation to an individual activity, is precisely that, at a particular time and place, no abstract regulatory standard properly addresses a potentially significant environmental impact. This may be because the regulator did not foresee a potential source of GHG emissions or potential ways of reducing it, or because it did not have the time or the resources needed to develop compre-hensive regulation. In these circumstances, EA can spur policy innovation in ways that may ultimately inform lawmakers and regulators.

C. Effects of the Decision

This last section assesses the effects of EA as a tool for climate change mitigation. The first subsection considers CA decisions' direct effects on the proposed activ-ities. It shows that, while many GHG-emitting activities are ultimately approved, CA procedures have caused some activities to be altered, or even to be set aside entirely, due to their documented climate impacts. The second subsection iden-tifies CAs' effects on climate change mitigation. It concludes that CA contributes to the mitigation of climate change, in particular by incentivizing proponents to mainstream the reduction of GHG emissions in proposed activities. The third

[198] *TrueNorth Energy Co*, Alberta Energy and Utilities Board, Decision (22 October 2002) 2002 CarswellAlta 2007 [231].
[199] Ibid [233].
[200] *Kearl Oil Sands Project*, Joint Review Panel (Canadian Environmental Assessment Agency and Alberta Energy and Utilities Board), Addendum to EUB Decision 2007-013: Additional Rationale for the Joint Review Panel's Conclusion on Air Emissions (6 May 2008), conclusion.
[201] See Fergus Green and Richard Denniss, 'Cutting with both Arms of the Scissors: The Economic and Political Case for Restrictive Supply-Side Climate Policies' (2018) 150 Climatic Change 73.

subsection discusses CAs' other potential effects on society, ranging from desirable outcomes such as the democratization of decision-making, to undesirable ones, in particular with regard to the burden that CA may place on proponents. The subsection concludes with a reflection on how CA frameworks can seek to minimize these unintended adverse impacts while reaping the benefits of an effective CA framework.

1. Effects on proposed activities

The effects of a CA procedure are not directly determined by the nature of the decision reached at the end of this procedure. On the one hand, a decision rejecting a proposed activity does not necessarily prevent the proponent from applying again, especially if the rejection is based on a procedural flaw that can easily be fixed. On the other hand, a decision approving a proposed activity does not necessarily mean that the activity will ultimately be implemented—especially if the decision imposes stringent conditions or if the publicity caused by the procedure has led investors to withdraw from the project. While rejection decisions tend to receive more attention in the public and academic discourse on CA,[202] the most effective decisions, from the perspective of climate change mitigation, may be those that allow the proposed activity to be implemented but ensure that this is done in the best possible conditions.

a) Implementation of the proposed activity

A CA requirement is mainly a requirement to gather information and to consider it, rather than a substantive barrier to the approval of a proposed activity.[203] As such, a CA requirement rarely prevents the implementation of a proposed activity. Many proposed activities are directly approved following a CA procedure. Other activities that were initially rejected may subsequently be approved, in particular when the rejection was based on a procedural flaw that could be rectified.[204]

An example of this is the decision of the District Court for the Southern District of California to reject the Imperial-Mexicali transmission lines project, in *Border Power Plant Working Group v Department of Energy*, on the ground, among other things, that the agency had not considered the carbon dioxide emissions from the

---

[202] See eg McGinness and Raff (n 112) 122–23 (positing that, in this context, '[a] clear demonstration of the influence of climate change litigation would be court refusal of a coal mine on the basis of its expected contribution to climate change or the imposition of conditions to limit such impact').

[203] See *Fundación Greenpeace Argentina v Estado Nacional*, Cámara Federal de Apelaciones de Mar Del Plata (Federal Court of Appeal of Mar del Plata), 3 June 2022 <https://climatecasechart.com/ non-us-case/greenpeace-argentina-et-al-v-argentina-et-al/> (approving fossil-fuel activities after the agency had complied with a previous order requiring the assessment of climate impacts).

[204] See text above at n 143. See also *Diné Citizens against Ruining Our Environment v US Office of Surface Mining Reclamation and Enforcement* (10th Cir 2016) 643 F Appx 799 (finding that the appeal proceedings were moot because the agency had already 'issued a Revised Environmental Assessment and Finding of No New Significant Impact as well as re-approved the permit revision').

operation of gas-fired power plants whose power would be transmitted by the electric lines.[205] Following the judicial decision, the proponent prepared a new, 1,198-page environmental impact statement. This statement predicted that the project would cause 1.3 Mt CO_2 per year, an amount that the statement characterized as 'negligible' in comparison with total US or global carbon dioxide emissions.[206] The Department of Energy then approved the project based on the assessment that the project's climate impact was 'expected to be negligible'.[207] After new judicial proceedings against the approval failed,[208] the project appears to have been implemented.[209]

Kearl Oil Sands, in Northern Alberta, is another project that had once been rejected but was ultimately approved and implemented. In *Pembina Institute for Appropriate Development v Canada*, the Federal Court of Canada quashed the approval of this project on the ground that the Joint Review Panel had not justified its conclusion that the project would not result in a significant climate impact.[210] The Panel reconvened and adopted an addendum to its original decision in which it explained that, while the project's GHG emissions would be 'considerable', its climate impact would not be 'significant'.[211] The Panel's reasoning was based on the assumption that a large share of the project's oil and gas would substitute for that from other projects and on the observation that the project's emission intensity would be lower than in those other projects.[212] The project was thus approved, and ultimately implemented,[213] without any new condition being imposed.

Yet another example of a project implemented in spite of an early decision against it is the Hazelwood coal mine and power plant. In *Australian Conservation Foundation v Latrobe City Council*, the Victoria Civil and Administrative Tribunal found that a Planning Panel had failed to comply with an implied statutory

---

[205] *Border Power Plant Working Group v Department of Energy* (SD California 2003) 260 F Supp 2d 997. See also *Baja California Power Inc*, EA-1391, Department of Energy, Finding of no significant impact (5 December 2001) <www.energy.gov/oe/articles/doe-environmental-assessment-ea-1391-presidential-permit-applications-baja-california>.

[206] *Imperial-Mexicali 230-kV Transmission Lines*, EIS-0365, Department of Energy, Final Environmental Impact Statement (17 December 2004) <www.energy.gov/nepa/articles/eis-0365-final-environmental-impact-statement>, ch 4, 58–59 (estimating that the project would increase the power plants' annual carbon dioxide emissions from 3.9 to 5.2 Mt).

[207] *Imperial-Mexicali 230-kV Transmission Lines*, PP-234-1, Department of Energy, Record of Decision and Floodplain Statement of Findings (18 April 2005), (2005) 70 Federal Register 21189, 21193.

[208] *Border Power Plant Working Group v Department of Energy* (SD California 2006) 467 F Supp 2d 1040.

[209] See Sempra Energy, '2008 Annual Report' (2009) <https://investor.sempra.com/static-files/74bbe168-75ca-46b2-852b-8d38e59c9bca>, 5 (reporting that the Mexicali power plant sells its electricity to a Californian energy company). The new transmission lines are visible on Google Map, satellite view, at 32°39'06"N 115°40'54"W (accessed 11 March 2024).

[210] *Pembina Institute for Appropriate Development* (n 133), [80].

[211] *Kearl Oil* (n 200).

[212] Ibid.

[213] See eg Imperial Oil Ltd, 2022 Report under the Extractive Sector Transparency Measures Act (23 May 2023) <www.imperialoil.ca/investors/investor-relations/annual-and-quarterly-reports-and-filings>, 3.

requirement to consider GHG emissions when assessing the potential environmental impacts of a planning decision necessary to permit an extension of the coal mine.[214] As a result, the Panel reconvened and estimated the project's climate impact.[215] In the Panel's estimation, the project could cause a global warming 'up to between 0.00009°C and 0.00027°C in 2030', which, in the Panel's view, was 'still very small, if taken in isolation to other emissions world wide'.[216] The Panel found that the project's unintended consequences, including its climate impact, were 'outweighed by the benefits to the State in terms of the significant contribution that Hazelwood Power Station will continue to make to Victoria's power supply, and the benefits to local economic activity, employment and social cohesion'.[217] As such, the Panel approved the mine, subject only to a formal recommendation that further investigation be carried out into potential ways to reduce some sources of GHG emissions.[218] The coal mining project was implemented and, following its completion, a rehabilitation of the site is now under way.[219]

A last example is the rail line project aimed at transporting coal from mines in Wyoming to power plants in Minnesota and South Dakota, whose approval was vacated by the Court of Appeal for the Eighth Circuit in *Mid States Coalition for Progress v Surface Transportation Board*, in part because the agency had not considered the foreseeable effect of the project on coal consumption.[220] Subsequently, the Surface Transportation Board reapproved the project based on a new environmental impact statement, asserted that the case was 'not the proper vehicle in which to address' concerns related to climate change,[221] and imposed no conditions on carbon dioxide emissions. The courts dismissed a new judicial challenge.[222] While the proponents ultimately abandoned the project, this was for financial reasons not directly related to the project's climate impact.[223]

---

[214] See *Australian Conservation Foundation v Latrobe City Council* (2004) 140 LGERA 100.

[215] *Hazelwood West Field EES La Trobe Planning Scheme Amendment C32*, Panel established under the Planning and Environment Act 1987, Final Report (March 2005) 1.

[216] Ibid, 4.

[217] Ibid, 243.

[218] Ibid, 246.

[219] See Melissa Davey, 'Hazelwood coal power station to close with loss of up to 1,000 jobs', *The Guardian* (3 November 2016); 'Hazelwood Rehabilitation Project' (*ENGIE* 2023) <www.hazelwoo drehabilitation.com.au/engie/hazelwood>.

[220] *Mid States Coalition for Progress v Surface Transportation Board* (8th Cir 2003) 345 F 3d 520. See also *Dakota, Minnesota and Eastern Railroad Corporation Construction into the Powder River Basin*, STB Finance Docket No 33407, Surface Transportation Board, Decision (28 January 2002) 2002 WL 121210 (containing no mention of the project's climate impact).

[221] *Dakota, Minnesota and Eastern Railroad Corporation Construction into the Powder River Basin*, STB Finance Docket No 33407, Surface Transportation Board, Decision (13 February 2006) 2006 WL 383507, at *12.

[222] *Mayo Foundation v Surface Transportation Board* (8th Cir 2006) 472 F 3d 545.

[223] See 'FRA Administrator Denies DM&E Powder River Basin Loan Application Citing Unacceptable Risk to Federal Taxpayers', *US Department of Transportation* (26 February 2007) <https://railroads.dot.gov/elibrary/fra-administrator-denies-dme-powder-river-basin-loan-appl ication-citing-unacceptable-risk> (reporting the rejection of a loan application for the project); Scott Deveau, 'CP Rail Takes $180M Hit as it Shelves Wyoming Powder River Expansion', *Financial Post* (3 December 2012)(noting the company's decision to shelve the project).

The frequent lack of effect of CA procedure is unsurprising. EA procedures aim at informing agencies, not at determining the decision that they ought to take. While agencies are expected to look at the information and views gathered throughout the CA with an open mind, they may nonetheless come to a reasonable conclusion in favour of a project. What is perhaps more concerning, however, is that some agencies do not appear willing to approach CA as anything more than a formality. For instance, when a Montana district court applied the Montana Environmental Policy Act as requiring the state to conduct CAs before deciding on proposed fossil-fuel projects,[224] the state pleaded for a stay of execution of the judgment pending appeal, among other things, by asserting that the judgment would make no difference to the plaintiffs' alleged harm because the state would 'continue to issue permits for "fossil fuel projects" regardless of whether this decision is stayed or not'.[225] This argument was a rare candid admission of a government's unwillingness to even consider information and views gathered through CA procedures before making a decision.

Governments that are unwilling to consider the climate impacts of proposed activities have occasionally sought to do away with any CA requirement. EA instruments sometimes allow governments to exempt an activity from the EA requirement in exceptional circumstances.[226] Alternatively, governments can adopt legislation approving the proposed activity to set aside the EA requirement. Thus, after the District Court for the District of Columbia vacated the sale of offshore oil and gas leases in the Gulf of Mexico on the ground that the NEPA review had failed to consider downstream emissions,[227] the Congress included in the Inflation Reduction Act of 2022 an instruction for the Secretary of the Interior to issue the said leases,[228] thus pre-empting the application of NEPA and allowing the leases to become effective.[229]

b) Alteration or cancellation of the proposed activity

Decisions taken on the basis of a CA procedure may seek to avoid or reduce climate impacts in two ways. First, CA decisions can prevent the implementation of the proposed activity, either through a formal rejection decision, or by imposing conditions so onerous as to precluding the implementation of the project. Second, CA decisions can impose conditions aimed at avoiding or (more frequently) reducing

---

[224] *Held* (Montana District Court 14 August 2023) (n 145).

[225] *Held v State of Montana* (Montana Sup Ct 16 January 2024) DA 23-0575 <https://climatecasech art.com/case/11091/> *5. See also *Held v State of Montana* (Montana District Court 21 November 2023) CDV-2020-307 <https://climatecasechart.com/case/11091/>.

[226] See eg Directive 2011/92 (n 12), consolidated as of 15 May 2014, art 2(4); Environmental Impact Assessment Ordinance (n 106) s 30.

[227] *Friends of the Earth v Haaland* (2022) (n 139) 139.

[228] Inflation Reduction Act of 2022, Pub L No 117-169, 136 Stat 1818, § 50264(b).

[229] See *Friends of the Earth v Haaland* (DC Cir 28 April 2023) 22-5036, 2023 WL 3144203 (concluding that, '[a]s a result of those instructions, this case is moot').

the activity's climate impact. Once these conditions are identified, their implementation is sometimes in the proponent's own best interest if, for instance, they save costs by reducing energy consumption. In most cases, however, these conditions impose a burden on the proponent, and their implementation depends on the existence of effective monitoring and enforcement mechanisms.

Some of the most prominent rejections of proposed activities on substantive grounds have taken place in the Australian coal sector. In 2019, the New South Wales Land and Environment Court, in a merit review, upheld the Minister's decision to refuse consent to the Rocky Hill coal mine project.[230] The proponent, *Gloucester*, decided not to appeal[231] and abandoned the project.[232] A few years later, the government of Queensland followed the Queensland Land Court's recommendation to reject Waratah's Galilee Coal mine and power plant project due to concern for its climate impact.[233]

Other proposed activities were abandoned by a decision of the applicant, but in a context where CA had led to increased public scrutiny. An example of this is the Thabametsi coal-fired power plant project. In *Earthlife Africa Johannesburg v Minister of Environmental Affairs*, the High Court of South Africa set aside the project's environmental authorization on the ground that the Minister of Environmental Affairs had not properly considered a climate impact assessment.[234] Following the judgment, the Minister reconsidered the CA report and decided to confirm the environmental authorization.[235] Yet, investors became increasingly concerned with the intense public scrutiny exacerbated by the court decision,[236] and the Thabametsi project was eventually cancelled.[237]

---

[230] *Gloucester* (n 127).

[231] Peter Hennam, 'Key Climate Ruling against Coal Mine Stands after Miner Declines to Appeal', *The Sydney Morning Herald* (8 May, 2019).

[232] Anne Keen, 'Gloucester Resources Limited Relinquishes Rocky Hill Mine Licence and is under New Ownership', *Gloucester Advocate* (8 February 2022).

[233] Department of Environment, Science and Innovation (n 147). See also *Waratah Coal* (2022) (n 87); Graham Readfearn, 'Clive Palmer-owned company withdraws appeal against ruling that coalmine would worsen climate crisis', *The Guardian* (12 February 2023)(reporting that the project proponent withdrew its appeal against the court's decision).

[234] *Earthlife Africa Johannesburg* (2017) (n 134).

[235] Minister of Environmental Affairs, *Thabametsi Coal-Fired Power Station*, Reconsideration of Appeal Decision, LSA 142346 (30 January 2018) <https://perma.cc/7CEL-JABB>, s 3.2.

[236] See 'Court orders government to disclose records on coal power plants' (*Centre for Environmental Rights*, 7 February 2019) <https://cer.org.za/news/court-orders-government-to-disclose-records-on-coal-power-plants> (reporting that Nedbank, First Rand and Standard Bank withdrew from the project); Mfuneko Toyana and Yuka Obayashi, 'Marubeni to pull out of S.African coal-fired power plant project', *Reuters* (11 November 2020) (reporting that Marubeni Corp and Korea Electric Power Corp withdrew); Jean Marie Takouleu, 'South Africa: Pretoria Cancels Thabametsi Coal-Fired Power Plant Project', *Afrik 21* (7 December 2020) (reporting 'the withdrawal of all investors, including private banks Standard Bank, FirstRand, Nedbank and Absa; as well as the Development Bank of South Africa, the Public Investment Commission and the country's Industrial Development Corporation', along with Korea Electric Power Corporation and Marubeni).

[237] See 'The Writing Is on the Wall for Coal-Fired Stations in South Africa', *ESI Africa* (4 June 2021); Isabella Kaminski, 'Climate campaigners take South Africa to Court over Coal Policy', *The Guardian* (19 November 2021); *Earthlife Africa Johannesburg v Minister of Environmental Affairs* (19 November

The Lamu Coal Power Station project, in Kenya, is another case in point. The project was proposed in 2013 and approved in 2016, prompting NGOs to appeal to the National Environmental Tribunal that same year.[238] In 2019, the National Environment Tribunal set the project's approval aside on the ground that its climate impact had not been assessed.[239] The proponent appealed.[240] Three years later, as the appeal proceedings were still ongoing, the Court rejected a motion of the NGO which it viewed as a delaying tactic.[241] Meanwhile, key investors have decided to leave the project consortium[242] and the project appears to have lapsed.[243]

At times, the proponent appeared to be pushed towards abandoning a proposed activity that would otherwise be rejected by competent authorities. The Frontier Oil Sands mine project, in Alberta, is a case in point. A Joint Review Panel had estimated the project's scope 1 and scope 2 emissions at 4.1 Mt CO_2 equivalent (CO_2e) per year,[244] which the Panel had described as a 'large source of greenhouse gas emissions.'[245] Nonetheless, the Panel had recommended that the governments of Canada and Alberta approve the project with conditions aimed at limiting these emissions.[246] Alberta's government had expressed strong support for the project,[247] but the position of the federal government was less clear.[248] Two days before the expected decision of the federal cabinet, the proponent pulled the plug, causing a reported C\$1.13 billion write-down.[249] While some questions had been raised about the project's economic viability,[250] Teck Resources' decision was explicitly

---

2020) <https://perma.cc/UJ4B-5534>. On the more diffuse effect on other fossil-fuel projects, see below at n 268.

[238] *Save Lamu v National Environmental Management Authority*, NET 196/2016, Notice of appeal (2016)   <www.accountabilitycounsel.org/wp-content/uploads/2017/11/NET-Notice-of-Appeal-Cover-Page-w_-NET-Stamp-and-Appeal.pdf>.

[239] *Save Lamu* (n 134) [154].

[240] *AMU Power Company v Lamu*, Environment and Land Court at Malindi, Appeal No 6 of 2019, Memorandum of Appeal (24 July 2019) <www.accountabilitycounsel.org/wp-content/uploads/2021/09/memorandum-of-appeal-amu-power-lamu-coal-net.pdf>.

[241] See *AMU Power Company v Lamu* [2022] KEELC 2999 (KLR) <http://kenyalaw.org/caselaw/cases/view/236406/> [38].

[242] See eg Victor Juma, 'General Electric abandons Lamu coal power plant deal in policy shift', *Business Daily* (25 September 2020); James Anyanzwa, 'Centum writes off \$16.3m invested in Lamu coal project', *The East African* (3 December 2022); Jevans Nyabiage, 'Are China's banks going cool on coal power plants in Africa?' *South China Morning Post* (5 July 2021).

[243] Sasha Kinney, 'Lamu Coal Power Plant', *The People's Map of Global China* (27 January 2022).

[244] *Teck Resources* (n 179) [889].

[245] Ibid [909].

[246] Ibid [4807], app 6 (recommendation 10), app 7 (recommendation 5.5).

[247] Marieke Walsh, 'Alberta Offers to Regulate Oil Sands Emissions Cap, Pending Frontier Approval', *The Globe and Mail* (21 February 2020).

[248] Tim Kiladze, 'Teck Resources Pledges Net-Zero Emissions by 2050, as Frontier Oil-Sands Project Decision Looms', *The Globe and Mail* (3 February 2020).

[249] Marieke Walsh and Robert Fife, 'Teck Resources Pulling Application for Frontier Oil Sands Mine', *The Globe and Mail* (24 February 2020).

[250] Andrew Coyne, 'For both Alberta and Indigenous Peoples, Now Is the Winter of our Disrespect', *The Globe and Mail* (14 February 2020).

justified by the lack of 'a constructive path forward for the project' in the absence of 'a framework ... that reconciles resource development and climate change' in Canada.[251]

Yet other proposed activities were approved with the imposition of conditions that could significantly reduce their climate impact.[252] Newhall Ranch, a large real-estate development project in California, is an example of such effective measures imposed through a CA procedure. The project was partly decertified following a decision of the California Supreme Court in *Center for Biological Diversity v Department of Fish and Wildlife*, which held that the agency had not properly supported its finding that the project would not have a significant climate impact.[253] The project was reapproved following a corrective environmental review. Yet, the new approval was subjected to a range of conditions aimed at ensuring the applicant's compliance with its 'voluntary commitment to achieve zero net GHG emissions'.[254] Among other things, the project proponent committed to ensure that every building achieves net-zero energy emissions, to install electric vehicle charging stations on-site and elsewhere in the county, to subsidize the purchase of private electric vehicles and electric buses, and to offset any emissions that could not be avoided through funding for emission-reduction or carbon-sequestration projects.[255]

2. Effects on climate change

A CA procedure may contribute to the mitigation of climate change by reducing the GHG emissions associated with the project, but it can also contribute to or, at times, hinder climate action in less direct and predictable ways. In this regard, while climate activists often seek to prevent the approval of GHG-intensive projects, this is not necessarily the most effective way of limiting or reducing global GHG emissions. For instance, blocking a new fossil-fuel production project would lead to increased production from other such projects, possibly in worse environmental conditions.[256] Similarly, Chin J, at the Supreme Court of California warned that blocking a new real estate project on the ground of its climate impact will likely push people to live and work somewhere 'far less green', where these people 'will be emitting business-as-usual amounts of greenhouse gases'.[257] Blocking a proposed

[251] Don Lindsay (President and Chief Executive Officer, Teck Resources Ltd), Letter to Minister Wilkinson (23 February 2020) <www.teck.com/news/news-releases/2020/teck-withdraws-regulatory-application-for-frontier-project>.

[252] Some examples of these measures were presented in subsection B.2 of this chapter.

[253] *Center for Biological Diversity v Department of Fish and Wildlife* (n 136) 226.

[254] *Newhall Ranch*, Final Actions and Supplemental Findings (n 195) 8.

[255] Ibid, 15–23.

[256] See discussion in Chapter IV, Section B.

[257] *Center for Biological Diversity v Department of Fish and Wildlife* (n 136) 254.

activity can also cause political discontent, leading to a political backlash, for instance if the government decides to bypass the CA procedure through legislation[258] or seeks to change the law in order to set aside any CA requirement.[259]

When taking stock of the implementation of the EU's EIA framework, the EU Commission observed that the improvement of a project design through an assessment of various reasonable alternatives 'can be seen as the main added value of the EIA process'.[260] The same likely holds true in relation to CA. The approval of the proposed activity with conditions aimed at limiting its climate impact can often be more favourable to the mitigation of climate change than the rejection of the project. While the implementation of the proposed activity avoids most of the unintended effects that could result from blocking the project, conditions on the approval of a proposed activity can reduce climate impacts. For instance, the revision of the Newhall Range project has avoided an estimated 16 Mt CO_2 over 30 years.[261]

A CA decision on a proposed activity may also have at least two types of diffuse effect, in two ways. First, each CA decision can act as a precedent for subsequent CA procedures, in the same jurisdiction and also possibly elsewhere.[262] For instance, the decision of the Land and Environment Court of New South Wales in *Gloucester*[263] influenced subsequent decisions by the New South Wales Independent Planning Commission[264] as well as the recommendation by the Land Court of Queensland in *Waratah Coal*,[265] among other things. Less formally, proponents might also consider previous decisions before seeking approval for a proposed activity, which might lead them to renounce to GHG-intensive activities or to seek ways to minimize climate impacts.[266] In California, the decision in *Center for Biological Diversity v Department of Fish and Wildlife* has certainly incentivized

---

[258] See above, text at n 228.

[259] See eg Montana Laws 2023, ch 703, § 1, effective 19 May 2023. See also New South Wales, *Parliamentary Debates*, Legislative Assembly, 24 October 2019, 1576 (Rob Stokes, Minister for Planning and Public Spaces).

[260] See generally European Commission, '35 years of EU Environmental Impact Assessment' (Publications Office of the European Union 2021) 7.

[261] See *Newhall Ranch Resource Management and Development Plan and Spineflower Conservation Plan*, California Department of Fish and Wildlife, Final Additional Environmental Analysis (12 June 2017), chapter 2 <https://nrm.dfg.ca.gov/FileHandler.ashx?DocumentID=145706>, 24 (table 2.3-3) (estimating unmitigated annual emissions at 526,103 t CO_2e). On the conditions imposed in this project, see above, text to n 195.

[262] See Jeff Collins, 'A Tale of 2 Housing Projects: Tejon Ranch and Newhall Ranch Developers Take Different Paths on Global Warming', *Angeles Daily News* (20 April 2021) <www.dailynews.com/2021/04/20/a-tale-of-2-housing-projects-tejon-and-newhall-ranch-developers-take-different-paths-on-global-warming/> (showing how the Newhall Range project influenced the CA on a subsequent real-estate project).

[263] *Gloucester* (n 127).

[264] See *Dartbrook Coal* (n 142) [73]; *Bylong Coal* (n 155) [817].

[265] *Waratah Coal* (2022) (n 87) [123], [711], [769], [773]–[774], [788], [1230].

[266] See generally Megan Jones and Angus Morrison-Saunders, 'Understanding the Long-Term Influence of EIA on Organizational Learning and Transformation' (2017) 64 Environmental Impact Assessment Review; Eskild H Nielsen, Per Christensen, and Lone Kørnøv, 'EIA Screening in Denmark: A New Regulatory Instrument?' (2005) 7 Journal of Environmental Assessment Policy & Management 35, 35.

other corporations to mainstream climate change mitigation in their projects or to concede to demands by environmental NGOs.[267]

Second, CA can raise public awareness and trigger political debates about climate change, climate impacts, and ways to address them. The conditions imposed on a proposed activity can prompt the development of new technologies or other innovative ways to reduce emissions, which could have a spillover effect on subsequent activities. CA decisions can also push investors to reconsider their support to GHG-intensive activities. Thus, the public scrutiny of the Thabametsi coal-fired power station led several institutional investors to distance themselves not just from this particular coal plant, but also from fossil-fuel projects more generally.[268] Further, CA may put pressure on governments to consider adopting general measures to reduce or limit GHG emissions,[269] or even test measures that governments could then adopt in abstract and general terms. Perhaps most importantly, CA forces public deliberations on infrastructural projects with long-term climate impacts, such as large fossil-fuel production project, power plants, and transportation infrastructure. Rather than simply blocking proposed activities, CAs could stir the development of alternative visions of what a long-term low-emission development could be.

3. Unintended effects

CA may have important unintended consequences. Identifying, predicting, and appraising climate impacts imposes a financial and social burden on the implementation of proposed activities, and litigation can cause further costs and delays. This burden can contribute to climate change mitigation by impeding

---

[267] See 'Settlement Agreement Reached in Centennial Lawsuit' (*Tejon Ranch*, 1 December 2021) <https://tejonranch.com/settlement-agreement-reached-in-centennial-lawsuit/> (reporting the adoption of a settlement agreement through which the proponent commits to take measures to limit the climate impact of its real-estate project, including by installing 30,000 chargers and supporting the purchase of 10,500 electric vehicles). See also *Center for Biological Diversity v Department of Fish and Wildlife* (n 136).

[268] Centre for Environmental Rights, 'Celebrating a Major Climate Victory: Court Sets Aside Approval for Thabametsi Coal Power Plant', (1 December 2020) <https://cer.org.za/news/celebrat ing-a-major-climate-victory-court-sets-aside-approval-for-thabametsi-coal-power-plant>   (noting that 'Nedbank, Absa, FirstRand and Standard Bank have ... released policies that constrain their future funding of coal power'). See also Nedbank Group credit policy, 'Energy Policy' (30 May 2023) <www. nedbank.co.za/content/dam/nedbank/site-assets/AboutUs/Information%20Hub/Integrated%20 Report/2023/Nedbank%20Energy%202023.pdf>; Absa Group, 'Summary Coal Financing Standard' (28 April 2023) <www.absa.africa/wp-content/uploads/2023/04/Coal-Financing-Standard-Summary-28-April-2023.pdf>; FirstRand, 'Policy on Thermal Coal Financing' (August 2019) <www. firstrand.co.za/media/society/risk/policy-on-thermal-coal-financing.pdf>;   Standard   Bank,   'Coal-Fired Power Finance Policy' (nd) <www.standardbank.com/static_file/StandardBankGroup/filedo wnloads/PolicySummary_CoalFiredPowerFinance.pdf>. On the Thabametsi coal plant project, see text above at n 234.

[269] See eg *TrueNorth Energy* (n 198) [233] (calling on the government of Alberta to 'continue to implement measures' on climate change mitigation in the fossil-fuel extraction sector).

GHG-intensive projects, but potentially at a disproportionate cost, for instance for energy security or economic development, in particular in developing countries. Thus, by impeding the development of new coal plants in South Africa, the decision of the High Court in *Earthlife Africa Johannesburg* has also aggravated a national energy security crisis, in a country where load shedding averages eight hours per day and, in some estimates, is 'costing 2–3% of GDP growth to the economy'.[270] Even when projects end up being implemented in spite of early setbacks, years of delays can cause significant loss for proponents and society in general.

Such concerns were discussed in *Held v State of Montana*, when the defendants suggested that a court-imposed CA requirement would hinder the implementation of certain activities and, thus, 'undermine Montana's energy system, increase costs to consumers, compromise grid reliability, or cause any other irreparable harms'.[271] The Montana District Court brushed these legitimate concerns aside on two grounds. First, it invoked a precedent according to which 'litigation burdens do not constitute irreparable harm'.[272] Yet, it is highly questionable that significant delays in the implementation of proposed activities of public importance could be repaired in any meaningful way, for instance when delays in the project's implementation could affect the enjoyment of human rights (as the economic consequence of widespread power cuts would certainly do). Second, the court asserted that 'Montana need not rely on fossil fuels to meet its energy needs'[273] and that the state could 'meet those needs by transitioning to renewable energy sources, which would have climate benefits, create jobs, reduce air pollution, save lives and costs from air pollution, and reduce energy costs for Montanans'.[274] However, asserting that all fossil-fuel activities can be delayed because they are not in the public interest pre-empts the conclusions of CA procedures as to the merits of each of these proposed activities. While some GHG-intensive activities may not be in the public interest, others certainly are, and delaying those activities that are in the public interest would generally be prejudicial to the collective good. In other words, asserting that all fossil-fuel activities can be delayed because they are not in the public interest pre-empts the conclusions of CA procedures as to the merits of each of these proposed activities.

For CA to be a cost-effective tool for climate change mitigation, the burden it imposes on society needs to be minimized. This can be done in a number of ways.

---

[270] The World Bank, 'South Africa: World Bank Backs Reforms to Advance Energy Security and Low Carbon Transition' (25 October 2023) <https://perma.cc/3ZEK-EPV6>. See also Ralph Mathekga, 'Southern Africa's energy crisis', *GIS* (31 August 2023) <www.gisreportsonline.com/r/southern-africa-energy/>. One could object that South Africa could solve its energy crisis by relying on clean energy— but the point is that South Africa has not done so, perhaps due to concerns about affordability, reliance, or practicality, and that, all things being equal otherwise, a different decision in *Earthlife Africa Johannesburg* would certainly have alleviated this crisis.

[271] *Held* (Montana District Court 21 November 2023) (n 225) *7.

[272] Ibid.

[273] Ibid, *3–*4.

[274] Ibid.

For instance, screening and scoping processes should ensure the assessment of all potentially significant climate impacts, but only of those impacts. The repeated assessment of the same climate impact should be avoided, for instance when an activity is subject to decisions by several agencies.[275] Thus, the EU Commission encourages Member States to implement coordinated or joint EIA procedures 'so as to avoid overlaps and redundancy'.[276] CA procedures imposed by different levels of government, for instance by federal and state governments, should ideally be coordinated to avoid duplication of work.[277]

A tiered approach can improve CA's efficiency by avoiding repeated consideration for similar strategic issues in successive projects.[278] Thus, a preliminary assessment of a nationally determined contribution (NDC), a long-term mitigation strategy, or similar policy documents can inform subsequent assessments of policies, programmes, plans, and ultimately projects. For instance, once a decision has been made to develop a particular source of electricity (eg fossil or renewable) at the strategic level, a project-based EIA can focus on the best way of implementing the project (eg minimizing the emission intensity) rather than on the nature of the project.

Many of the costs and delays associated with CAs result from disputes about the existence, contents, and modalities of CA requirements. As CA becomes more firmly grounded in national practice, its implementation should be streamlined with clearer and more detailed rules. In particular, these rules should define thresholds of significance that could be used at the screening and scoping stages, as well as methodologies to structure the appraisal of significance at the decision-making stages. Further, these rules should clearly define the scope of the GHG emissions that are to be assessed, including scope 2 and scope 3 emissions overseas. Public debates, in turn, should be directed to assessing the merits of the proposed activity and ways to reduce its climate impact rather than on general information about climate change and the need to mitigate it.

---

[275] See eg Conseil d'État (State Council) (6ème–5ème chambres réunies), 10 February 2022, 455465, ECLI:FR:CECHR:2022:455465.20220210 [4]–[5]; *Sierra Club v Federal Energy Regulatory Commission* (n 129) 1379, 1380 (Brown CJ dissenting) ('when the occurrence of an indirect environmental effect is contingent upon the issuance of a license from a separate agency, the agency under review is not required to address those indirect effects in its NEPA analysis').

[276] Commission guidance document on streamlining environmental assessments conducted under Article 2(3) of the Environmental Impact Assessment Directive [2016] OJ C 273/1 para 2. See also Directive 2011/92 (n 12), consolidated as of 15 May 2014, art 2(3).

[277] See discussion in Chapter II, subsection C.2(d).

[278] Elizabeth A Masden and others, 'Cumulative Impact Assessments and Bird/Wind Farm Interactions: Developing a Conceptual Framework' (2010) 30 Environmental Impact Assessment Review 1, 6. See also Meeting of the Parties to the Convention on Environmental Impact Assessment in a Transboundary Context and Meeting of the Parties to the Protocol on Strategic Environmental Assessment, Decision VII/7–III/6, 'Development of a Strategy and an Action Plan for the Future Application of the Convention and the Protocol', ECE/MP.EIA/23/Add.1–ECE/MP.EIA/SEA/7/Add.1 (19 September 2017) para 9.

Conclusion

This chapter has shown that, under certain conditions, CA can prompt governments and society to make better decisions—that is, decisions that are based on a fuller understanding of the climate impact associated with a proposed activity and of the way to avoid or reduce this impact. At times, CA procedures can prevent the implementation of a proposed activity, either because a government denies the necessary approval, or because increased public scrutiny leads to the withdrawal of investors. More commonly, these procedures result in changes to some aspects of the proposed activity, for instance when agencies impose conditions aimed at avoiding or reducing GHG emissions.

The effectiveness of EA as a tool for climate change mitigation depends not only on the content of the relevant law, but also on the context in which it is implemented. CA procedures may encourage productive political deliberations, especially in democratic and open societies. Constructive comments from the public may point to alternatives to or alterations of the proposed activity with the potential of avoiding or reducing GHG emissions, which the proponent and the government may consider. On the other hand, there is a risk that CA procedures are implemented as a mere formality. Whether information and ideas that emerge during the CA can affect the final outcome depends, among other things, on a culture of public deliberation and on the political accountability of a government to a people interested in the mitigation of climate change.

VI

Conclusion

This book has shown that environmental assessment (EA) has widely been used as a tool for climate change mitigation. Chapter II has documented developments in multiple national, subnational, and supranational jurisdictions across the world that led to the identification, adoption, and clarification of climate assessment (CA) requirements. Chapter III has identified three types of methodologies to appraise the significance of climate impacts at various stages of an EA procedure. Chapter IV has taken stock of the debates on the assessment of indirect climate impacts and argued that, in order to ensure that decisions are well informed, any potentially significant climate impact caused by the activity ought to be assessed. Lastly, Chapter V has shown how CA can help society to achieve better decisions based on more careful consideration of more complete information.

Beside documenting the developments taking place around the world and analysing common issues, this book aims at identifying and evaluating potential solutions. A central conclusion from this study, in this regard, is that applying EA as a tool for climate change requires a rethink of EA frameworks, reflecting a shifting emphasis from the local to the global, and from direct and concrete effects to more diffuse and abstract ones. The implementation of a CA requirement must be guided by specific rules, for instance to determine how significance is to be appraised, how indirect effects are to be assessed, and how public participation is to be encouraged. As such, while CA requirements frequently arise from judicial or administrative interpretations of existing EA instruments, legislative or regulatory reforms play an essential role in defining the content of this requirement.

The previous chapters have identified a number of good practices and policy recommendations. Some of these approaches have been widely implemented throughout the world, while others are yet to be adopted in most jurisdictions. Ten of the main recommendations are summarized in the following.

1. The climate impact of a proposed activity should be assessed as part of the EA of this activity whenever the impact is likely to be significant. Although a given activity would only ever have an extremely small effect on the global climate system as a whole, this could nonetheless be considered significant given the extraordinarily broad set of ecological, economic, and social risks

Environmental Assessment as a Tool for Climate Change Mitigation. Benoit Mayer, Oxford University Press.
© Benoit Mayer 2024. DOI: 10.1093/oso/9780198939184.003.0006

unfolding globally and in the long-term. This assessment of significant climate impacts is justified by the conclusion that EA can be an effective tool for climate change mitigation, in particular when society and agencies are willing to take climate impacts seriously. Finding that an activity would cause large amounts of greenhouse gas (GHG) emissions provides decision-makers with potentially actionable conclusions: it may cast doubt on the merits of the activity or justify alterations aimed at avoiding or reducing these emissions.

2. Possible redundancies with other policies and measures on climate change mitigation are not ordinarily a good reason to exclude climate impacts from the scope of EA procedures. In most cases, EA is useful to complement other policies and measures on climate change mitigation. In particular, CA can be an effective tool for climate change mitigation even in jurisdictions that have established a cap-and-trade mechanism, in spite of the risk of a 'waterbed effect'. This is because CA procedures often consider sources of GHG emissions taking place beyond the sectorial, temporal, and geographical scope of cap-and-trade mechanisms.

3. At the screening and scoping stages of an EA procedure, the potential significance of climate impacts should be determined based on clear, convenient, and consistent criteria. These criteria could involve quantitative thresholds of magnitude expressed in terms of a level of GHG emissions (at present, typically around 100 kt CO_2 equivalent per year). But these criteria may also include activity-specific thresholds, the application of which would not require a preliminary calculation of GHG emissions, such as the generation capacity of a power plant, the additional traffic expected from a road project, or the area of forest affected by a land conversion activity.

4. Indirect climate impacts (eg scope 2 and scope 3 emissions) should be assessed, not *in spite of* the difficulties of predicting these indirect climate impacts, but precisely *because of* these difficulties. The exclusion of significant sources of indirect emissions from the scope of an EA procedure should only be justified based on robust evidence that these sources of emissions are not significantly affected by the proposed activity. Some degree of market substitution is likely to take place in relation to some indirect climate impacts, but market substitution is not ordinarily 'perfect'. As such, a preliminary assumption should be that any new fossil-fuel production project would increase the global fossil fuel consumption and global GHG emissions.

5. Extraterritorial sources of GHG emissions should also be considered as part of an EA when these sources are likely to be significant. The location of a source of GHG emissions does not affect its climate impact. A state is entitled (and might even be required) to take measures within its territory that are intended to reduce GHG emissions from sources situated in the territory of other states, for instance by limiting the exportation of fossil fuels. These

measures, however, must respect the sovereignty of other states, including their right to development and the human rights of their population.

6. A CA procedure needs to be carried out for activities with a significant indirect climate impact, even if they do not have any on-site environmental impacts. It is regrettable that, even in some of the jurisdictions with the most advanced CA law and practice, some activities without major on-site environmental impact escape any CA requirement because of the lack of consideration for indirect climate impacts at the screening stage. Such is perhaps most obviously the case of cryptocurrency-mining facilities, whose high electricity consumption can cause large amounts of GHG emissions from public utility companies, but which have seldom been subjected to any sort of scrutiny.

7. Interested citizens and organizations should be encouraged to participate in CA procedures. The absence of directly affected population should not prevent or restrict the role of the public in expressing concerns about climate impacts and proposing ways to avoid or reduce these impacts. On the other hand, comments should not merely seek to repeat well-known information about the existence and consequences of climate change; rather, they should focus on the merits of the proposed activity and on possible ways to avoid or reduce climate impacts in the circumstances of the activity under consideration.

8. At the substantive decision-making stages, the significance of climate impacts and the merits of the proposed activity should be appraised transparently, fairly, and consistently across projects, based on methods outlined by statutes, regulations, or guidance documents. Some existing approaches rely on benchmarks such as emission inventories or emission-reduction goals identified at a scale relevant to the proposed activity. Other approaches build on economic valuation tools such as estimates of the social cost of GHG emissions or of the willingness of a society to pay to avoid such emissions (ie the 'shadow price' of GHG emissions). A further distinction can be drawn between empirical approaches, which rely on the level of climate ambitions embedded in existing policies (eg the value of an existing carbon tax), and normative ones, based on goals that society aspires to achieve by enhancing its ambition over time (eg the level at which a carbon tax should be if it were to be consistent with global temperature goals). The political debate on the significance of climate impacts and on the merits of a proposed activity is best informed by a combination of several approaches.

9. Conditions imposed on proposed activities should follow a mitigation hierarchy, seeking first to avoid, then to minimize, and lastly to offset climate impacts. These conditions should be defined in such a way as to ensure that their effective implementation can be monitored. When some conditions are repeatedly imposed on all similar activities, lawmakers and regulatory agencies should consider adopting them as general and abstract standards.

10. Governments should regularly review the implementation of EA frameworks in order both to enhance their effectiveness as a tool for climate change mitigation and to reduce the burden on proponents and society. In doing so, governments should experiment with new approaches to the many issues faced when implementing CA, draw lessons from their own experience, and share this experience with other governments. While CA has yielded some tangible and important results, there remains room for improvement in every jurisdiction.

It is hoped that this book will prompt further research on the role of EA as a tool for climate change mitigation. For instance, the significance of many past, ongoing, and future national developments needs to be better understood and evaluated, both in their national context and from a global comparative perspective. Further policy analysis could suggest more innovative approaches to common issues in CA, including (but not limited to) the appraisal of significance, the assessment of indirect climate impacts, and the roles of the public, agencies, and courts in decision-making processes. As states intend to deploy negative emission technologies at scale in the coming decades, researchers could also contemplate whether EA could be a useful tool to enhance sinks and reservoirs of GHGs, including in relation to activities that are not necessarily net sources of GHG emissions. Lastly, a more theoretical reflection could assess how applying EA as a tool for climate change mitigation may have affected EA law and practice, including when this tool is applied to traditional environmental impacts.

More generally, it is hoped that this book has shown the potential for comparative legal research in climate law. As states around the world are striving to reducing their GHG emissions, they are all experimenting with similar rules and policies— framework legislation, carbon budgets, cap-and-trade mechanisms, carbon taxes, subsidies to renewable energies, technical regulations, tort liability, state-ownership policies, public information disclosure, etc. In doing so, they are likely to face similar issues, and innovative approaches adopted in some jurisdictions could be of interest to governments in others. Comparative legal scholarship has a crucial role to play in documenting these rules and policies, identifying the policy issues they face, and evaluating potential solutions to these issues, in ways that can help governments devise effective climate change mitigation strategies.

References

Monographs

Bardach, E, *A Practical Guide for Policy Analysis* (7th edn, Sage 2023)

Barrett, C, and Therivel, R, *Environmental Policy and Impact Assessment in Japan* (2nd edn, Routledge 2019)

Bates, GM, *Environmental Law in Australia* (LexisNexis Butterworths 2016)

Benidickson, J, *Environmental Law* (Irwin Law 2013)

Boyle, A, and others, *Birnie, Boyle, and Redgwell's International Law and the Environment* (4th edn, OUP 2021)

Craik, N, *The International Law of Environmental Impact Assessment: Process, Substance and Integration* (CUP 2008)

Dias, C, *Getting Heard but not Listened to: An Analysis of Public Participation in Environmental Impact Assessment (EIA) in Brazil* (Lexington 2021)

Dobson, N, *Extraterritoriality and Climate Change Jurisdiction: Exploring EU Climate Protection under International Law* (Bloomsbury 2021)

Donnelly, A, Dalal-Clayton, B, and Huges, R, *A Directory of Impact Assessment Guidelines* (2nd edn, Russel 1998)

Dunn, W, *Public Policy Analysis: An Integrated Approach* (6th edn, Routledge 2017)

Eccleston, C, *Environmental Impact Assessment: A Guide to Best Professional Practices* (CRC Press 2011)

Fisher, E, Lange, B, and Scotford, E, *Environmental Law: Text, Cases, and Materials* (2nd edn, OUP 2019)

Gibson, R, and others, *Sustainability Assessment: Criteria and Processes* (Routledge 2005)

Glasson, J, and Therivel, R, *Introduction to Environmental Impact Assessment* (5th edn, Routledge 2019)

Harrop, O, and Nixon, A, *Environmental Assessment in Practice* (Taylor & Francis 2005)

Holder, J, *Environmental Assessment: The Regulation of Decision Making* (OUP 2005)

Holling, CS, *Adaptive Environmental Assessment and Management* (Wiley 1978)

Hutley, N, *A Social Cost of Carbon for the ACT: Prepared for the ACT Government* (ACT Government 2021)

Hyman, E, and others, *Combining Facts and Values in Environmental Impact Assessment: Theories and Techniques* (Routledge 2018)

IPCC Core Writing Team and others (eds), *Climate Change 2023: Synthesis Report* (WMO & UNEP 2023)

Langlet, D, and others, *EU Environmental Law and Policy* (OUP 2016)

Lyster, R, and others, *Environmental and Planning Law in New South Wales* (5th edn, Federal Press 2021)

Mayer, B, *International Law Obligations on Climate Change Mitigation* (OUP 2022)

McDonald, M, and others, *Environmental Impact Assessment Guide to: Climate Change Resilience & Adaptation* (IEMA, 2020) 1

Morrison-Saunders, A, *Advanced Introduction to Environmental Impact Assessment* (Edward Elgar 2018)

Rowland, W, *The Plot to Save the World: The Life and Times of the Stockholm Conference on the Human Environment* (Clarke 1973)

Sabel, C, and Victor, D, *Fixing the Climate: Strategies for an Uncertain World* (Princeton University Press 2022)

Samuelson, P, and Nordhaus, W, *Economics* (19th edn, McGraw-Hill Irwin 2010)

Sandbach, F, *Environment, Ideology, and Policy* (Allanheld, Osmun & Co 1980)

Stern, P, and others, *Public Participation in Environmental Assessment and Decision Making* (National Academies Press 2008)

Taylor, S, *Making Bureaucracies Think: The Environmental Impact Statement Strategy of Administrative Reform* (Stanford University Press 1984)

Therivel, R, and Partidário, MR, *The Practice of Strategic Environmental Assessment* (Earthscan 1996)

Wood, C, *Environmental Impact Assessment: A Comparative Review* (2nd edn, Routledge 2013)

Wouters, W, Hardie-Boys, N, and Wilson, C, *Evaluating Public Input in National Park Management Plan Reviews: Facilitators and Barriers to Meaningful Participation in Statutory Processes* (New Zealand Department of Conservation 2011)

Zweigert, K, *Introduction to Comparative Law* (3rd revised edn, Clarendon 1998)

Edited Volumes

Bastmeijer, CJ, and Koivurova, T (eds), *Theory and Practice of Transboundary Environmental Impact Assessment* (Brill 2007)

Biswas, A, and Agarwal, SBC (eds), *Environmental Impact Assessment for Developing Countries* (Elsevier 1992)

Blakey, J, and Franks, D (eds), *Handbook of Cumulative Impact Assessment* (Edward Elgar 2021)

Doelle, M, and Sinclair, A (eds), *Impact Assessment in Transition: A Critical Review of the Canadian Impact Assessment Act* (Irwin Law 2021)

Eggleston, S, and others (eds), *2006 IPCC Guidelines for National Greenhouse Gas Inventories* (IGES 2006)

Farber, D, and Peeters, M (eds), *Climate Change Law* (Edward Elgar 2016)

Faure, M (ed), *Elgar Encyclopedia of Environmental Law* (Edward Elgar 2016)

Fischer, T, and González, A (eds), *Handbook on Strategic Environmental Assessment* (Edward Elgar 2021)

Fisher, D (ed), *Research Handbook on Fundamental Concepts of Environmental Law* (Edward Elgar 2016)

Fisher, E, and Preston, B (eds), *An Environmental Courtin Action* (Hart 2022)

Hanna, K (ed), *Routledge Handbook of Environmental Impact Assessment* (Routledge 2022)

Holder, J, and McGillivray, D (eds), *Taking Stock of Environmental Assessment: Law, Policy and Practice* (Routledge 2007)

Kamm, F, *Intricate Ethics: Rights, Responsibilities, and Permissible Harm* (OUP 2006)

Kevin, H, and Arnold, L (eds), *An Introduction to Environmental Impact Assessment* (Routledge 2020)

Lees, E, and Vinuales, J (eds), *The Oxford Handbook of Comparative Environmental Law* (OUP 2019)

Masson-Delmotte, V, and others (eds), *Climate Change 2021: The Physical Science Basis. Working Group I Contribution to the Sixth Assessment Report of the Intergovernmental Panel on Climate Change* (CUP 2021)

Masson-Delmotte, V, and others (eds), *Global Warming of 1.5°C: An IPCC Special Report* (IPCC 2019)

Mayer, B, and Zahar, A (eds), *Debating Climate Law* (CUP 2021)

Monateri, PG (ed), *Methods of Comparative Law* (Edward Elgar 2012)

O'Riordan, T, and Hey, R (eds), *Environmental Impact Assessment* (Saxon House 1976)

Oliver-Smith, A, and Hoffman, S (eds), *The Angry Earth: Disaster in Anthropological Perspective* (Routledge 1999)

Paehlke, R, and Torgerson, D (eds), *Managing Leviathan: Environmental Politics and the Administrative State* (2nd edn, University of Toronto Press 2019)

Pearson, L, Harlow, C, and Taggart, M (eds), *Administrative Law in a Changing State: Essays in Honour of Mark Aronson* (Hart 2008)

Peters, A (ed), *Max Planck Encyclopedias of International Law* (OUP 2024)

Petts, J (ed), *Handbook of Environmental Impact Assessment* (Wiley-Blackwell 1999)

Pörtner, HO, and others (eds), *Climate Change 2022: Impacts, Adaptation and Vulnerability. Working Group II Contribution to the Sixth Assessment Report of the Intergovernmental Panel on Climate Change* (CUP 2022)

Rajamani, L, and Peel, J (eds), *The Oxford Handbook of International Environmental Law* (OUP 2021)

Rayfuse, R (ed) *Research Handbook on International Marine Environmental Law* (Edward Elgar 2015)

Reimann, M, and Zimmermann, R (eds), *The Oxford Handbook of Comparative Law* (2nd edn, OUP 2019)

Salmon, P, and Grinlinton, D (eds), *Environmental Law in New Zealand* (Thomson Reuters 2015)

Sanford, R, and Holtgrieve, D, *Environmental Impact Assessment in the United States* (Routledge 2023)

Schmidt, M, and others (eds), *Standards and Thresholds for Impact Assessment: Environmental Protection in the European Union* (Springer 2008)

Shukla, P, and others (eds), *Climate Change 2022: Mitigation of Climate Change. Working Group III Contribution to the Sixth Assessment Report of the Intergovernmental Panel on Climate Change* (CUP 2022)

Sindico, F, and Mbengue, MM (eds), *Comparative Climate Change Litigation: Beyond the Usual Suspects* (Springer 2021)

Sindico, F, Switzer, S, and Qin, T (eds), *The Transformation of Environmental Law and Governance* (Edward Elgar 2021)

Stocker, T, and others (eds), *Climate Change 2013: The Physical Science Basis* (CUP 2013)

Sullivan, T (ed.), *Environmental Law Handbook* (24th edn, Bernan Press 2019)

Tabatabaei, M, and Aghbashlo, M (eds), *Biofuel* (Springer 2019)

Therivel, R, and Wood, G (eds), *Methods of Environmental and Social Impact Assessment* (4th edn, Taylor & Francis 2017)

Journal Articles

Agrawala, S, and others, 'Incorporating Climate Change Impacts and Adaptation in Environmental Impact Assessments: Opportunities and Challenges' (2012) 4 Climate and Development 26

Alomari, M, and Heffron, R, 'Environmental Impact Assessment: A Middle Eastern Experience' (2021) 33 Journal of Environmental Law 309

Andresen, S, 'The Paris Agreement and Its Rulebook in a Problem-Solving Perspective' (2019) 9 Climate Law 122

Antoniazzi, CT, 'Strengthening the Complaint Mechanisms of Multilateral Climate Funds and Carbon Markets: A Critical Step Towards a Human Rights-Based Green Transition' (2023) 32 Review of European, Comparative & International Environmental Law 173

Armeni, C, and Lee, M, 'Participation in a Time of Climate Crisis' (2021) 48 Journal of Law & Society 549

Arnstein, S, 'A Ladder of Citizen Participation' (1969) 35 Journal of the American Institute of Planners 216

Asheim, GB, and others, 'The Case for a Supply-Side Climate Treaty' (2019) 365 Science 325

Aydos, E, and others, 'Rocky Hill: A Legal Breakthrough in the Consideration of Climate Change and Social Impacts of Coal Mines' (2020) 14 Carbon & Climate Law Review 98

Bartlett, R, and Kurian, P, 'The Theory of Environmental Impact Assessment: Implicit Models of Policy Making' (1999) 27 Policy & Politics 415

Baxter, W, Ross, W, and Spaling, H, 'Improving the Practice of Cumulative Effects Assessment in Canada' (2001) 19 Impact Assessment & Project Appraisal 253

Bell-James, J, and Collins, B, ' "If We Don't Mine Coal, Someone Else Will": Debunking the "Market Substitution Assumption" in Queensland Climate Change Litigation' (2020) 37 Environmental & Planning Law Journal 167

Bell-James, J, and Ryan, S, 'Climate Change Litigation in Queensland: A Case Study in Incrementalism' (2016) 33 Environmental & Planning Law Journal 515

Bilgin, A, 'Analysis of the Environmental Impact Assessment (EIA) Directive and the EIA Decision in Turkey' (2015) 53 Environmental Impact Assessment Review 40

Blakley, J, and Russell, J, 'International Progress in Cumulative Effects Assessment: A Review of Academic Literature 2008–2018' (2022) 65 Journal of Environmental Planning & Management 186

Bodansky, D, 'The Paris Climate Change Agreement: A New Hope?' (2016) 110 American Journal of International Law 288

Boyle, A, 'Human Rights and the Environment: Where Next?' (2012) 23 European Journal of International Law 613

Browne, J, and others, 'A Guide to Policy Analysis as a Research Method' (2019) 34 Health Promotion International 1032

Burger, M, and Wentz, J, 'Downstream and Upstream' (2017) 41 Harvard Environmental Law Review 110

Burke, M, Hsiang, S, and Miguel, E, 'Global Non-Linear Effect of Temperature on Economic Production' (2015) 527 Nature 235

Burton, P, 'Power to the People? How to Judge Public Participation' (2004) 19 Local Economy 193

Castaneda, J, 'The World Bank Adopts Environmental Impact Assessments' (1992) 4 Pace Yearbook of International Law 241

Castro, F, 'Canada's Climate Change Mitigation Commitments and the Role of the Federal Impact Assessment Act' (2020) 33 Journal of Environmental Law & Practice 211

Cattino, M, and Reckien, D, 'Does Public Participation Lead to More Ambitious and Transformative Local Climate Change Planning?' (2021) 52 Current Opinion in Environmental Sustainability 100

Chen, Y, Wang Y, and Zhang Z, 'Suggestions to Response to Climate Change by Environmental Impact Assessment Mechanisms Innovation' (2016) 41 Environment & Sustainable Development 17 (in Chinese)

Cho, HS, 'An Overview of Korean Environmental Law' (1999) 29 Environmental Law 501

Christopher, C, 'Success by a Thousand Cuts: The Use of Environmental Impact Assessment in Addressing Climate Change' (2008) 9 Vermont Journal of Environmental Law 549

Clark, B, 'Improving Public Participation in Environmental Impact Assessment' (1994) 20 Built Environment 294

Clausen, A, 'An Evaluation of the Environmental Impact Assessment System in Vietnam: The Gap between Theory and Practice' (2011) 31 Environmental Impact Assessment Review 136

Connelly, S, and Richardson, T, 'Value-Driven SEA: Time for an Environmental Justice Perspective?' (2005) 25 Environmental Impact Assessment Review 391

Cooper, L, and Sheate, W, 'Cumulative Effects Assessment: A Review of UK Environmental Impact Statements' (2002) 22 Environmental Impact Assessment Review 415

Craik, N, and Gu, K, 'Strategic Environmental Assessment in Marine Areas beyond National Jurisdiction: Implementing Integration' (2022) 37 International Journal of Marine & Coastal Law 189

Crockett, A, 'Addressing the Significance of Greenhouse Gas Emissions under CEQA: California's Search for Regulatory Certainty in an Uncertain World' (2011) 4 Golden Gate University Environmental Law Journal 203

Dales, J, 'Death by a Thousand Cuts: Incorporating Cumulative Effects in Australia's Environment Protection and Biodiversity Conservation Act' (2011) 20 Pacific Rim Law & Policy Journal 149

De Castro, FV, 'Canada's Climate Change Mitigation Commitments and the Role of the Federal Impact Assessment Act' (2020) 33 Journal of Environmental Law & Practice 211

Dehm, J, 'Coal Mines, Carbon Budgets and Human Rights in Australian Climate Litigation: Reflections on *Gloucester Resources Ltd v Minister for Planning and Environment*' (2020) 26 Australian Journal of Human Rights 244

Doelle, M, and Majekolagbe, A, 'Meaningful Public Engagement and the Integration of Climate Considerations into Impact Assessment' (2023) 101 Environmental Impact Assessment Review (article#107103) 4

Dwyer, G, '"Market Substitution" in the Context of Climate Litigation' (2022) 12 Climate Law 1

Eccleston, C, 'Applying the Significant Departure Principle in Resolving the Cumulative Impact Paradox: Assessing Significance in Areas That Have Sustained Cumulatively Significant Impacts' (2006) 8 Environmental Practice 241, 244

Eccleston, C, 'Assessing Cumulative Significance of Greenhouse Gas Emissions: Resolving the Paradox – the Sphinx Solution' (2010) 12 Environmental Practice 105

Ehrlich, A, and Ross, W, 'The Significance Spectrum and EIA Significance Determinations' (2015) 33 Impact Assessment & Project Appraisal 87

Enriquez-de-Salamanca, A, and others, 'Environmental Impacts of Climate Change Adaptation' (2017) 64 Environmental Impact Assessment Review 87

Eskander, S, and Fankhauser, S, 'Reduction in Greenhouse Gas Emissions from National Climate Legislation' (2020) 10 Nature Climate Change 750

Esty, D, 'Revitalizing Environmental Federalism' (1996) 95 Michigan Law Review 570

Fisher, E, 'Law and Energy Transitions: Wind Turbines and Planning Law in the UK' (2018) 38 Oxford Journal of Legal Studies 528

Foley, M, and others, 'The Challenges and Opportunities in Cumulative Effects Assessment' (2017) 62 Environmental Impact Assessment Review 122

Formby, J, 'The Australian Government's Experience with Environmental Impact Assessment' (1987) 7 Environmental Impact Assessment Review 207

Fowler, R, 'Environmental Impact Assessment: What Role for the Commonwealth? – An Overview' (1996) 13 Environmental & Planning Law Journal 246

Fragnière, A, 'Climate Change and Individual Duties' (2016) 7 WIREs Climate Change 798

Gao, Q, 'Mainstreaming Climate Change into the EIA Procedures: A Perspective from China' (2017) 10 International Journal of Climate Change Strategies & Management 342

Garcia, B, Foerster, A, and Lin, J, 'Net Zero for the International Shipping Sector? An Analysis of the Implementation and Regulatory Challenges of the IMO Strategy on Reduction of GHG Emissions' (2021) 33 Journal of Environmental Law 85

Gerrard, M, 'Climate Change and the Environmental Impact Review Process' (2008) 22 Natural Resources & Environment 20

Giglio, S, Maggiori, M, and Stroebel, J, 'Very Long-Run Discount Rates' (2015) 130 Quarterly Journal of Economics 1

Glover, J, and Scott-Taggart, M, 'It Makes no Difference whether or Not I Do It' (1975) 49 Proceedings of the Aristotelian Society (Supplementary Volumes) 171

Glucker, A, and others, 'Public Participation in Environmental Impact Assessment: Why, Who and How?' (2013) 43 Environmental Impact Assessment Review 104

Green, F, and Denniss, R, 'Cutting with both Arms of the Scissors: The Economic and Political Case for Restrictive Supply-Side Climate Policies' (2018) 150 Climatic Change 73

Green, J, 'Does Carbon Pricing Reduce Emissions? A Review of ex-post Analyses' (2021) 16 Environmental Research Letters (article #43004) 1–17

Hands, S, and Hudson, M, 'Incorporating Climate Change Mitigation and Adaptation into Environmental Impact Assessment: A Review of Current Practice within Transport Projects in England' (2016) 34 Impact Assessment & Project Appraisal 330

Hartley, N, and Wood, C, 'Public Participation in Environmental Impact Assessment: Implementing the Aarhus Convention' (2005) 25 Environmental Impact Assessment Review 319

He, X, 'Integrating Climate Change Factors within China's Environmental Impact Assessment Legislation: New Challenges and Developments' (2013) 9 Law, Environment & Development Journal 50

He, X, 'Mitigation and Adaptation through Environmental Impact Assessment Litigation: Rethinking the Prospect of Climate Change Litigation in China' (2021) 10 Transnational Environmental Law 413

Hegmann, G, and Yarranton, GA, 'Alchemy to Reason: Effective Use of Cumulative Effects Assessment in Resource Management' (2011) 31 Environmental Impact Assessment Review 484

Hein, JF, and Jacewicz, N, 'Implementing NEPA in the Age of Climate Change' (2020) 10 Michigan Journal of Environmental & Administrative Law 1

Hepburn, C, and others, 'The Economics of the EU ETS Market Stability Reserve' (2016) 80 Journal of Environmental Economics & Management 1

Hill, J, 'Comparative Law, Law Reform and Legal Theory' (1989) 9 Oxford Journal of Legal Studies 101

Hofferth, S, 'Significant Impacts under NEPA: The Social Cost of Greenhouse Gases as a Tool to Mitigate Climate Change' (2022) 11 Michigan Journal of Environmental & Administrative Law 333

Hollander, R, 'Rethinking Overlap and Duplication: Federalism and Environmental Assessment in Australia' (2010) 40 Publius: The Journal of Federalism 136

Holt, S, and McGrath, C, 'Climate Change: Is the Common Law up to the Task?' (2018) 24 Auckland University Law Review 10

Howie, P, and Atakhanova, Z, 'Assessing Initial Conditions and ETS Outcomes in a Fossil-Fuel Dependent Economy' (2022) 40 Energy Resources Reviews 1

Hulme, M, 'Climate Emergency Politics Is Dangerous' (2019) 36 Issues in Science & Technology 23

Iyer, C, and Thomas, N, 'A Critical Review on Regional Connectivity Scheme of India' (2020) 48 Transportation Research Procedia 47

Jarke-Neuert, J, and Perino, G, 'Energy Efficiency Promotion Backfires under Cap-and-Trade' (2020) 62 Resource & Energy Economics (article #101189) 1–21

Jin, Z, 'Environmental Impact Assessment Law in China's Courts: A Study of 107 Judicial Decisions' (2015) 55 Environmental Impact Assessment Review 35

Jiricka, A, and others, 'Consideration of Climate Change Impacts and Adaptation in EIA Practice: Perspectives of Actors in Austria and Germany' (2016) 57 Environmental Impact Assessment Review 78

Jones, M and Morrison-Saunders, A, 'Understanding the Long-Term Influence of EIA on Organisational Learning and Transformation' (2017) 64 Environmental Impact Assessment Review 131

Jones, M, and Morrison-Saunders, A, 'Making Sense of Significance in Environmental Impact Assessment' (2016) 34 Impact Assessment & Project Appraisal 87

Joseph, C, and others, 'Improving Cumulative Effects Assessment: Alternative Approaches Based upon an Expert Survey and Literature Review' (2023) 41 Impact Assessment & Project Appraisal 162

Kakonge, J, 'Environmental Impact Assessment in Sub-Saharan Africa: The Gambian Experience' (2006) 24 Impact Assessment & Project Appraisal 57

Kamau, J, and Mwaura, F, 'Climate Change Adaptation and EIA Studies in Kenya' (2013) 5 International Journal of Climate Change Strategies and Management 152

Kang, S, and others, 'A Review of Black Carbon in Snow and Ice and Its Impact on the Cryosphere' (2020) 210 Earth-Science Reviews (article #103346) 1–12

Kaufman, N, and others, 'A Near-Term to Net Zero Alternative to the Social Cost of Carbon for Setting Carbon Prices' (2020) 10 Nature Climate Change 1010

Kaupa, C. 'Is It Still Permissible under EU Law to Issue New Permits for Oil and Gas Extraction?' (2024) 33 Review of European, Comparative & International Environmental Law 236

Kellman, B, 'NEPA Review of Climate Change' (2016) 46 Environmental Law Reporter News & Analysis 10378

Kennedy, W, 'Environmental Impact Assessment in the Federal Republic of Germany' (1980) 1 Environmental Impact Assessment Review 92

Kersten, CM, 'Rethinking Transboundary Environmental Impact Assessment' (2009) 34 Yale Journal of International Law 173

Kim, K, and Murabayashi, DHL, 'Recent Developments in the Use of Environmental Impact Statements in Korea' (1992) 12 Environmental Impact Assessment Review 295

Kim, KT, and Kim, I, 'The Significance of Scope 3 GHG Emissions in Construction Projects in Korea: Using EIA and LCA' (2021) 9 Climate 33

Knox, J, 'The Myth and Reality of Transboundary Environmental Impact Assessment' (2002) 96 American Journal of International Law 291

Knox, J, and Voigt, C, 'Introduction to the Symposium on Jacqueline Peel and Jolene Lin: Transnational Climate Litigation: The Contribution of the Global South' (2020) 114 AJIL Unbound 35

Knutti, R, and others, 'A Scientific Critique of the Two-Degree Climate Change Target' (2016) 9 Nature Geoscience 13

Kotzampasakis, M, and Woerdman, E, 'The Legal Objectives of the EU Emissions Trading System: An Evaluation Framework' *Transnational Environmental Law* (forthcoming)

Krasnova, M, 'Legal Problems in the Implementation of the Environmental Impact Assessment in Ukraine: A Critical Review' (2021) 4 Grassroots Journal of Natural Resources 91

Kuramochi, T, and others, 'Greenhouse Gas Emission Scenarios in Nine Key Non-G20 Countries: An Assessment of Progress toward 2030 Climate Targets' (2021) 123 Environmental Science & Policy 67

Kurasaka, H, 'Japanese Environmental Impact Assessment Law: Before and after' (2001) 27 Built Environment 16

Lamb, W, and others, 'A Review of Trends and Drivers of Greenhouse Gas Emissions by Sector from 1990 to 2018' (2021) 16 Environmental Research Letters (article #073005) 1–31

Lamboll, R, and others, 'Assessing the Size and Uncertainty of Remaining Carbon Budgets' (2023) 13 Nature Climate Change 1360

Landau, M, 'Redundancy, Rationality, and the Problem of Duplication and Overlap' (1969) 29 Public Administration Review 346

Lawal, A, Bouzarovski, S, and Clark, J, 'Public Participation in EIA: The Case of West African Gas Pipeline and Tank Farm Projects in Nigeria' (2013) 31 Impact Assessment & Project Appraisal 226

Lawrence, D, 'Impact Significance Determination: Designing an Approach' (2007) 27 Environmental Impact Assessment Review 730

Lazarus, M, and van Asselt, H, 'Fossil Fuel Supply and Climate Policy: Exploring the Road Less Taken' (2018) 150 Climatic Change 1

Lazarus, S, 'The Equator Principles at Ten Years' (2014) 5 Transnational Legal Theory 417

Le Quéré, C, and others, 'Drivers of Declining CO_2 Emissions in 18 Developed Economies' (2019) 9 Nature Climate Change 213

Ledda, A, and others, 'Integrating Adaptation to Climate Change in Regional Plans and Programmes: The Role of Strategic Environmental Assessment' (2021) 91 Environmental Impact Assessment Review (article #106655) 1–9

Lee, M, and others, 'Public Participation and Climate Change Infrastructure' (2013) 25 Journal of Environmental Law 33

Lee, N, and Wood, C, 'Environmental Impact Assessment in the European Economic Community' (1980) 1 Environmental Impact Assessment Review 287

Lightfoot, T, 'Climate Change and Environmental Review: Addressing the Impact of Greenhouse Gas Emissions under the Minnesota Environmental Policy Act' (2010) 36 William Mitchell Law Review 1068

Liu, D, and others, 'Contribution of International Photovoltaic Trade to Global Greenhouse Gas Emission Reduction: The Example of China' (2019) 143 Resources, Conservation & Recycling 114

Lostarnau, C, and others, 'Stakeholder Participation within the Public Environmental System in Chile: Major gaps between Theory and Practice' (2011) 92 Journal of Environmental Management 2470

Maamoun, N, 'The Kyoto Protocol: Empirical Evidence of a Hidden Success' (2019) 95 Journal of Environmental Economics &Management 227

Maphanga, T, and others, 'The State of Public Participation in the EIA Process and its Role in South Africa: A case of Xolobeni' (2023) 105 South African Geographical Journal 277

Marzuki, A, 'A Review on Public Participation in Environmental Impact Assessment in Malaysia' (2009) 3 Theoretical & Empirical Researches in Urban Management 126

Masden, E, and others, 'Cumulative Impact Assessments and Bird/Wind Farm Interactions: Developing a Conceptual Framework' (2010) 30 Environmental Impact Assessment Review 1

Mauhs, F, 'Cumulative Impact Analysis in NEPA Climate Assessments' (2022) 39 Pace Environmental Law Review 211

Mayembe, R, and others, 'Integrating Climate Change in Environmental Impact Assessment: A Review of Requirements Across 19 EIA Regimes' (2023) 869 Science of the Total Environment (article #161850) 1–13

Mayer, B, 'International Law Obligations Arising in Relation to Nationally Determined Contributions' (2018) 7 Transnational Environmental Law 251

Mayer, B, 'Climate Assessment as an Emerging Obligation under Customary International Law' (2019) 68 International & Comparative Law Quarterly 271

Mayer, B, 'Environmental Assessments in the Context of Climate Change: The Role of the UN Economic Commission for Europe' (2019) 28 Review of European, Comparative & International Environmental Law 82

Mayer, B, 'Climate Change Mitigation as an Obligation under Human Rights Treaties?' (2021) 115 American Journal of International Law 409

Mayer, B, 'Temperature Targets and State Obligations on the Mitigation of Climate Change' (2021) 33 Journal of Environmental Law 585

Mayer, B, 'Attribution Science and the Fate of Climate Litigation' (2022) 13 Global Policy 831

Mayer, B, 'Climate Change Mitigation as an Obligation under Customary International Law' (2023) 48 Yale Journal of International Law 105

Mayer, B, 'Prompting Climate Change Mitigation through Litigation' (2023) 72 International & Comparative Law Quarterly 233

Mayer, B, 'The "Highest Possible Ambition" on Climate Change Mitigation as a Legal Standard' (2024) 73 International and Comparative Law Quarterly 285

Mayer, B, and Ding, Z, 'Climate Change Mitigation in the Aviation Sector: A Critical Overview of National and International Initiatives' (2023) 12 Transnational Environmental Law 14

McGinness, V, and Raff, M, 'Coal and Climate Change: A Study of Contemporary Climate Litigation in Australia' (2020) 37 Environmental & Planning Law Journal 87

Meazell, E, 'Super Deference, the Science Obsession, and Judicial Review as Translation of Agency Science' (2011) 109 Michigan Law Review 733

Mehling, M, 'The Comparative Law of Climate Change: A Research Agenda' (2015) 24 Review of European, Comparative and International Environmental Law 341

Minx, J, and others, 'A Comprehensive and Synthetic Dataset for Global, Regional, and National Greenhouse Gas Emissions by Sector 1970–2018 with an Extension to 2019' (2021) 13 Earth System Science Data 5213

Montz, B, and Dixon, J, 'From Law to Practice: EIA in New Zealand' (1993) 13 Environmental Impact Assessment Review 89

Mora, C, and others, 'Bitcoin Emissions Alone Could Push Global Warming above 2°C' (2018) 8 Nature Climate Change 931

Morgan, R, 'Environmental Impact Assessment: The State of the Art' (2012) 30 Impact Assessment & Project Appraisal 5

Murcott, MJ, and Vinti, C, 'The Judge-Made "Duty" to Consider Climate Change in South Africa' (2024) 16 Journal of Human Rights Practice 125

Nelson, R, 'Breaking Backs and Boiling Frogs: Warnings from a Dialogue between Federal Water Law and Environmental Law' (2019) 42 University of New South Wales Law Journal 1179

Nelson, R, and Shirley, LM, 'The Latent Potential of Cumulative Effects Concepts in National and International Environmental Impact Assessment Regimes' (2022) 12 Transnational Environmental Law 150

Nenasheva, M, and others, 'Legal Tools of Public Participation in the Environmental Impact Assessment Process and their Application in the Countries of the Barents Euro-Arctic Region' (2015) 1 Barents Studies: Peoples, Economies and Politics 13

Nielsen, E, Christensen, P, and Kørnøv, L, 'EIA Screening in Denmark: A New Regulatory Instrument?' (2005) 7 Journal of Environmental Assessment Policy & Management 35

Nordhaus, W, 'Climate Change: The Ultimate Challenge for Economics' (2019) 109 American Economic Review 1991

Normile, D, 'Indonesia's Utopian New Capital May Not Be as Green as It Looks' (2022) 375 Science 479

O'Dea Brill, M, 'Assessing the Scope of the National Environmental Policy Act: Recent Attempts by Environmentalists to Add Climate Change Considerations into NEPA Review' (2014) 54 Natural Resources Journal 409

O'Faircheallaigh, C, 'Public Participation and Environmental Impact Assessment: Purposes, Implications, and Lessons for Public Policy Making' (2010) 30 Environmental Impact Assessment Review 19

Ohsawa, T, and Duinker, P, 'Climate-Change Mitigation in Canadian Environmental Impact Assessments' (2014) 32 Impact Assessment & Project Appraisal 222

Olsen, ASH, and Hansen, AM, 'Perceptions of Public Participation in Impact Assessment: A Study of Offshore Oil Exploration in Greenland' (2014) 32 Impact Assessment & Project Appraisal 72

Omenge, P, Makindi, S, and Obwoyere, G, 'Public Participation in Environmental Impact Assessment and its Substantive Contribution to Environmental Risk Management: Insights from EIA Practitioners and Other Stakeholders in Kenya's Renewable Energy Sub-Sector' (2019) 8 Energy & Sustainability 133

Palerm, J, and Aceves, C, 'Environmental Impact Assessment in Mexico: An Analysis from a "Consolidating Democracy" Perspective' (2004) 22 Impact Assessment & Project Appraisal 99

Panovics, A, 'The Aarhus Convention Model' [2016] Hungarian Yearbook of International Law & European Law 251

Parenteau, P, 'The Atmosphere as A Global Public Good' (2023) 16 Journal of Law and Public Policy, St Thomas U 217

Parker, K, 'Litigating at the Source: Attributing Climate Change Impacts to Coal Mines' (2020) 37 Environmental & Planning Law Journal 67

Pearce, D, 'The Social Cost of Carbon and its Policy Implications' (2003) 19 Oxford Review of Economic Policy 362, 363

Peel, J, 'Issues in Climate Change Litigation' (2011) 5 Carbon & Climate Law Review 15

Peel, J, 'The *Living Wonders* Case: A Backwards Step in Australian Climate Litigation on Coal Mines' (2024) 36 Journal of Environmental Law 125

Peel, J, and Lin, J, 'Transnational Climate Litigation: The Contribution of the Global South' (2019) 113 American Journal of International Law 679

Preston, B, 'Contemporary Issues in Environmental Impact Assessment' (2020) 37 Environmental & Planning Law Journal 423

Prieur, M, 'Le respect de l'environnement et les études d'impact' (1981) 6 Revue juridique de l'environnement 103

Quaas, J, and others, 'Robust Evidence for Reversal of the Trend in Aerosol Effective Climate Forcing' (2022) 22 Atmospheric Chemistry & Physics 1221

Raff, M, 'Ten Principles of Quality in Environmental Impact Assessment' (1997) 14 Environmental & Planning Law Journal 207

Ramanathan, V, and Carmichael, G, 'Global and Regional Climate Changes Due to Black Carbon' (2008) 1 Nature Geoscience 221

Reitze, A, 'Dealing with Climate Change under the National Environmental Policy Act' (2018) 43 William & Mary Environmental Law & Policy Review 173

Rendall, M, 'Carbon Leakage and the Argument from no Difference' (2015) 24 Environmental Values 535

Rissman, J, and others, 'Technologies and Policies to Decarbonize Global Industry: Review and Assessment of Mitigation Drivers through 2070' (2020) 266 Applied Energy (article #114848) 1–34

Roberts, AE, 'Traditional and Modern Approaches to Customary International Law: A Reconciliation' (2001) 95 American Journal of International Law 757

Roeck, M, and Drennen, T, 'Life Cycle Assessment of Behind-the-Meter Bitcoin Mining at US Power Plant' (2022) 27 International Journal of Life Cycle Assessment 355

Roesler, S, 'Agency Reasons at the Intersection of Expertise and Presidential Preferences' (2019) 71 Administrative Law Review 491

Rose, A, '*Gray v Minister for Planning*: The Rising Tide of Climate Change Litigation in Australia' (2007) 29 Sydney Law Review 725

Rosendahl, KE, 'EU ETS and the Waterbed Effect' (2019) 9 Nature Climate Change 734

Ruhl, JB, 'Climate Change and the Endangered Species Act: Building Bridges to the No-Analog Future' (2008) 88 Boston University Law Review 1

Rushovich, N, 'Climate Change and Environmental Policy: An Analysis of the Final Guidance on Greenhouse Gas Emissions and the Effects of Climate Change in National Environment Policy Act Reviews' (2018) 27 Boston University Public Interest Law Journal 327

Ryall, A, 'Enforcing the Environmental Impact Assessment Directive in Ireland: Evolution of the Standard of Judicial Review' (2018) 7 Transnational Environmental Law 515

Rydin, Y, and others, 'Local Voices on Renewable Energy Projects: The Performative Role of the Regulatory Process for Major Offshore Infrastructure in England and Wales' (2018) 23 Local Environment 565

Rydin, Y, Lee, M, and Lock, S, 'Public Engagement in Decision-Making on Major Wind Energy Projects' (2015) 27 Journal of Environmental Law 139

Schäfer, S, 'Decoupling the EU ETS from Subsidized Renewables and Other Demand Side Effects: Lessons from the Impact of the EU ETS on CO_2 Emissions in the German Electricity Sector' (2019) 133 Energy Policy (article #110858) 1–12

Scott, J, 'The Geographical Scope of the EU's Climate Responsibilities' (2015) 17 Cambridge Yearbook of European Legal Studies 92

Selmi, D, 'The Judicial Development of the California Environmental Quality Act' (1984) 18 UC Davis Law Review 197

Serletis, A, Timilsina, G, and Vasetsky, O, 'Interfuel Substitution in the United States' (2010) 32 Energy Economics 737

Shapovalova, D, 'Arctic Petroleum and the 2°C Goal: A Case for Accountability for Fossil-Fuel Supply' (2020) 10 Climate Law 282

Shapovalova, D, 'Climate Change and Oil and Gas Production Regulation: An Impossible Reconciliation?' (2023) 26 Journal of International Economic Law 817

Shen, Q, Pan, Y, and Feng, Y, 'The Impacts of High-Speed Railway on Environmental Sustainability: Quasi-Experimental Evidence from China' (2023) 10 Humanities and Social Sciences Communications (article #719) 1–19

Shepherd, A, and Bowler, C, 'Beyond the Requirements: Improving Public Participation in EIA' (1997) 40 Journal of Environmental Planning & Management 725

Siddik, MAB, and others, 'The Environmental Footprint of Data Centers in the United States' (2021) 16 Environmental Research Letters 1

Smets, H, 'Le principe de non-discrimination en matière de protection de l'environnement' (2000) 4 Revue européenne de droit de l'environnement 3

Spaling, H, 'Cumulative Effects Assessment: Concepts and Principles' (1994) 12 Impact Assessment 231

Sprain, L, 'Paradoxes of Public Participation in Climate Change Governance' (2016) 25 Good Society 62

Steele, J, 'Participation and Deliberation in Environmental Law: A Problem-Solving Approach' (2001) 21 Oxford Journal of Legal Studies 415

Suškevičs, M, and others, 'Public Participation in Environmental Assessments in the EU: A Systematic Search and Qualitative Synthesis of Empirical Scientific Literature' (2023) 98 Environmental Impact Assessment Review 106944

Tekayak, D, 'An Overview of Environmental Impact Assessment in Turkey: Issues and Recommendations' (2014) 13 Ankara Avrupa Çalışmaları Dergisi 133

Tenney, A, Kværner, J, and Gjerstad, KI, 'Uncertainty in Environmental Impact Assessment Predictions: The Need for Better Communication and More Transparency' (2006) 24 Impact Assessment & Project Appraisal 45

Tol, R, 'The Economic Impacts of Climate Change' (2018) 12 Review of Environmental Economics & Policy 7

Unruh, G, 'Escaping Carbon Lock-In' (2002) 30 Energy Policy 317

Unruh, G, 'Understanding Carbon Lock-In' (2000) 28 Energy Policy 817

Van Hoecke, M, 'Methodology of Comparative Legal Research' [2015] Law & Method 11

Vanclay, F, 'International Principles for Social Impact Assessment' (2003) 21 Impact Assessment & Project Appraisal 5

Vanhala, L, 'The Comparative Politics of Courts and Climate Change' (2013) 22 Environmental Politics 447

Voigt, C, 'The First Climate Judgment before the Norwegian Supreme Court: Aligning Law with Politics' (2021) 33 Journal of Environmental Law 697

Wagner, W, 'The Science Charade in Toxic Risk Regulation' (1995) 95 Columbia Law Review 1613

Wang, P, and others, 'Estimates of the Social Cost of Carbon: A Review Based on Meta-Analysis' (2019) 209 Journal of Cleaner Production 1494

Ward, R, and Hong, DU, 'Environmental Impact Assessment in North Korean Environmental Law: Origins, Evolution, and a Comparative Analysis' (2021) 48 Ecology Law Quarterly 38

Watkins, J, and Durning, B, 'Carbon Definitions and Typologies in Environmental Impact Assessment: Greenhouse Gas Confusion?' (2012) 30 Impact Assessment & Project Appraisal 296

Webster, E, 'The Planning Regime and Climate Change Mitigation' (2022) 4 Conveyancer and Property Lawyer 415

Weiner, K, 'NEPA and State NEPAs: Learning from the Past, Foresight for the Future' (2009) 39 Environmental Law Reporter News & Analysis 10675

Weisbach, D, and Sunstein, C, 'Climate Change and Discounting the Future: A Guide for the Perplexed' (2008) 27 Yale Law & Policy Review 433

Wentz, JA, 'Draft NEPA Guidance Requires Agencies to Consider both GHG Emissions and the Impact of Climate Change on Proposed Actions' (2015) 26 Environmental Law in New York 57

Wiklund, H, 'Why High Participatory Ideals Fail in Practice: A Bottom-Up Approach to Public Nonparticipation in EIA' (2011) 13 Journal of Environmental Assessment Policy & Management 159

Wishnie, LG, 'NEPA for a New Century: Climate Change & the Reform of the National Environmental Policy Act' (2008) 16 New York University Environmental Law Journal 628

Woloszyn, W, 'Evolution of Environmental Impact Assessment in Poland: Problems and Prospects' (2004) 22 Impact Assessment & Project Appraisal 109

Wood, C, and Dejeddour, M, 'Strategic Environmental Assessment: EA of Policies, Plans and Programmes' (1992) 10 Impact Assessment 3

Wu, H, and Zhang, Y, 关于中国将气候变化因素融入环境影响评价的探讨 ('Discussion of China's Integration of Climate Change Factors into Environmental Impact Assessment') (2011) 33 Environmental Pollution and Control 91

Wu, Y, and Ren, J, 规划环评中纳入气候变化因素的现状调查与分析 ('Survey and Analysis of the Status Quo of the Climate Change Factors in Strategic Environmental Assessments') [2014] 中国环境科学学会学术年会 (Annual Meeting of the Chinese Society of Environmental Sciences) 1192

Xiong, X, and others, 'Aviation and Carbon Emissions: Evidence from Airport Operations' (2023) 109 Journal of Air Transport Management (article #102383) 1–9

Yang, Y, 'Reformed Environmental Impact Assessment in China: An Evaluation of its Effectiveness' (2020) 11 Journal of Environmental Protection 889

Yang, Y, and others, 'Integrating Climate Change Factor into Strategic Environmental Assessment in China' (2021) 89 Environmental Impact Assessment Review (article #106585) 1–8

Yi, J, and Hacking, T, 'Incorporating Climate Change into Environmental Impact Assessment: Perspectives from Urban Development Projects in South Korea' (2011) 21 Procedia Engineering 907

Index